Applications of DIFFERENTIAL THERMAL ANALYSIS in Cement Chemistry

Applications of DIFFERENTIAL THERMAL ANALYSIS in Cement Chemistry

V. S. Ramachandran, M. Sc., D. Phil.

Senior Scientific Officer
Central Building Research Institute
Roorkee, India

CHEMICAL PUBLISHING COMPANY, INC • NEW YORK • 1969

Applications of Differential Thermal Analysis in Cement Chemistry

ISBN: 978-0-8206-0024-6

Chemical Publishing Company:
www.chemical-publishing.com
www.chemicalpublishing.net

Printed in the United States of America

Dedicated to
Henry Le Chatelier (1850-1936),
as much a pioneer in thermal analysis
as in cement chemistry

Foreword

In this book, the first of its kind, the author ably summarizes and competently discusses information scattered through the literature in many languages, and presents a balanced review of the use of DTA in studies on the chemistry of cement and related materials. The selectively prepared bibliography presented with respect to each chapter on the logical basis of the various technical fields constitutes a useful aid and guide to further reading.

A particularly valuable aspect of the work is the systematic way in which the interrelationships and interdependence of the essentially complimentary techniques of DTA, X-ray diffraction, TGA, and chemical and physical techniques are stressed. This is an important aspect that is so often overlooked in publications of this type.

This pleasing and concise but critical presentation and appraisal of the state of current knowledge will prove valuable to the research worker, student, and technologist concerned with use of DTA in the investigation, production, and processing of mineral products.

August 1967
Pretoria

T. L. Webb
Director, South African National Building Research Institute

Preface

Many of the physicochemical phenomena associated with the formation and behaviour of cementitious compounds are so complex that they have defied unequivocal interpretation. Application of techniques such as X-ray diffraction, optical and electron microscopy, and infrared analysis has helped unravel many complicated problems, but it has also created new ones. By and large each technique has enabled a fresh approach to many intricate problems and has helped in narrowing down the gap in the existing knowledge. Compared with many other techniques Differential Thermal Analysis (DTA) is simpler, more easily adaptable, and yields quicker and sometimes very important results and hence has attracted great attention in recent years.

The DTA technique has been widely applied in clay mineralogy and only since a decade and a half has it been adopted as an important tool in cement chemistry. Within this period it has attained so much recognition, as evidenced by the phenomenal growth of publications, that it is difficult to conceive of a cement laboratory without the facilities of a DTA apparatus.

The results pertaining to the application of DTA to cementitious materials are widely scattered and hence the Director, Division of Building Research, National Research Council of Canada, assigned me the task of preparing a comprehensive and critical paper on the subject—which I presented at the 42nd meeting of the Highway Research Board held in 1964 in Washington. The reaction of cement chemists during and subsequent to the meeting was so encouraging that the Chemical Publishing Co., New York, felt the need of a book incorporating the existing knowledge on the application of DTA to cementitious materials.

It is needless to say that a sound knowledge of the DTA technique and of cement chemistry is essential for a correct interpretation of differential thermograms. Hence necessity was felt for the inclusion of (i) the treatment of the DTA technique and its applicability, (ii) chemistry of cementitious materials, and (iii) the application of DTA to inorganic cements. The book treats the essentials of the first two and lays emphasis on the third.

The book, as the title suggests, is intended primarily for those engaged in research on cement chemistry and other allied fields. However, ideas developed in the use of DTA to cement chemistry problems could be profitably applied to any other branch of study.

No attempt has been made to provide an exhaustive bibliography, but the list of references given at the end of each chapter or sub-chapter should serve as a reasonably good guide for further reading on the topics covered.

I would like to express my gratitude to Prof. Dinesh Mohan, Director, Central Building Research Institute, Roorkee, for his constant encouragement and permission to write the book. I acknowledge my thankfulness to Dr. T.L. Webb, Director, South African National Building Research Institute, for his valuable comments on the book. Mr. P.L.De was kind enough to translate some of the important German publications of relevance to this book, which I gratefully acknowledge. Thanks are also due to Mr. P.S. Shankar, Scientist, Publication Directorate, Council of Scientific and Industrial Research, New Delhi, for checking the manuscript, and to Mr. Krishan Lal for his accurate typing of the manuscript in the final form. To my wife Vasundhara, who devotedly undertook the task of typing the draft, assisting me in reading the proof, and preparation of indexes, I owe my especial thanks.

I am grateful to authors, publishers of journals, and organizations who permitted me to reproduce figures and tables from their publications.

August 1967 **V. S. Ramachandran**

Contents

Acknowledgments

The author gratefully acknowledges permission from the following publishers to use figures and tables from periodicals and books.

1. E.I. du Pont de Nemours & Co., U.S.A.
2. Building Research Station, U.K.
3. Society of Chemical Industry, U.K.
4. Academia Publishing House of the Czechoslovak Academy of Sciences
5. Macmillan & Co., Ltd., U.K.
6. Cement and Concrete Association, U.K.
7. Zhurnal Prikladnoi Khimii, U.S.S.R.
8. American Mineralogist, U.S.A.
9. The British Ceramic Society, U.K.
10. McGraw-Hill Book Co., U.S.A.
11. Hungarian Academy of Sciences, Hungary
12. Reinhold Book Division, U.S.A.
13. John Wiley & Sons, U.S.A.
14. Pergamon Press Ltd., U.K.
15. U.S. Department of Commerce, National Bureau of Standards
16. The South African Chemical Institute
17. Plenum Publishing Corporation, U.S.A.
18. Concrete Publications Ltd., U.K.
19. The American Ceramic Society
20. Rock Products, Maclean-Hunter Publishing Corporation, U.S.A.
21. Highway Research Board, U.S.A.
22. Tonindustrie-Zeitung und Keramische Rundschau, W. Germany
23. Nature, Macmillan (Journals) Ltd., U.K.
24. Central Road Research Institute, India
25. Commonwealth Scientific, Industrial and Research Organization, Australia
26. Council of Scientific & Industrial Research, India
27. American Concrete Institute
28. National Institute of Sciences of India
29. American Society for Testing and Materials, Materials Research

and Standards
30. Academic Press, U.K.
31. Indian Standards Institution
32. *Analytica Chimica Acta*, Elsevier Publishing Co., Netherlands
33. Revue Des Materiaux De Construction, Societe d'Editions Scientifiques, France.

ABBREVIATIONS USED IN THE TEXT

DTA	:	Differential thermal analysis
DTG	:	Differential thermogravimetric analysis
TGA	:	Thermogravimetric Analysis
MTA	:	Mass Spectro thermal analysis
DDC	:	Dynamic differential calorimetry
EGA	:	Effluent gas analysis
DDTA	:	Derived Differential thermal analysis
c.e.c.	:	Cation exchange capacity
DSC	:	Differential scanning calorimeter

Nomenclature for Cement Components

$C = CaO$	$S = SiO_2$	$F = Fe_2O_3$	$A = Al_2O_3$
$T = TiO_2$	$\bar{S} = SO_3$	$H = H_2O$	$B = BaO$
$M = MgO$	$N = Na_2O$	$K = K_2O$	

Thus, $C_4AH_{10} = 4CaO \cdot Al_2O_3 \cdot 10H_2O$

$C_4A_3\bar{S} = 4CaO \cdot 3Al_2O_3 \cdot SO_3$

I. Differential Thermal Analysis Technique

A substance subjected to thermal treatment may undergo physical or chemical change in dimensions, magnetic susceptibility, decomposition, crystalline transition, etc. The method of thermal analysis determines these changes as a function of temperature. The effect of heat treatment on physicochemical changes in a substance may be studied by three methods. The substance may be heated for a long period at a predetermined temperature until the equilibrium is established. This method is used in phase equilibria studies. The quasistatic method involves heating a substance at known intervals of temperature for a few hours. By far the quickest method is to study the substance by heating it continuously at a uniform rate. This forms the dynamic method and is very widely used.

Table I.1 shows some of the thermal analysis techniques used for measuring various parameters. Often a single method may not yield useful results unless other related thermal techniques are applied as adjuncts.

Enthalpy Changes

In a physical change the chemical nature of the transformed material will be same as that of the starting material. Typical examples of physical change are fusion, crystalline modification and destruction, sublimation, etc. The rhombic → rhombohedral transformation in $AgNO_3$ at 160°C and the fusion of m-dinitrobenzene at 90°C are physical changes. In a chemical change the thermal treatment may rupture the chemical bonds to form new compounds. Decomposition, oxidation, and chemisorption are all examples of chemical changes. The decomposition of $CaCO_3$ to CaO and CO_2 at 925°C and the chemisorption of CO on Ni catalyst are attended by a chemical change.

Every substance has a certain energy content. The energy content of the substance after a thermal change is different from that of the starting material; this difference may manifest itself as heat energy. The heat contents of $Mg(OH)_2$, MgO, and $H_2O(g)$ are respectively 223.4, 145.76, and 57.826 kcal/mole. The decomposition of $Mg(OH)_2$ at about 385°C is represented by the equation

TABLE I.1
Thermal Methods of Analysis [Wendlandt (51)]

Technique	*Physical parameter measured as function of temperature*	*Instrument employed*
Thermogravimetric analysis	Change in mass	Thermobalance
Differential thermogravimetric analysis	First derivative of change in mass	Thermobalance or differential thermobalance
Differential thermal analysis	Temperature difference between sample and reference material	DTA apparatus
Derivative differential thermal analysis	First derivative of temperature difference	DTA apparatus
Differential scanning calorimetry analysis	Heat change supplied to sample	Differential calorimeter
Specific heat measurement	Specific heat	Differential calorimeter
Gas evolution analysis; linear pyrolysis	Gas thermal conductivity	Thermal conductivity cell
Pyrolysis	Pyrolysis fragments	Gas chromatograph; mass spectrometer; infrared spectrophotometer; and others
Thermoluminescence; oxyluminescence	Light emission	Photomultiplier tube in thermoluminescence apparatus
Dilatometry	Change in volume	Dilatometer
Electrical conductivity	Change in electrical resistance	Resistance bridge
High-temperature X-ray diffraction	Change in d-spacing	X-ray diffractometer
Thermometric titrimetry	Temperature change vs. time or volume of titrant	Thermometric titrimeter
Thermal analysis	Temperature vs. time or heat content	Calorimeter; thermal analysis apparatus
Dynamic reflectance spectroscopy	Reflectance of sample	Spectroreflectometer

$$Mg(OH)_{2(s)} \longrightarrow MgO_{(s)} + H_2O_{(g)}$$

The difference in the heat contents of the reactant $Mg(OH)_2$ and the products MgO and H_2O is 19.814 kcal, the amount of heat that would be absorbed dur-

TABLE I.2
Physical and Chemical Changes and Their Character

	Thermal effect	
Type of reaction	*Endothermic*	*Exothermic*
Physical		
Absorption	−	+
Adsorption	−	+
Crystalline transition	+	+
Desorption	+	−
Fusion	+	−
Second-order transition	+	−
Sublimation	+	−
Vaporization	+	−
Chemical		
Chemisorption	−	+
Decomposition	+	−
Dehydration	+	−
Desolvation	+	−
Oxidation in gaseous atmosphere	−	+
Oxidative degradation	+	+
Redox reaction	+	+
Reduction in gaseous atmosphere	+	+
Solid-state reaction	+	+

[By permission, *Handbook of Analytical Chemistry*, Meites, L (Editor), McGraw Hill Book Co., New York, 1963.]

ing the decomposition of 1 mole of $Mg(OH)_2$.

Reactions attended by the absorption of heat are 'endothermic' while those that evolve heat are 'exothermic'. Table I.2 gives the physical and chemical changes that occur during thermal treatment and also the endothermal or exothermal character of each of these changes.

Development of Differential Thermal Analysis

As described earlier, when a substance is gradually heated at a predetermined rate, at a particular temperature the substance may undergo a thermal change that manifests as an endothermal or exothermal change. The temperature of the thermal change, its intensity, and endo- or exothermal nature, being characteristic of a particular substance under study, enable identification.

In 1887, in his very significant paper entitled *De l'action de la chaleur sur les argiles*, Le Chatelier (*1*) devised for the first time a heating-curve method to measure temperature as a function of time. This method was successfully employed to investigate clays and other minerals.

In earlier work, thermal analysis consisted of heating gradually a platinum

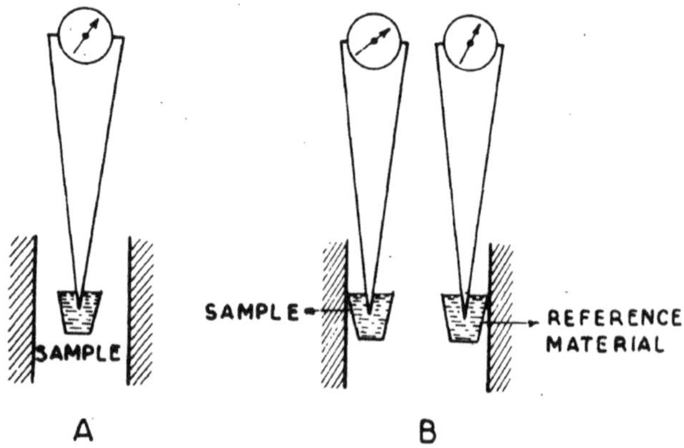

Fig. I.1: Thermocouple assembly.

crucible containing the sample with a thermocouple embedded in it (Fig. I.1). The temperature was determined by a galvanometer at definite intervals of time. A time-temperature curve thus plotted was compared with that obtained with the crucible containing no sample. A comparison showed differences that could be attributed to thermal reactions. Improvement could be effected by using two crucibles, one containing the sample and another a thermally inert material, i.e. a material showing no thermal reactions. Two thermocouples were used, one in each of the crucibles as is shown in Fig.I.1.

A typical time-temperature plot is shown in Fig. I.2. As the sample is heated uniformly, a linear curve results until point A is reached. From A onwards the time-temperature curve shows a deviation from the linearity and at point B it becomes steeper until point C is reached. At D the curve becomes steeper again and regains linearity at point F. This curve shows two transformations at A-B-C and D-E-F. The change at A-B-C is caused by the absorption of heat and that at D-E-F is caused by the evolution of heat. It can be seen that the time-temperature method cannot be sensitive because a single thermocouple that measures higher temperatures has also to measure very small decreases or increases in temperatures, and in addition the peaks are not sharp.

A sensitive method of following the thermal reactions was invented by Roberts-Austen (*2*, *3*) in 1899 and is popularly known as Differential Thermal Analysis (DTA). In this method two crucibles were used, one to house the sample and the other a reference material that does not undergo any thermal change in the temperature range of study. Of the two thermocouples

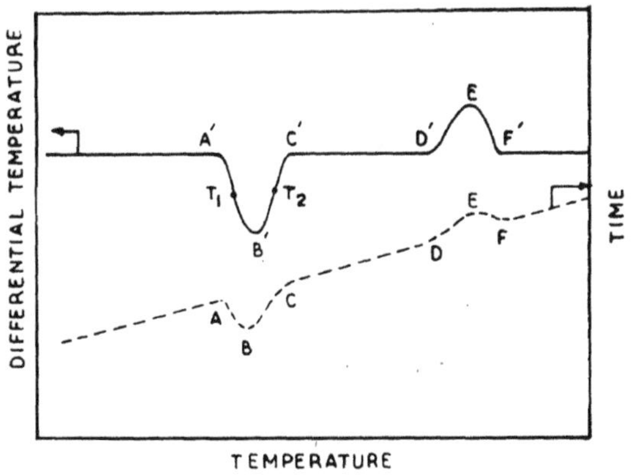

Fig. I.2: Comparison of time–temperature curve with differential curve.

used one was embedded in the sample and the other in the reference material. They were connected in such a way that the electromotive force developed by one opposed that developed by the other (Fig. I.3).

The galvanometer connected to the differential thermocouple shows no deflection if the temperature in the sample and the reference is the same. If a reaction occurs, for an interval of time the temperature in the sample decreases or increases with respect to the temperature of the reference material. The galvanometer would show deflection, the direction of which depends on whether the reaction occurs with an absorption or evolution of heat. After the reaction is over, the temperature in both the crucibles is substantially same.

The differential thermal curves are plotted with the temperature of the reference material on the X-axis and the differential temperature on the Y-axis. Thermal curves showing exothermic effects are shown as upward peaks whereas the endothermic effects are shown as downward peaks with respect to the base line. Accordingly, a thermal reaction with a hump is exothermic and that with a valley is endothermic. It is to be noted that in DTA it is not the differential temperature that is plotted but only the difference in temperature. Because of its long usage the word 'differential' continues to be used in all publications.

A typical differential thermogram containing an endothermal effect A'-B'-C' and an exothermal effect D'-E'-F' is shown in Fig. I.2. A comparison with the time-temperature curve clearly indicates the sharper nature of the DTA peaks. The reaction starts at a temperature A' and is completed at a temperature T_2 somewhere between B' and C'. In some of the earlier work it

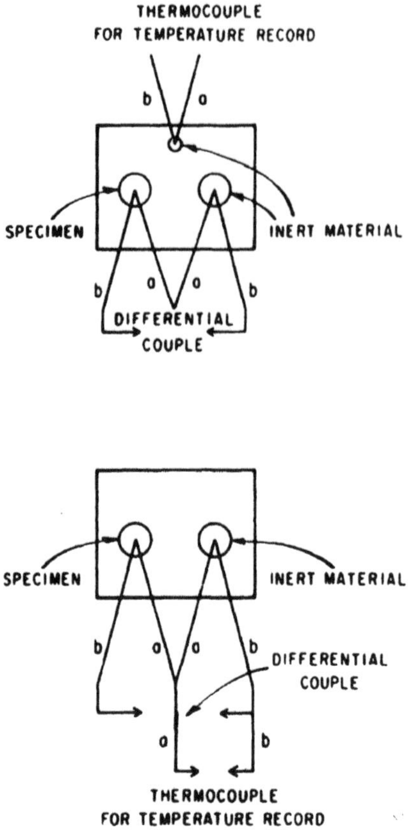

Fig. I.3: Differential thermocouple connection.

was implied that the reaction ends at a temperature C′. The peak temperature B′ denotes the temperature at which the absorption of heat is equal to the heat energy supplied, and does not refer to the temperature at which the reaction rate is maximum. It may in fact occur at a temperature T_1 prior to B′.

The DTA technique was mainly used in the field of metallurgy until 1913 when Fenner's work brought out the importance of this method in other fields (*4*). Orcel (*5*, *6*) later demonstrated the utility of this technique for investigating clays and since then DTA has become an invaluable tool in clay mineralogy.

In the past three decades many refinements have been incorporated into the DTA equipment. The technique has established itself in diverse fields as a very useful qualitative and quantitative method. Very valuable data have been obtained by using correlative techniques such as X-ray diffraction, infrared, TGA, dilatometry, microscopy, chemical analysis, gas analysis, surface area, and calorimetry.

Development of multipurpose equipment incorporating DTA with one or more of other techniques has further widened its scope. Many of these are available commercially. For details of multipurpose equipment, reference should be made to the publications shown below:

1. DTA–TGA (*7-18*)
2. DTA–Effluent Gas Analysis (EGA) (*19-22*)
3. DTA–TGA–EGA (*23*)
4. DTA–Mass Spectroscopy (*24*)
5. DTA–TGA–DTG (*25-30*)
6. DTA–Visual Observation (*31*)
7. DTA–Electrical Conductivity (*32*, *33*)
8. DTA–Heat Capacity–Thermal Conductivity–Thermal Diffusivity (*34*, *35*)
9. DTA–Resistivity (*36*)
10. DTA–Derivative Differential Thermal Analysis (*37*)
11. DTA–TGA–Radioactive Emission (*38*)
12. DTA–Dilatometer (*39-41*)
13. DTA–Modular Calorimeter (*42*)
14. DTA–TGA–Dilatometer (*43-45*)
15. DTA–X-ray (*46*).

Of the multipurpose equipment, the DTA-TGA combination seems to be the most widely applied. TGA gives the weight changes that occur in a body at different temperatures and in DTA this might reflect as endothermal and exothermal effects. It is possible that a phase transformation shows an inflexion in the DTA but may fail to show any weight change in TGA. An example is given in Fig. I.4, comparing the DTA, DTG, and TGA behaviours of $Ca(NO_3)_2 \cdot 4H_2O$. Inflexions in DTA generally correspond to those in DTG (*47*).

Equipment for Differential Thermal Analysis

Essentially the DTA assembly consists of a sample holder to place the sample, thermocouples for measuring temperatures, a furnace to heat the sample, a program controller to heat the sample at a uniform rate, and a recorder for registering the temperature difference between the sample and the reference material. A schematic diagram of the DTA assembly is shown in Fig. I.5.

Sample Holder. Sample holders varying in size and shape and made of different materials are in use. The most important prerequisite is that the

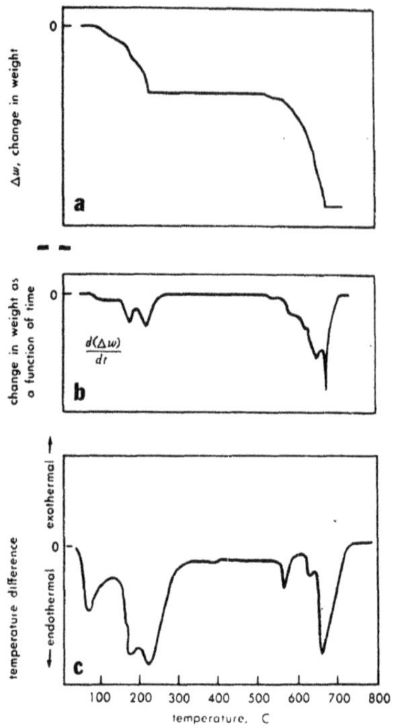

Fig. I.4: (a) TGA, (b) DTG, (c) DTA of Ca $(NO_3)_2 \cdot 4H_2O$ (*47*) [from McGraw-Hill *Encylopaedia of Science & Technology*, vol. 13, p. 557, 1960, by permission of McGraw-Hill Book Co.].

sample holder should not itself undergo any thermal transformation and should not interact with the sample. In cement research a sample holder made of alumina, quartz, or glass is to be avoided because the material may react with the cement.

Usually the sample holder contains two cavities, one for housing the sample and the other for the reference material. Sometimes a third hole is used for determining the temperature of the block. The optimum size of the block and the cavities is determined by the ratio of the volume of the block to the volume of the cavities. A large ratio promotes better stabilization of heat and acts as an insulator for heat flow between the two materials. The smaller ratio reduces the absorption of heat by the block and increases the sharpness of the peak. A multiple sample holder designed to contain

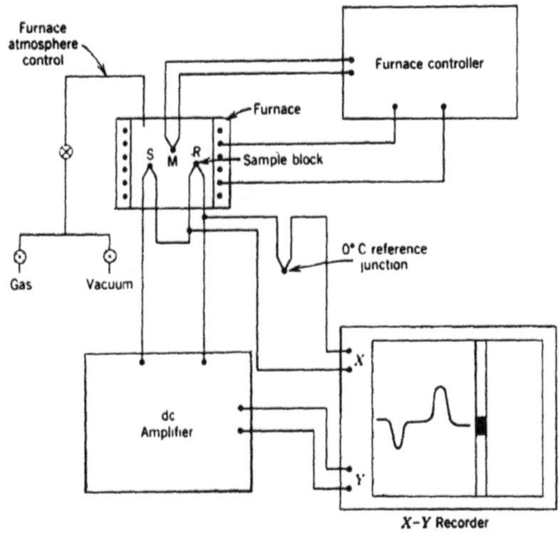

Fig. I.5: Typical differential thermal analysis equipment. (S) sample thermocouple, (R) reference thermocouple, (M) monitor thermocouple (*51*).

more cavities can be used to carry out simultaneous DTA on six samples.

Several materials have been used to make the sample holder and these include sintered alumina, fused quartz, vycor glass, porcelain, silica, zirconia, beryllia, graphite, pyrophylite, nickel, platinum block or crucible, silver, aluminum, copper, inconel, stainless steel, brass, platinum–platinum +10% rhodium, and palladium-stainless steel. It is possible to use a sample holder of platinum which may also serve as a thermocouple (*48*). Sometimes the sample holder may be dispensed with if the samples are of large size. In the DTA studies of Grim and Johns (*49*) two whole bricks were used and into each of these a thermocouple was introduced. The sample holder is also not necessary in case pressed pellets are used (*50*).

It is a common practice in many laboratories to cover the sample holder with a lid for effecting equal temperature distribution.

The various shapes of sample holders used for DTA are illustrated in the publications of Wendlandt (*51*) and Mackenzie (*52*).

Thermocouples. The temperature measuring devices include thermistors, resistance thermometers, and thermocouples. The thermocouple is by far the most widely used temperature measuring device. It measures the differential temperature as well as the temperature of the reference material or of the furnace. The choice of the thermocouple is dictated mainly by the maximum temperature required, the sensitivity and its resistance to attack by the sample or the effluents. The emf of the thermocouple should be almost linear with respect to the changing temperature. For sub-zero temperatures an iron–constantan thermocouple may be used; a copper–constantan thermocouple is used from −185 to 370°C. A chromel–alumel thermocouple could be used up to 1200°C. Thermocouples such as platinum–platinum +10 or 13% rhodium, platinum +5% rhodium–platinum +20% rhodium, iridium–60% iridium +40% rhodium, and gold +40% palladium–platinum +10% rhodium are satisfactorily used up to 1400°C. For temperatures in excess of 1400°C use of thermocouples made of tungsten, molybdenum, rhenium, tantalum, niobium, and their alloys is recommended. The tungsten –molybdenum thermocouple may be used up to 2200°C (*53*). The tungsten –graphite thermocouple has been used up to 2400°C and tungsten–tungsten +26% rhenium thermocouple, up to 2850°C. A graphite–tantalum carbide thermocouple is recommended for temperatures up to 3000°C (*54*). Goton's patent (*55*) pertains to the use of a graphite–borated graphite thermocouple for temperatures up to 3000°C. Wisely (*56*) has reviewed the thermocouples used for high temperature applications.

Sometimes a protective cover is necessary if the thermocouple is liable to attack by any chemical. Hence, shields of molybdenum, zirconium nitride, or tungsten may have to be used in such cases. Thermocouples connected in series form what is known as a 'thermopile'—this method may be used to increase the electromotive force of a thermocouple.

Table I.3 gives the characteristics of four thermocouples commonly used in DTA. (*57*).

The accuracy of the DTA interpretation is very much dependent on the precision with which the temperature can be determined. The quantitative DTA requires calibration with substances that have known latent heats of fusion. Various substances have been suggested for calibration and some of them are shown in Table I.4.

Furnace. Furnaces are either of vertical or horizontal type. They are usually tubular and heated by resistance elements. Infrared radiation and high frequency r.f. induction are other methods of heating. For low temperature applications the sample block is cooled by air or nitrogen flowing through a coil.

Various types of heating elements are used for temperatures up to 2800°C. Table I.5 gives the types of heating elements that may be used for winding DTA furnaces.

Special types of furnaces have been designed to provide a constant flow of gases such as CO_2, O_2, N_2, H_2, He, N_2, or A at atmospheric or higher pressures. High vacuum DTA equipment is also available.

Temperature Controller. Reproducibility of thermograms is possible only

TABLE I.3

Characteristics of Some Commonly Used Thermocouples in Differential Thermal Analysis

	Polarity and identification of conductors		*Approximate limits of error*		
Thermocouple	*Positive*	*Negative*	*Temp. range °C*	*Standard error*	*Recommended temperature range, °C*
Chromel–Alumel	Chromel (non-magnetic)	Alumel (magnetic)	−18 to +280 280 to 1260	±2°C ±0.75%	0—1250 (useful in oxidizing atmospheres)
Copper–Constantan	Copper (yellow)	Constantan (silvery)	−185 to −60 −100 to −60 − 60 to +95 95 to 370	— ±1°C ±0.8°C ±0.75%	−185 to 370°C (useful at very low temperatures)
Iron–Constantan	Iron (magnetic)	Constantan (non-magnetic)	−18 to +275 275 to 760	±2°C +0.75%	0—760°C (used in reduced atmosphere only)
Platinum–Platinum +10% Rhodium	Platinum+ Rhodium (slightly harder)	Platinum (slightly softer)	0 to 540 540 to 1480	+3°C ±5%	0—1650°C

[By permission, McGraw Hill Book Company, *Handbook of Analytical Chemistry*. Meites, L. (Ed.), 1963.]

TABLE I.4
Data for Temperature and Enthalpic Calibrations of Differential Thermal Analysis Systems

Temperature, °C	*Substance*	*Transformation*	*Heat of fusion* (cal/g)
32	NH_4NO_3	T	—
85	NH_4NO_3	T	—
90	m-dinitrobenzene	F	24.7
117	o-dinitrobenzene	F	32.3
122	benzoic acid	F	33.9
125	NH_4NO_3	T	—
128	KNO_3	T	—
147	AgI	T	—
160	$AgNO_3$	T	—
170	NH_4NO_3	F	—
212	$AgNO_3$	F	16.7
300	$KClO_4$	T	—
307	AgCl	F	—
314	$NaNO_3$	F	—
327	Pb	F	—
432	Ag_2SO_4	T	—
552	AgI	F	—
573	SiO_2 (quartz)	T	—
583	K_2SO_4	T	—
642	$NaMo_2O_4$	T	—
652	Ag_2SO_4	F	—
687	$NaMo_2O_4$	F	—
804	NaCl	F	—
961	Ag (from Ag_2SO_4)	F	—
1069	K_2SO_4	F	—

[By permission, McGraw Hill book Co., *Handbook of Analytical Chemistry*, Meites, L. (Ed.)., 1963.]

when the sample is heated at a linear rate. Heating rates ranging from 0.1 to 300°C/min have been used by different workers. In the manual method heating is controlled by an autotransformer manipulated at required intervals. In the automatic method the voltage input to the heating element is controlled by a variable transformer through a synchronous motor. A more accurate method consists of a feed back circuit controlled by a thermocouple inserted into the furnace tube. In low temperature DTA a liquid refrigerant is passed through a heat exchanger to cool the primary liquid circulating through the coil surrounding the sample holder.

Normally the substances are studied as they are heated uniformly from room temperature up to 1000°C or more. In many instances a steady cooling of the sample from the high temperature gives valuable results.

Recorder. Various types of recorders are in use for registering the differ-

TABLE I.5
Temperature Limits for Furnace Resistance Elements [Wendlandt (51)]

Element	*Approximate temperature, °C*
Nichrome	1000
Chromel A	1100
Molybdenum	1200
Tantalum	1330
Kanthal	1350
Platinum	1400
Globar	1500
Platinum—10% Rhodium	1500
Platinum—20% Rhodium	1500
Kanthal Super	1600
Rhodium	1800
Tungsten	2800

TABLE I.6
Commercial DTA Equipment Including Typical Apparatus Used in Various Laboratories

Firm	*Country*
Linseis Prüfgerätelbau	Germany
Stegmüller-siebnlist Apparatus	Germany
de Bruijn–van der Marel Apparatus	Holland
Metripex	Hungary
Rigaku–Denki Co.	Japan
Shimadzu Manufacturing Co.	Japan
Bolton & Co., Ltd.	U.K.
Gallenkamp & Co., Ltd.	U.K.
Mecaulay Institute Apparatus	U.K.
American Instrument Co.	U.S.A.
Apparatus Manufacturers	U.S.A.
Brinkmann Instruments Inc.	U.S.A.
Du Pont de Nemours & Co.	U.S.A.
Eberbach Corporation	U.S.A.
Harrop Precision Furnace Co.	U.S.A.
Leeds and Northrup Co.	U.S.A.
Mettler Instruments Corporation	U.S.A.
Perkin-Elmer Corporation	U.S.A.
Robert L. Stone & Co.	U.S.A.
Schuco Scientific	U.S.A.
Technical Equipment Corporation	U.S.A.
Tem-Press Research Inc.	U.S.A.
The Wheelco Instrument Co.	U.S.A.
Kaurkovskii Apparatus	U.S.S.R.
Teitelbaum-Berg Apparatus	U.S.S.R.

ential temperature. In earlier studies the differential temperature was either read off by a sensitive galvanometer or was automatically recorded on a photographic paper attached to a rotating drum. Most of the commercial recorders available today use electronic potentiometric recorders that simultaneously record the differential as well as the temperature of the reference material. Often a dc amplifier is used to amplify the differential signal before feeding to the recorder. Six point recorders that automatically analyse six samples are also available. Portable DTA apparatus has also been designed for exploration work in the field.

Commercial DTA equipment is available incorporating different types of features. Table I.6 gives a list of some of the manufacturers of DTA equipment and also some of the apparatus used in established laboratories.

Differential Thermal Curves

In the different thermal method, calcined alumina is usually used as an inert or reference material. The temperature difference between the sample and the reference material is zero if the sample does not undergo any thermal change. When a thermal change occurs there is either an absorption or evolution of heat during a temperature interval. As a consequence of this the temperature of the sample becomes lower or higher than that of the reference material. An emf is thus set up, the intensity of which depends on the nature of the change and the amount of the sample. Thermograms are obtained such that the temperature of the sample is plotted or recorded on the X-axis while the differential temperature is shown on the Y-axis.

Types of Thermograms. If both the cavities of the sample holder are filled with an inert sample such as calcined alumina, a straight line would be recorded (Fig. I.6A). If a sample in one of the cavities is replaced with another sample that does not undergo any thermal reaction, a gradient line might result (Fig. I.6B). It is possible that a sample exhibits a thermal reaction with the resultant product having similar thermal properties as the reference material. In such a case a curve of type C or D may be registered. The base line shift is common when the sample has a different thermal property from the inert material (Fig. I.6E and I.6F). Sometimes two thermal effects occur very close to each other giving a doublet as in Fig. I.6G. It is not uncommon that two reactions occur at about the same temperature, resulting in obscuring of the peak with lesser heat of reaction. The curve H refers to an exothermal effect caused by an organic impurity. The vacuum DTA of the sample (curve I) would show a small endothermal effect which had been obscured by the exothermal effect in curve H.

In practice, complex thermal curves are obtained and it needs no emphasis that the interpretation of thermograms should be done with extreme caution.

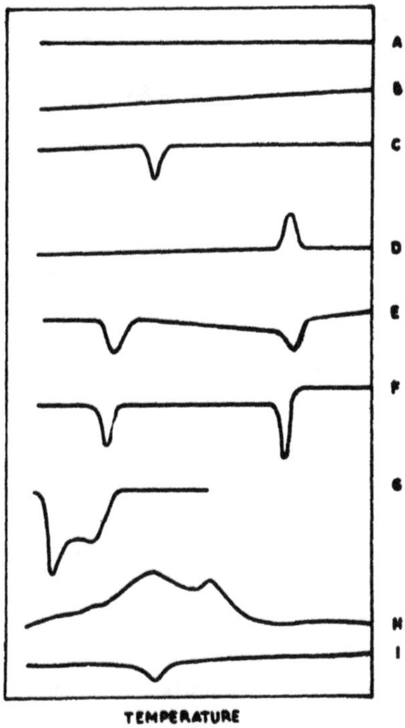

Fig. I.6: Typical thermograms.

Designations Used to Define the Thermal Change

Differential thermal curves are defined in terms of their endo- or exothermal nature and the temperatures at which they occur. In DTA literature the thermal changes are defined in terms of characteristic temperature, peak temperature, temperature range of the peak, peak width, peak amplitude, and peak area.

Characteristic Temperature. The characteristic temperature denotes the temperature at which the curve deviates from the base line to give an endothermic or exothermic peak. Often it is not easy to determine this temperature. Hence, three methods have been suggested to define and determine this temperature (*58*). In the first method the temperature can be directly determined, provided the thermal curve is sharp (Fig. I.7A, point X_1). In the second method an exatrapolated line from the linear portion of the endothermal effect cuts the base line AB at X_2, which becomes the characteristic tem-

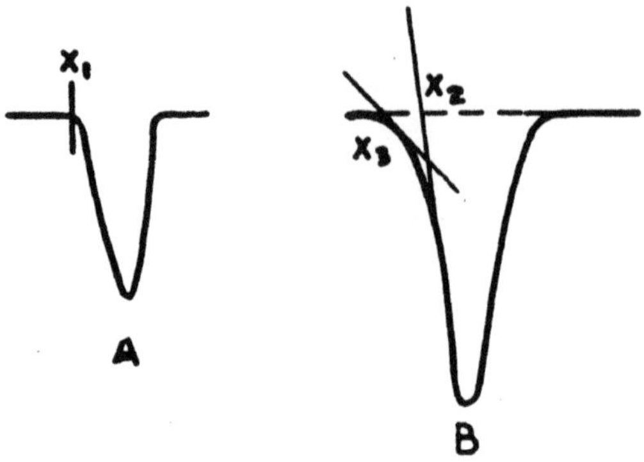

Fig. I.7: Determination of characteristic temperature.

perature. In the third method a line is drawn at an angle of 45° to the base line so as to be a tangent to the curve at the commencement of the endothermic effect. The point X_3 on the base line forms the characteristic temperature. The characteristic temperature should form a better basis to define the temperature of the reaction, but because of the difficulties associated with its determination, comparison is usually based on peak temperatures.

Peak Temperature. This is the temperature of the apex of the endothermic or exothermic effect. In the DTA studies all curves are invariably defined in terms of peak temperatures as they are more easily determinable than the characteristic temperature.

It is also usual practice to give the peak range of a reaction, and this is the range of temperature between the point at which the deviation of the curve occurs and the point at which the curve just touches the base line after the reaction.

Peak Width. This is the distance between the points at which the curve starts and ends. In Fig. I.8 the distance AB is the peak width. Peak width is not always easy to determine and certain extrapolations are made. In the method adopted by Webb (*59*) the linear parts of the curve on both sides of the peak intersect AB at C and D. Then CD is defined as the peak width. Two alternative methods have been suggested by Webb to determine the peak widths of curves that do not return to the base line after the reaction. In the first method the extrapolated lines from the two linear parts of the curve intersect the line AB drawn as a continuation of the base line (peak width =1). In the second method the extrapolated lines intersect the linejoining the points of commencement and ending of the curve (peak width =2).

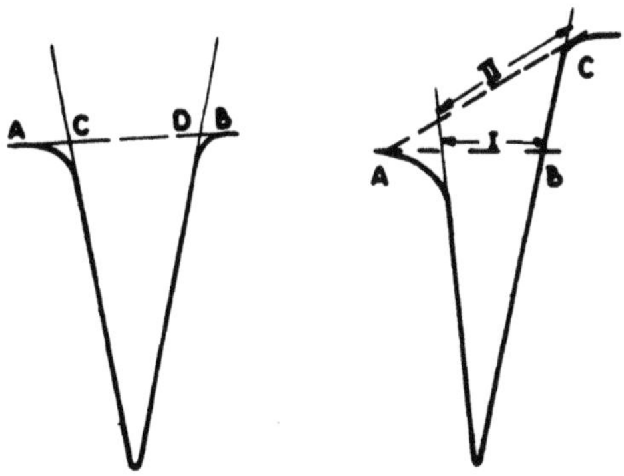

Fig. I.8: Determination of peak width.

Peak Amplitude. In a thermogram exhibiting a sharp peak the peak amplitude is directly proportional to the heat developed in the reaction. The peak amplitude is the height of the peak from the base line. If there is no base line shift the height X_1 of the vertical line CD is the peak amplitude [Fig. I.9 (I)].

Two methods have been suggested to determine the peak amplitude for thermal effects showing a base line shift. In the first method the amplitude is the vertical line from the peak to the base line AB [X_2, Fig. I.9 (II)]. In the second method the peak height X_3 is the distance between C and the line DE

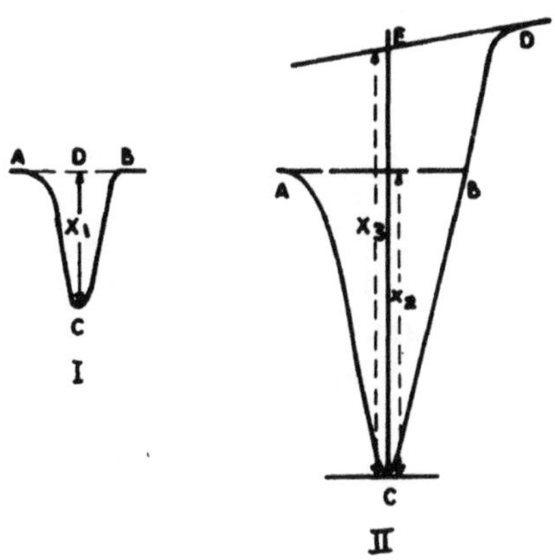

Fig. I.9: Determination of the peak amplitude.

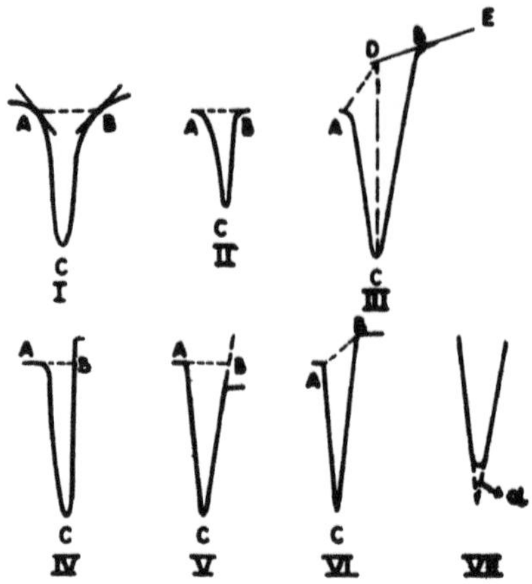

Fig. I.10: Methods of delineation of peak area.

drawn as an extrapolation of the drifted base line.

Peak Area. The area enclosed by the peak has to be accurately determined for quantitative work. Norton (*60*) was perhaps the first to attempt a quantitative DTA of high alumina clays. In his method two tangents were drawn on both sides of the peak and a straight line AB joined the points of tangency [Fig. I.10 (I)]. The area enclosed by ABC formed the peak area. In the method of Speil *et al* (*61*) the line AB was obtained by joining two points where the curve starts and ends. The delineation of the peak area for a curve showing a considerable base line shift is described by Berg (*62*). In this method the extrapolated line from BE meets the vertical line drawn from the peak at D. The point A represents the temperature at which the curve deviates from the base line. The space enclosed by ADBC is the peak area [Fig. I.10 (III)]. The other methods of delineating peak area of the curves that do not return to the base line are illustrated in Fig. I.10 (curves IV, V and VI). They are self-explanatory.

Dean (*63*) has suggested determining the cosecant of the peak angle α[Fig. I.10 (VII)].

The peak area can be obtained with a planimeter or by weighing the cut out area of the curve drawn on transparent paper.

Factors Influencing Results

Some workers have expressed doubts concerning the reliability of the DTA technique for qualitative as well as quantitative analysis. This is because not infrequently the thermogram of a substance obtained in one laboratory may not be exactly reproducible in another. Extensive studies over the past two decades have resulted in a better understanding of the various factors influencing the thermograms. These studies along with the development of refined techniques have considerably increased the reliability and applicability of DTA to diverse fields. The phenomenal growth of publications on DTA in recent years is in itself evidence of its popularity. A knowledge of the factors influencing the thermal curves is essential for a correct interpretation of the DTA results.

The variability in the nature of the thermograms can be traced to the instrument or to the sample. The instrumental factors include: (a) sample holder—type, shape, and size; (b) furnace; (c) thermocouple—diameter, bead size, and positioning; (d) heating rate; (e) sensitivity of the recording system.

Variations in the sample may be due to: (a) method of drying; (b) amount and diluent; (c) particle size and crystallinity; (d) packing density; (e) thermal conductivity and heat capacity; (f) swelling or shrinkage; (g) inert standard.

Sample Holder. Metallic and ceramic holders are in common use. The metallic holder, with a higher heat conductivity, gives raise to less intense peaks than the ceramic holder. Metallic holders decrease the gradient between the sample and the furnace or the block, thus minimizing the base line shift (*64*). Spurious peaks are induced by the block and this can be eliminated by proper grounding. In reactions involving gas evolution—such as the decomposition of $Ca(OH)_2$ or $CaCO_3$—the linear part of the downsweep may be shifted to a higher temperature if a porous alumina holder is used in place of a nickel holder. This is due to easier escape of the gaseous decomposition products through the pores of the alumina holder (*65*, *66*). It is important that the sample holder should not react with the sample. Boron may react with sugar if borosilicate glass is used as a holder in the DTA of sugars (*67*). Small sample holders, though useful in eliminating thermal gradients, give small peaks and hence are not recommended. The sample holder should be placed symmetrically in the sample block and the sample block itself should be positioned symmetrically in the furnace, otherwise drift in the base line is liable to occur.

Lids are used to cover both the sample and the inert material to obtain a uniform temperature distribution. Very close fitting of the lid does not provide free access to the effluent gases and hence should be avoided. A close fitting may be advantageous to minimize oxidation effects. The size and shape of the specimen holders influence the curve but have a negligible effect if the holder is relatively larger than the sample well.

Furnace. The peaks are influenced to a great extent if the nature and pressure of the furnace atmosphere are not kept uniform for a particular set of experiments. In vacuum or at normal atomspheric conditions, thermal curves of phase transformations are not affected. However, in reactions attended by gas evolution, vacuum heating shifts the peaks to lower temperatures. Oxidation reactions are eliminated if the DTA is carried out in vacuum or in an atmosphere of nitrogen, argon or helium. At higher pressures certain gases retard the decomposition reactions. In presence of CO_2 under high pressure the peak due to $CaCO_3$ decomposition is shifted to a higher temperature. Dehydroxylation peak of kaolinite can also be raised to a higher temperature at high water vapor pressures.

Wire wound furnaces may exhibit spurious thermal effects due to stray currents and these may be eliminated by using rod elements. The furnace of smaller diameter has a lower temperature gradient and should be preferred to one of larger diameter.

Thermocouples. Conduction through the thermocouple wire is an important means of transfering the heat of reaction. Cole and Rowland (*68*) in their studies on vermiculites observed an abnormal endothermic peak when the differential thermocouple was arranged radially or when the amount of sample was less than a particular amount. This spurious effect could be eliminated by covering the sample with a lid or by arranging the thermocouple along the axis of the sample holder.

As far as practicable, the thermocouple bead embedded in the sample as well as in the inert material should be of equal size to prevent the base line shift. A larger bead gives a more pronounced thermal peak (*69*) but it possesses a larger thermal capacity.

The diameter of the thermocouple wire is also an important factor affecting results. Thicker wires have better thermal conductivity but lower electrical resistance. As these are two opposing factors, compromise has to be made in choosing the optimum diameter of the wire. The heat leakage through the wires may be large enough to reduce the peak area by more than 50% of the theoretical value (*70*). Faster deterioration is the factor that prevents the use of too thin a wire.

Nonplacement of the thermocouple bead at the centre of the sample is often the cause of base line drift. The thermocouple should be symmetrically located with respect to the walls of the cavity, as well as to the top and bottom of the sample mass.

It is not uncommon for a thermocouple to be attacked by the sample. Under such conditions not only unusual effects appear but also the thermocouples undergo quick deterioration. In the DTA of BaO and MnO_2 the thermocouples need replacement because of chemical attack (*71*, *72*). A protective shield or a coating of the thermocouple with Al_2O_3 may prevent the chemical attack on the thermocouple.

Rate of Heating. Generally, a slower rate of heating results in the broadening of the peaks and lowering of the peak temperature. The area of the peak is, however, not affected, but the work of Barrall and Rogers (*73*) seem to indicate that the peak area in a mixture of $AgNO_3$ and carborundum increases as the rate of heating is increased. Higher rates of heating may coalesce two peaks occurring adjacent to each other. Hence a rate of heating of 10 ± 1°C/min is recommended for clay mineral studies.

DTA, being a dynamic method, requires a constant rate of heating for reproducible results. Much work has been done on the effect of rate of heating on the variation of the endothermal and exothermal effects in kaolinite. Langer and Kerr (*74*) found that the endothermal peak of dehydration and the exothermal peak of phase change increase as the rate of heating is increased from 10°C/min to 50°C/min. A plot of log heating rate bears a linear relationship to the peak temperature for both the endothermic dehydration and the exothermic phase change peaks. In Table I.7 are given the variation of peak temperature at different rates of heating and the equation of the lines obtained in the plot of log heating rate against peak temperature.

Sensitivity. Unless the sensitivity of the recording system is high, reactions with small enthalpic changes may be unregistered. General shapes of curves may also change, depending on the sensitivity. A mixture may exhibit several peaks, greatly differing in intensity from one another. In such cases, if a low sensitivity is chosen to ensure registration of all the large peaks on the recording paper, there is a possibility of missing the smaller peaks.

Drying of the Sample. Uniform drying of the sample is advisable for obtaining reproducible thermal effects. Too severe a drying may eliminate the endothermic peak caused by the desorption of water at low temperatures. Air dried samples of clays give varying results because of the fluctuations in the humidity conditions. Hence epuilibration for at least four days in vacuo over a saturated solution of $Mg(NO_3)_2 \cdot 6H_2O$ is suggested for clay studies (*52*).

Amount of Sample. Ideally, the amount of sample should be such that it is in the form of a small sphere around the thermocouple bead. But this is not practicable, and amounts in the range 0.2-0.5g are usually used. Excess sample size results in broader peaks, larger peak areas, shift in peak temperatures, and overlapping effects. If the sample is filled in the cavity beyond a particular limit of height it has no influence on the thermal peak because it is situated far away from the influence zone of the thermocouple (*75*).

A sample showing a large base line shift due to shrinkage or spurting requires an inert material as a diluent. Dilution also reduces the possibility of the reaction products sticking to the thermocouple or to the specimen holder.

TABLE I.7
Effect of Heating Rate on the Peak Temperature in Kaolinite [Langer and Kerr (74)]

		Temperature of dehydration (Td) as related to experimental variables, °C/min (β)					
		10	*20*	*30*	*40*	*50*	*Heating rate and Td*
Curve set	(1)	535	555	568	574	589	$Td = 460.8 \pm 73.0 \log_{10}\beta$
	(2)	549	564	572	584	596	$Td = 448.5 + 87.9 \log_{10}\beta$
	(3)	561	584	596	601	608	$Td = 496.3 \pm 66.2 \log_{10}\beta$
	(4)	549	566	572	589	596	$Td = 480.5 + 66.3 \log_{10}\beta$
Average		549±6	570±6	580±7	587±7	597±4	
		Temperature of phase change (Tp) as related to experimental variables, °C/min (β)					
		10	*20*	*30*	*40*	*50*	*Heating rate and Tp*
Curve set	(1)	1013	1023	1035	1040	1045	$Tp = 973.6 + 40.7 \log_{10}\beta$
	(2)	1018	1031	1036	1040	1049	$Tp = 985.5 + 34.7 \log_{10}\beta$
	(3)	1013	1023	1033	1040	1035	$Tp = 975.5 + 37.6 \log_{10}\beta$
	(4)	1015	1028	1033	1047	1049	$Tp = 963.0 + 50.4 \log_{10}\beta$
Average		1015±2	1025±2	1034±1	1041±3	1044±3	

The presence of impurities in small amounts may sometimes influence the curve to a substantial extent. A typical example is the presence of minute amounts of NaCl in dolomite which changes the ratios of areas in the thermograms over wide limits.

Particle Size and Crystallinity. The finer the particle size of the sample, the lower the temperature of the reaction due to larger surface area. The particle size has a considerable influence on the peak temperature and peak area, though in some materials it may have little effect. In the study of pyrites of particle size varying in the range 200-2μ, the peak temperature decreased by about 100°C (*76*).

Extensive studies have been made on the effect of particle size on the thermal peaks but the results are conflicting. Increasing fineness may sometimes show unusual effects such as development of new peaks and disappearance of the normal peaks owing to the destruction of the lattice itself. The particle size should be as uniform as possible so that a uniform density of packing is achieved. For most studies, grinding of the sample to pass through 100 mesh sieve may be recommended.

The differences in thermal curves attributed in many instances to the particle size may in fact be due to the lower degree of crystallinity. Grinding for prolonged periods may destroy the crystallinity and decompose $CaCO_3$ (*77*). Grinding of dolomite for 30 min in a vibratory mill reduces the particle size to 0.06-0.09μ and almost completely eliminates the $MgCO_3$ decomposition peak (*78*). The effect of the degree of crystallinity can be better understood by crushing $CaCO_3$ to any required particle size without destroying the crystallinity. A comparison of such a sample with that crushed normally shows that the former does not have any effect on the peak (*79*).

Packing Density. Nonuniform packing causes modification of the curve and change in the intensity of peaks and in the base line. As it is not easy to control the degree of packing, variability in thermal curves may often be traced to this factor.

A loose packing containing air pockets tends to impede heat conduction and decreases the peak intensity. Too much compaction on the other hand will retain the evolved gases from the sample, thus hindering further decomposition (*80, 81*). A uniform packing to $\pm 1\%$ reproducibility can be achieved by careful tamping by hand (*82*). Specially designed tools have been suggested for obtaining a uniform packing density (*83*).

Thermal Conductivity and Heat Capacity. The heat capacity, thermal conductivity, and packing density are interdependent variables. The coefficient of thermal diffusivity of the sample is almost always different from the inert material, and this causes a drift in the base line. Even if the sample has the same thermal characteristics as the inert material it is apt to change after a transformation. Hence it is practically difficult to eliminate this factor. One method to partly eliminate this is to use a sample containing a large amount of diluent. Too large an amount of the diluent would decrease the thermal effects. If a diluent has a high heat capacity lower peak areas will result (*84*).

Shrinkage and Swelling. Many samples exhibit shrinkage or swelling during heating or following a transformation. These change the thermal characteristics of the sample and cause base line drift. Spurious peaks have been observed in samples which recede from the thermocouple after an excessive shrinkage (*85*). An air gap is developed between the sample mass and the walls of the cavities during shrinkage. The air gap acts as a thermal insulator. Some samples may partly fuse and evolve a gas rapidly. This will result in an expansion and spurting and the thermocouple in the sample well may get exposed, causing considerable base line shift. This is often averted by using a diluent.

Inert Standard. The inert or the reference material should not undergo any thermal reaction in the temperature range of study. It is impossible to find a reference material that will have exactly the same thermal characteristics as the reactive sample over the entire temperature range. As far as as possible the reference material should have thermal characteristics not much different from those of the sample.

Calcined alumina is widely used as a reference material. Calcined alumina does not prevent the base line shift, though it could be minimized by grinding it to the same particle size as that of the sample or by using it as a diluent. If it could be adjusted such that the mass × specific heat of the sample is equal to the mass × specific heat of the inert material, base line shift could be eliminated to a great extent. It is possible that the so-called inert material may react with the sample. Calcined alumina has been found to be an unsuitable diluent with benzoic acid (*86*). Special circumstances require the use of inert standards such as copper oxide, quartz, calcined clay, magnesium oxide, sodium chloride, and potassium chloride.

The effects of some of the factors described above on the base line shift, peak area, amplitude, etc. have been treated theoretically by Arens (*87*), Eriksson (*88*, *89*), Murray and White (*90*), Kronig and Snoodjik (*91*), Smyth (*92*), Soule (*93*), Sewell (*94*, *95*), and Tsang (*96*).

REFERENCES

1. Le Chatelier, H., *Bull. Soc. Franc. Mineral., 10*, 204, 1887.
2. Roberts-Austen, W.C., *Proc. Inst. Mech. Engrs.* (*London*), 1899.
3. Roberts-Austen, W.C., *Metallographist, 2*, 186, 1899.

4. Fenner, C.N., *Am. J. Sci., 4th Series, 36*, 331, 1913.
5. Orcel, J., *Congr. Intern. Mines, Met. Geol. Appl.* 7 Session, Paris, *1*, 359, 1935.
6. Orcel, J., and Caillere, S., *Compt. Rend., 197*, 774, 1933.
7. Krüger, J.E., and Bryden, J.G., *J. Sci. Instr., 40*, 178, 1963.
8. Powell, D.A., *J. Sci. Instr., 34*, 225, 1957.
9. Reisman, A., *Anal. Chem., 32*, 1567, 1960.
10. Notz, K.J., and Jaffe, H.H. (quoted by Murphy, C.B., *Anal. Chem., 32*, 168, 1960).
11. Splitek, R., *Hutnicke listy, 13*, 697, 1958.
12. Papailhau, J., *Bull. Soc. Franc. Mineral Crist., 82*, 367, 1959.
13. Kassner, B., and Hartwig, K., *Tonind.-Z. Keram. Rundschau, 84*, 471, 1960.
14. Kissinger, H.E., and Newman, S.B., *High Polymers, Analytical Chemistry of Polymers*, Interscience, New York, 1962.
15. Piece, R., *Schwiez Mineral Petrog. Mitt., 41*, 303, 1961.
16. Wallace, R.W., and Ruh, E., *Bull. Am. Ceram. Soc., 44*, 156, 1965.
17. Charsley, E.L., and Redfern, J.P., *Intern. Symp. Thermal Anal.*, London, Paper B-3, 1965.
18. Khristianov, A.S., and Korovyatnikov, G.F., *Zavadsk. Lab., 30*, 495, 1964.
19. Teitelbaum, B.Y., and Berg, L.G., *Zh. Anal. Khim., 8*, 152, 1953.
20. Murphy, C.B., *Anal. Chem., 30*, 867, 1958.
21. Ayres, W.M., and Bens, E.M., *Anal. Chem., 33*, 568, 1961.
22. Wendlandt, W.W., *Anal. Chim. Acta., 27*, 309, 1962.
23. Blazek, A., *Hutnicke listy, 12*, 1096, 1957.
24. Langer, H.G., and Gohlke, R.S., *Anal. Chem., 35*, 1301, 1963.
25. Paulik, F., Erdey, L., and Gal, S.Z., *Anal. Chem., 163*, 321, 1958.
26. Paulik, F., Paulik, J., and Erdey, L., *Z. Anal. Chem., 160*, 241, 1958.
27. Paulik, F., Paulik, J., and Erdey, L., *U.S. Pat.* 3,045,472, July 24, 1962.
28. Erdey, L., Paulik, F., and Paulik, J., *Acta Chim. Acad. Sci., Hungary, 10*, 61, 1956.
29. *Idem, Nature, 174*, 885, 1954.
30. Sato, H., Otsuka, R., Iwata, S., and Tsuchimoto, K., *Waseda Univ. Bull. Sci. Eng. Res. Lab. No. 27*, 26, 1964.
31. Hogan, V.D., and Gordon, S., *Anal. Chem., 32*, 573, 1960.
32. Berg, L.G., and Burmistrova, N.P., *Z. Neorgn. Khim., 5*, 676, 1960.
33. Budnikov, P.P., Gorshkov, V.S., and Titoskaya, V.T., *Stroitel Materialy, 6*, 30, 1960.
34. Yagfarov, M.S., *Dokl. Akad. Nauk SSSR, 127*, 615, 1959.
35. Yagfarov, M.S., *Russ. J. Inorg. Chem., 6*, 1236, 1961.
36. Chiu, J., *Du Pont Thermogram, 2*, June 1965.
37. Freeman, E.S., and Edelman, D., *Anal. Chem., 31*, 624, 1959.
38. Bussiere, P., Claudel, T., Renouf, J., Trambouze, Y., and Prettre, M., *J. Chim. Phys., 58*, 668, 1961.
39. Barrall, E.M., Porter, R.S., and Johnson, J.F., *Anal. Chem., 36*, 2316, 1964.
40. Lehmann, H., and Gatzke, H., *Tonind.-Z. Keram. Rundschau, 80*, 7, 1956.
41. Pearce, J.H., and Mardon, P.G., *J. Sci. Instr., 36*, 457, 1959.
42. Anon, *Du Pont Thermogram, 2* (3), June 1965.
43. Koehler, E., *Bull. Soc. Franc. Ceram., 38*, 3, 1958.
44. Keler, E.K., and Kuznetsov, A.K., *Dokl. Akad. Nauk SSSR, 88*, 1031, 1953.
45. Avetikov, V.G., and Zinko, E.L., *Steklo i keram., 15*, 29, 1958.

46. Wendlandt, W.W., *Proc. 1st Toronto Symp. Thermal Analysis, Toronto*, p. 101, 1965.
47. *McGraw-Hill Encyclopaedia of Science and Technology, Vol. 13*, p. 557, 1960.
48. Herold, P.G., and Planje, T.J., *J. Am. Ceram. Soc.*, *31*, 20, 1948.
49. Grim, R.E., and Johns, W.D., *J. Am. Ceram. Soc.*, *34*, 71, 1951.
50. Walton, J.D., *J. Am. Ceram. Soc.*, *38*, 438, 1955.
51. Wendlandt, W.W., *Thermal Methods of Analysis*, Interscience Publishers, New York, 1964.
52. Mackenzie, R.C. (Ed.), *Differential Thermal Analysis of Clays.* Mineralogical Society, London, 1957.
53. Budnikov, P.P., and Tresvyatskii, S.G., *Ogneuopory*, *20*, 166, 1955.
54. Brewer, L., and Zavitsanos, P., *J. Phys. Chem. Solids*, *2*, 284, 1957.
55. Goton, R.P., *U.S. Pat. 3,084,534*, April 9, 1963.
56. Wisely, H.R., *Ceramic Age*, *66*, 15, 1955.
57. Meites, L. (Editor), *Handbook of Analytical Chemistry*, *McGraw-Hill Book Co., Inc., New York*, 1963.
58. Murray, J.A., Fischer, H.C., and Shade, R.W., *Proc. Nat. Lime Assoc.*, *48*, 73, 1950.
59. Webb, T.L., *D.Sc. Thesis.* Univ. of Pretoria, South Africa, 1958.
60. Norton, F.H., *J. Am. Ceram. Soc.*, *23*, 281, 1940.
61. Speil, S., Berkelheimer, L.S., Pask, J.A., and Davis, B., *U.S. Bur. Mines, Tech. Paper 664*, 1945.
62. Berg, L.G., *C.R. Acad. Sci.*, *URSS*, *49*, 648, 1945.
63. Dean, L.A., *Soil Sci.*, *63*, 95, 1947.
64. Carthew, A.R., *Aust. J. Inst. Tech.*, *9*, 23, 1953.
65. Webb, T.L., *Nature*, *174*, 686, 1954.
66. Mackenzie, R.C., *Nature*, *174*, 177, 1954.
67. Segal, L., and Stanonis, D.J., *Anal. Chem.*, *35*, 1750, 1963.
68. Cole, W.F., and Rowland, N.M., *Am. Mineralogist*, *46*, 304, 1961.
69. Hauser, R.E., *Thesis*, New York College of Ceramics, Alfred University, 1953.
70. Boersma, S.L., *J. Am. Ceram. Soc.*, *38*, 281, 1955.
71. Levin, E., and McMurdie, H.F., *J. Am. Ceram. Soc.*, *32*, 99, 1949.
72. Krishna Prasad, N.S., and Patel, C.C., *J. Indian Inst. Sci.*, *36A*, 23, 1954.
73. Barrall, E.M., and Rogers, L.B., *Anal. Chem.*, *34*, 1101, 1962.
74. Langer, M., and Kerr, P.F., *Du Pont Thermogram*, *3*, 1, 1966.
75. De Josselin De Jong, G., *J. Am. Ceram. Soc.*, *40*, 42, 1957.
76. Polizzotti, C., and Grillone, M.D., *Ann. Chim.* (*Rome*), *53*, 1476, 1963.
77. Tanaka, T., *Minerals Process*, *5*, 31, 1964.
78. Barta, R., and Bruthans, Z., *Proc. 6th Conf. Silicate Ind.*, *Budapest*, 1963.
79. Bayliss, P., *Nature*, *201*, 1019, 1964.
80. Foldvari-Vogl, M., and Koblenz, V., *Foldt. Kozl.*, *88*, 453, 1960.
81. Foldvari-Vogl, M., and Kliburszky, B., *Acta Geol. Acad. Sci.*, *Hungary*, *5*, 187, 1958.
82. McConnel, D., and Earley, J.W., *J. Am. Ceram. Soc.*, *34*, 183, 1951.
83. Breger, I.A., and Whitehead, W.L., *Fuel*, *30*, 247, 1950.
84. Barrall, E.M., and Rogers, L.B., *Anal. Chem.*, *34*, 1106, 1962.
85. Markowitz, M.M., and Boryta, D.A., *J. Phys. Chem.*, *64*, 1711, 1960.
86. Morita, H., and Rice, H., *Anal. Chem.*, *27*, 336, 1955.

87. Arens, P.L., *A Study of the DTA of Clays and Clay Minerals*, Gravehage, Wageningen, Netherlands, 1951.
88. Eriksson, E., *Lantbr Högsk Ann.*, *19*, 127, 1953.
89. Eriksson, E., *ibid*, *20*, 117, 1953.
90. Murray, P., and White, J., *Trans. Brit. Ceram. Soc.*, *48*, 187, 1949.
91. Kronig, R., and Snoodjik, F., *App. Sci. Res., Hague*, *A3*, 27, 1951.
92. Smyth, H.T., *J. Am. Ceram. Soc.*, *34*, 221, 1951.
93. Soule, J.L., *J. Phys. Radium*, *13*, 516, 1952.
94. Sewell, E.C., *Clay Minerals Bull.*, *2*, 233, 1955.
95. Sewell, E.C., *Res. Note, Building Res. Note, U.K.*, 1955, 1956.
96. Tsang, N.F., *Handbook of Differential Thermal Analysis*, Smothers and Yao Chiang (Ed.), Chemical Publishing Co., New York, 1966.

II. Theory and Typical Applications

II.1. THEORETICAL CONSIDERATIONS

In the theoretical interpretation of the differential thermal curve, major attention has been directed to finding a relationship between the peak area and other parameters. Speil *et al* (*1*) were the first to present a simplified theory showing that the peak area is proportional to the mass of the reacting material. This was considerably modified by Kerr and Kulp (*2*) but several misprints in this paper make the treatment somewhat difficult to follow. Tsang (*3*) derived a relationship by taking into consideration some factors not considered before. Tsang's treatment is discussed below with the help of an endothermal effect in Fig. II.1.1.

The endothermic effect representing a cooling effect begins at *a* and continues to *b*, at which point the rate of heat absorption by the transformation is equal to the rate of heat transmission into the sample. After this temperature the rate of heat absorption by the sample continues to decrease more rapidly than the flow of heat into the block. The reaction stops at some point

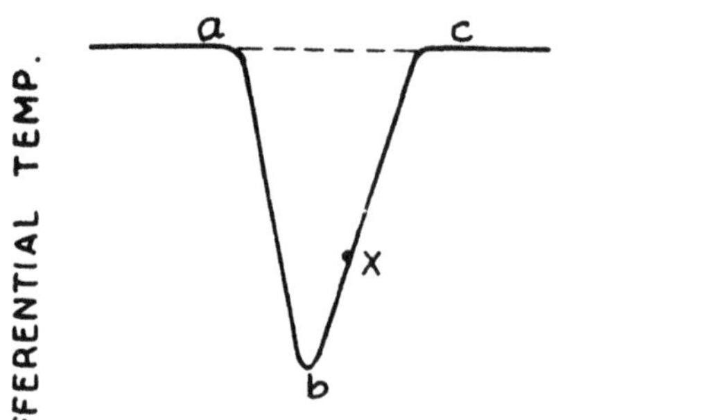

Fig. II.1.1: An endothermal curve.

X between b and c.

If the specimen is considered isolated, an endothermic reaction would cause a lowering in temperature $\Delta\theta$ given by

$$-\Delta\theta = \frac{m\Delta H}{m_s C_s} \tag{1}$$

where m=mass of reactive component in sample (g)
ΔH=heat of reaction (cal/g)
m_s=mass of sample (g)
C_s=mean specific heat of sample.

For any point between a and c such as x, the changes in heat content for the sample can be expressed as

$$\underbrace{m\int_a^x \frac{dH}{dt}dt}_{A} + \underbrace{gk_s\int_a^x (T_w - T_s)dt}_{B} = \underbrace{m_s C_s(T_s - T_0)}_{C} \tag{2}$$

and for the inert material

$$\underbrace{gk_r\int_a^x (T_w - T_r)dt}_{B'} = \underbrace{m_r C_r(T_r - T_0)}_{C'} \tag{3}$$

where t=time
m=mass of reactive component in sample
m_s=total mass of sample
m_r=total mass of inert material
dH=the decrement of heat absorbed from the sample by the endothermic reaction in time interval dt
g=geometric shape constant (same for both sample and inert experimentally)
k_s=thermal conductivity of sample
k_r=thermal conductivity of inert material
T_w=temperature of nickel block at t=X
T_s=temperature at center of sample at $t=x$
T_r=temperature at center of inert material at $t=x$
T_0=temperature at center of sample or inert material at $t=a$
C_s=mean specific heat of sample between $t=a$ and $t=x$
C_r=mean specific heat of inert material between $t=a$ and $t=x$.

These equations neglect the mode of variation of T_r, thus imposing that $C_r - C_s$ is negligible relative to C_s.

Factor A defines the quantity of heat added or subtracted from the sample due to the reaction. Factor B in equation (2) defines the heat absorbed by the sample. $A+B=C$ because at any time x along the DTA curve the amount of heat used in raising the temperature of the sample must be equal to the amount brought in by flow from the sample block plus the amount added or subtracted by the reaction.

In the inert material, factor A does not exist, so the heat of factor B' in equation (*3*) must be equal to the heat used in raising the temperature of the inert material C':

$$C_r = C_s + \delta C$$

and

$$k_r = k_s + \delta k$$

Experimentally, m_r is made essentially equal to m_s within the error of measurement. Therefore, subtracting equation (*3*) from equation (*2*) and rearranging gives

$$m\int_a^x \frac{dH}{dt}dt - gk_s\int_a^x (T_r - T_s)dt - g\delta k\int_a^x (T_w - T_r)dt$$
$$= m_s[C_s(T_s - T_r) - \delta C(T_r - T_0)] \qquad (4)$$

Let $T_r - T_s = y$, the temperature indicated by the differential thermocouple. Equation (*4*) can be considerably simplified by assuming that the terms containing $(T_w - T_r)$ and either δc or δk are small in comparison with other terms. By using a and c as integration limits:

$$m\int_a^c \frac{dH}{dt}dt - gk_s\int_a^c y\ dt = m_sC_s(T_s - T_r) \qquad (5)$$

at C, but in most cases

$$(T_s - T_r) = 0 \qquad (6)$$

at C.

Therefore, to a close approximation, the total heat of reaction is given by

$$m\int_a^c \frac{dH}{dt}dt = m\Delta H$$

and

$$\Delta H = \frac{gk_s}{m}\int_a^c y\ dt \qquad (7)$$

This expression shows that the area enclosed by the curve *abc* is proportional to the mass of the reacting material.

The geometric shape constant for a cylindrical specimen holder is given as $4\pi l$ and for the spherical shape it is equal to $8\pi l$, where l=length of specimen holder (*4*).

Boersma (*5*) considered a nickel block of high thermal conductivity and derived the equation

$$\int_{t_2}^{t_1} \theta_0 dt = \frac{qa^2}{4\lambda}$$

where t_1 and t_2 refer to the beginning and end of the peak, q is the heat of reaction per unit volume, θ_0 is the differential temperature, a is the radius of the cavity housing the sample, and λ is the thermal conductivity of the sample.

In the equation derived by Yagfarov and Berg (*6*), the geometric shape constant is given by

$$\frac{2\pi l}{ln(d_2/d_1)}$$

where l=length of the specimen
d_2=outer diameter of the specimen
d_1=internal diameter of the specimen.

The total heat of the reaction is given by

$$Q = \frac{2\pi l\lambda}{ln(d_2/d_1)} \int_{t_1}^{t_2} \Delta t dt$$

where λ represents the thermal conductivity of the specimen. This equation is very similar to the equation of Kerr and Kulp.

Speil's theory was extended by Vold (*7*) and the final equation takes the form

$$\frac{dy}{dt} + A(y - y_s) = \frac{\Delta H}{C_s} \cdot \frac{df}{dt}$$

where C_s=heat capacity of specimen
f=fraction of specimen active material transformed
$y=T_r-T_s$
$$y_s = \frac{(C_s-C_r)dT_r/dt+(T_w-T_s)\delta k+(T_0-T_s)\delta\alpha}{k_r+\alpha_r+2\sigma}$$
C_r=heat capacity of inert material
k_s=furnace wall to specimen heat conductivity
k_r=furnace wall to reference material heat conductivity
α_s=specimen to outside conductivity
α_r=reference material to outside conductivity
t=time
T_0, T_r, T_s, T_w=temperature of outside, reference material, specimen and

furnace wall

$$\delta k = k_r - k_s$$

$$\delta a = a_r - a_s$$

$$A = \frac{k_r + a_r + 2\sigma}{C_s}$$

In Speil's theory σ, a_r, a_s are all zero and Vold's equation reduces to

$$\frac{m\Delta H}{gk_s}\int_a^c y\ dt$$

In addition to the above, more rigorous theories have been proposed by Eriksson (*8*) and Sewell (*9*). Sewell's theory derives the peak area as follows:

$$\text{Peak Area} = \int_{t_0}^{t_2}\{(\Delta T_0) - (\Delta T)\}\,dt = W(P_1) - W(P_2)$$

The two functions $W(P_1)$ and $W(P_2)$ take into account heating rate and time. The peak area in this equation is shown as a function of the volume of the reacting material.

Arens (*25*) attempted to derive a general equation containing many parameters but its accuracy has been questioned by Mackenzie and Farmer (*10*) and Tsang (*3*).

Wittels (*11*) presented an empirical formula in which

$$Q = \frac{A}{\tan\left[\dfrac{lnR(R - c)}{m}\right]}$$

where A=area (in.^{2})

R=rate of heating (°C/min)

Q=calories

and m and c are slope and intercept of the straight line obtained by plotting $\tan^{-1}(A/Q)/ln\ R$ against R. The equation holds good for heating rates 10-30°C/min.

The theory of DTA has also been discussed by Smyth (*12*), Soule (*13*), Nagasawa (*14*), Allison (*15*), Murray and White (*16*), Kissinger (*17*), Deeg (*18*), Ziegler (*19*), Borchardt and Daniels (*20*), De Jong (*21*), Sewell (*22*), Eyraud and Goton (*23*), Foldvari-Vogl and Klibursky (*24*), and Tateno (*26*).

REFERENCES

1. Speil, S., Berkelheimer, L.H., Pask, J.A., and Davis, B., *U.S. Bur. Mines. Tech. Papers*, *664*, 1945.
2. Kerr, P.F., and Kulp, J.L., *Am. Mineralogist*, *33*, 387, 1948.

3. Tsang, N.F., *Handbook of Differential Thermal Analysis*, Smothers, W.J. and Yao Chiang (Ed), Chemical Publishing Co. Inc., New York, pp. 90-123, 1966.
4. Kronig, R., and Snoodijk, F., *Appl. Sci. Res., A3*, 27, 1953.
5. Boersma, S.L., *J. Am. Ceram. Soc., 38*, 281, 1955.
6. Yagfarov, M.S., and Berg, L.G., *Izv. Kazan. Fil. Akad. Nauk SSSR, Ser. Khim. No. 3*, 31, 1957.
7. Vold, M.J., *Anal. Chem., 21*, 683, 1949.
8. Eriksson, E., *LantbrHogsk. Ann., 19*, 127, 1953; *20*, 117, 1953; *21*, 189, 1954.
9. Sewell, E.C., and Honeyborne, D.B., *The Differential Thermal Investigation of Clays*, R.C. Mackenzie (Ed), p. 65, Mineralogical Society, London, 1957.
10. Mackenzie, R.C., and Farmer, V.C., *Clay Minerals Bull., 1*, 262, 1952.
11. Wittels, M., *Am. Mineral, 36*, 760, 1951.
12. Smyth, H.T., *J. Am. Ceram. Soc., 34*, 221, 1951.
13. Soule, J.L., *J. Phys. Radium, 13*, 516, 1952.
14. Nagasawa, K., *J. Earth Sci., Nagoya Univ., 1*, 156, 1953.
15. Allison, E.B., *Clay Minerals Bull., 2*, 242, 1955.
16. Murray, P., and White, J., *Clay Minerals Bull., 2*, 255, 1955.
17. Kissinger, H.E., *J. Res. Nat. Bur. Std., 57*, 217, 1956; *Anal. Chem., 29*, 1702, 1957.
18. Deeg, E., *Ber. Dtsch. Keram. Ges., 33*, 321, 1956.
19. Ziegler, G., *ibid, 33*, 363, 1956.
20. Borchardt, H.J., and Daniels, F., *J. Am. Chem. Soc., 79*, 41, 1957.
21. De Jong, G. de J., *J. Am. Ceram. Soc., 40*, 42, 1957.
22. Sewell, E.C., *Clay Minerals Bull., 2*, 233, 1955.
23. Eyraud, C., and Goton, R., *Compt. Rend., 240*, 423, 1955.
24. Foldvari-Vogl, M., and Kliburszky, B., *Acta Geol. Acad. Sci., 2*, 215, 1954.
25. Arens, P.L., *Dissertation, Wageningen*, 1951.
26. Tateno, J., *Trans. Faraday Soc., 62*, 1885, 1966.

II.2. TYPICAL APPLICATIONS

The technique of DTA has found wide application, sometimes in unexpected fields. The following list, which is not exhaustive, indicates the diversity of fields in which DTA has been applied with considerable success.

1. Agronomy
2. Analytical chemistry
3. Atomic fuels
4. Carbohydrate chemistry
5. Catalysis
6. Cellulose (textiles)
7. Cement chemistry
8. Ceramics
9. Coal technology
10. Criminology
11. Dentifrice
12. Explosives
13. Genesis study
14. Greases
15. Glasses
16. Gypsum
17. Lime
18. Metallurgy
19. Mining
20. Oils and Fats
21. Petroleum (exploration)
22. Phase equilibrium
23. Pigments
24. Polymers
25. Proteins
26. Radioactive minerals
27. Refractories
28. Rocket fuels
29. Soaps
30. Soils
31. Synthetic fibers
32. Waxes
33. Wood

It is evident that any treatment of DTA based on the above list of applications tends to be arbitrary, overlapping, and perhaps confusing. Hence, typical applications of DTA can be conveniently treated under the broad fields of study, *viz:* (1) Inorganic Chemistry, (2) Organic Chemistry, and (3) Physical Chemistry.

Studies on DTA are widely scattered in the literature in the form of research papers, reviews (*1-18*), books or chapters in books (*19-34*, *111*), and in conference proceedings (*35-38*). These references provide a sufficiently exhaustive treatment of DTA.

In addition to the above publications, Mackenzie (*39*) has prepared DTA index cards containing 2400 references, of which 60% pertains to the minerals, 20% to inorganic compounds, and the rest to organic materials. The cards give the following data:

1. Name of mineral or compound
2. Literature references
3. Classification code for minerals
4. Endothermic or exothermic peak temperatures and their relative

intensities, including temperature range

5. Reference to other relevant cards.

Recently, Sadtler Research Laboratories (*40*) prepared a collection of 2000 DTA thermograms comprising 450 commercial compounds, 50 pharmaceuticals and steroids, 1000 organic compounds, and 360 inorganic compounds.

In this chapter a few typical applications of DTA are given with a view to indicating the diversified fields in which this method has been employed. This should serve as a background to a study of cements, possibly also as a basis for further research into the unexplored branches of cement chemistry.

Inorganic Chemistry

Coal Mine Dusts. Continual inhalation of very fine silica dusts by workers in the coal mining industry causes a lung incapacitating desease known as *silicosis.* The pathogenicity depends on the particle size and amount of soluble silica (*41*). DTA has been successfully employed by Dempster and Ritchie (*42*) to estimate the quartz content in mine dusts.

The quantitative DTA results show that with increasing fineness of the dust the amount of quartz decreases. For instance, the dust with a surface area of 9350 cm^2/g has 67.0% quartz, while at 17,550 cm^2/g the value de-decreases to 50.0%. If the silica dust is etched with hydrofluoric acid the resultant material shows a quartz content of 100%. The fine particles have a surface layer of non-quartz silica that fails to exhibit an endothermic effect at 573°C for the $\alpha \rightarrow \beta$ transition of quartz. Hence by knowing the total amount of silica in the dust, DTA would be able to determine the nonquartz silica, by difference.

Soil Color. An interesting application of DTA pertains to the study of causes for the dark color in certain soils in India (*43*) known as black cotton soils. The soil shows two broad exothermic effects at about 325 and 410°C, representing a stepwise oxidation of organic matter. The dark color could be removed by treatment with 10% sodium hexametaphosphate at boiling point, with hydrogen peroxide or potassium dichromate at 80°C, or by heat treatment at 400°C. Any treatment that practically eliminates the exothermic peak at 410°C removes the darkness. A treatment that eliminates only the effect at 325°C fails to remove the dark color. This leads to the conclusion that in the clay part of the organic matter is loosely bound and is oxidized at 325°C. The other part responsible for darkness is more rigidly bound and is oxidized at a temperature of 410°C.

Dehydration Mechanism. Application of two techniques becomes necessary for studying the mechanism of certain reactions. The mechanism of dehydration of $BeSO_4 \cdot 4H_2O$ has not yet been resolved. DTA shows four endothermic effects for $BeSO_4 \cdot 4H_2O$, the peaks representing dehydration and/or phase transformations. A simultaneous analysis of volatile gases by Mass Spectrometric Thermal Analysis (MTA) is essential in resolving the DTA peaks. By a combination of DTA and MTA (*44*) the endothermic peak at 210°C may be ascribed to the reaction

$$BeSO_4 \cdot 2H_2O \longrightarrow BeSO_4 \cdot H_2O + H_2O$$

and that at 275°C to the reaction

$$BeSO_4 \cdot H_2O \longrightarrow BeSO_4 + H_2O$$

The four endothermic peaks represent dehydration in four steps. The decomposition of $BeSO_4$ to BeO and SO_3 results in a peak at 750°C.

Nitrides in Steel. Presence of several types of nitrides influences the physical and mechanical properties of steel. Techniques such as optical and electron microscopy, X-ray and electron diffraction, the electron microprobe, and chemical analysis have been used with partial success. By a combination of DTA–EGA techniques it is possible to analyse the compounds that can be extracted from steel (*45*). The DTA and EGA curves have similar characteristics. Only 2 mg of the material is needed to study the nitrides of vanadium, titanium, niobium, boron, and aluminum in steel.

Kiln Efficiency. The efficiency of a kiln used for making structural clay products is defined as a percentage relationship between the heat energy essential for the process and the energy actually supplied. The theoretical amount of heat supplied to the clay comprises (i) the latent heat of mechanical and combined water, (ii) the heat required to decompose carbonates, and (iii) the heat required to complete other irreversible reactions. Ramachandran and Majumdar (*46*) applied DTA for determining the heat of dehydration of brick clays.

Potassium sulfate undergoing an endothermic reaction at 585°C with a heat energy of 12.3 cal/g was used as a standard for calibration. In Fig. II.2.1 the thermal peak areas for various percentages of K_2SO_4 are shown. A straight line relationship was obtained by plotting the potassium sulfate

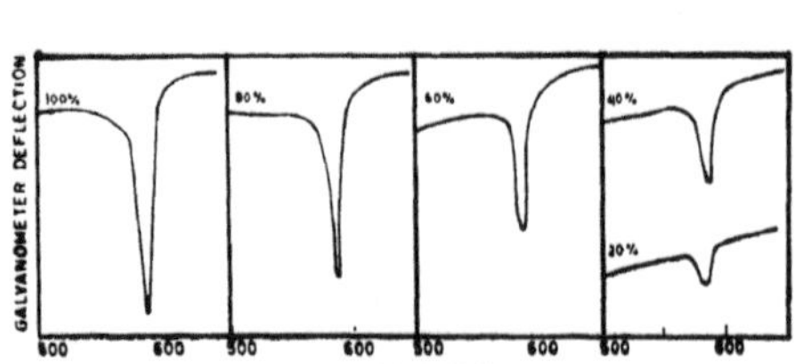

Fig. II.2.1: Effect of potassium sulfate concentration on endothermal area (*46*).

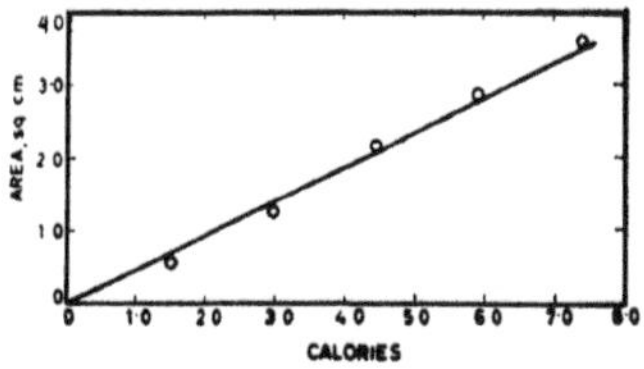

Fig. II.2.2: Area under endothermal peak vs. concentration of potassium sulfate (in cal) (*46*).

content against peak area (Fig. II.2.2.) Using this calibration, values of heats of heats of dehydration of six brick clays and a bentonite were determined.

In the method adopted by Noble (*47*) the heat associated with combined water was calculated by assuming the residual ignition loss, *i.e.* total ignition loss—[mechanically held water+carbon (organic)+carbon dioxide], to be proportional to the true clay substance. This method is not accurate because even for the same type of clay mineral procured from different sources the heat values vary, and it cannot be applied to mixtures of clay minerals.

Table II.2.1 gives the heats of dehydration determined by DTA. The values are slightly higher after the organic matter in the clay was destroyed by H_2O_2. The heat values calculated on the basis of the residual ignition loss of clays are not linear with respect to the values calculated by DTA because the clays do not contain the same type of clay mineral.

Types of Water in a Porous System. Water in a porous system may be bound with different energies. If a porous system containing water is slowly cooled, freezing of water occurs at different temperatures depending on the state of water. The solid material and its pore size seem to influence the state of water in the system. A low temperature adiabatic calorimeter may be used to study the types of water, but is often cumbersome and time consuming. The low temperature DTA quickly reveals the types of water present in a porous system. Interesting features emerge in the DTA of clay minerals carried out in the temperature range 0° to −195°C (*48*). Montmorillonite exhibits three peaks at −4, −7, and −20 to −30°C due respectively to freezing of mechanically-held water, adsorbed water, and water present in the interlayer. Other minerals such as attapulgite, nontronite, vermiculite, and endellite also give three peaks. Kaolinite and illite give a single peak at −2 to −5°C corresponding to the freezing of the mechanically-held water.

Prediction of Firing Behavior. DTA is a powerful tool for predicting the firing behaviour of clay products. Some of the practical examples provided by Everhart and Van Der Beck (*49*) are indicative of this.

TABLE II.2.1
Residual Ignition Loss and Heat of Dehydration of Clays [Ramachandran and Majumdar (46)]

Clay	*Moisture, % (air-dry basis)*	*Ignition loss, %*	*CO_2, %*	*Organic matter, %*	*Residual ignition loss, %*	*Heat of dehydration, cal/g*	*Heat of dehydration, cal/g (after oxidation with H_2O_2)*
S-229	6.75	17.83	3.32	0.65	7.11	8.9	9.7
S-230	6.75	13.96	2.93	0.50	3.78	14.1	15.6
S-231	4.89	19.91	5.99	0.40	8.63	16.0	18.3
S-232	6.66	14.15	0.47	1.10	5.92	31.7	34.3
S-233	8.06	19.94	4.27	1.50	6.11	10.0	12.1
S-234	5.03	10.54	0.47	1.30	3.74	21.5	23.9
Bentonite (B)	10.29	15.80	—	—	5.51	69.0	70.0

Two clays designated A and B were used as raw materials in a plant contained organic matter. This was shown as exothermic effect in thermograms. Clay A exhibited a wide exothermic effect extending up to 760°C while clay B showed the same effect extending up to 650°C. In the plant the clays were pre-fired to 440-540°C to oxidize the organic carbon. On the basis of DTA results the temperature of oxidation was raised to 700°C, resulting in a saving of time and in an improvement in the first quality ware.

Black coring and popping in clay products are sometimes caused by iron sulfide. Iron sulfide either in the form of marcasite or pyrite shows exothermic peaks in the range 440-540°C. If an exothermic effect is noticed in this temperature range, caution should be exercised in the firing procedure by having longer periods of firing and a supply of excess air.

The presence of magnesium and calcium carbonates is shown by peaks in the range 760-815°C. If these are present, a short vitrification range and a high shrinkage may result.

An endothermic effect below 200°C and at 500-700°C, followed by an endothermic effect at about 900°C and an exothermic peak immediately thereafter, signifies illite or montmorillonite. A clay containing these clay minerals may show high plasticity, high shrinkage, and low vitrification range.

An intense endothermic peak at about 600°C and a sharp exothermic peak at about 950°C are indicative of kaolinite. The kaolinitic clay is refractory, with low plasticity and wide vitrification range.

Clays may also show endothermic reactions between 300 and 500°C due to the presence of hydroxides of aluminum or iron. Ferric oxide has a fluxing action, whereas aluminum oxide increases the refractoriness and decreases the shrinkage.

Presence of quartz is indicated by a reversible thermal effect at about 575°C. Quartz rduces the plasticity and shrinkage of clay.

Deterioration Aspects. As discussed above DTA is a useful tool in predicting the firing behavior of clay products. This technique can also be applied to investigage the causes leading to the deterioration of a structural clay product in use. Honeyborne (*4*) applied the technique of DTA to study the causes of deterioration of a brick, and he could also estimate the temperature to which the brick had been fired. The results are presented in Fig. II.2.3. Curve A refers to the thermogram of a clay obtained from a locality where the deterioration of bricks under question had taken place. Curves B, C, and D are the thermograms of the clay fired to 600°, 800° and 1000°C. A comparison shows that the thermogram of the decayed brick is very similar to that of the clay heated to 800°C. The decayed brick, in addition, shows two low temperature endothermic peaks for the presence of calcium sulfate. These results suggest that the bricks had been fired to 800°C perhaps delibe-

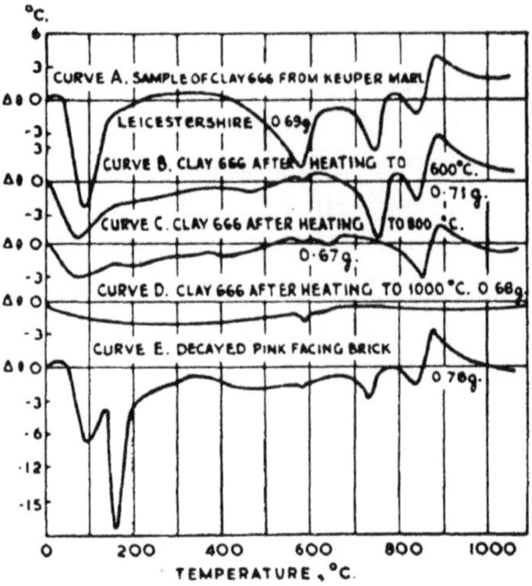

Fig. II.2.3: Comparison of an underfired facing brick with a clay fired to various temperatures (*4*) [Crown Copyright, reproduced by permission of Her Majesty's Stationery Office].

rately to realize a pink color. The deterioration must have occurred due to the crystallization of calcium sulfate in the pores of the poorly bonded brick.

Low Temperature Applications. The bulk of the work on DTA pertains to the study of materials from ambient to temperatures of the order of 1200°C. Compared with this, progress at sub-zero temperatures is inadequate, though it has great promise. Small energy changes associated with the first order transition in an Mn–Ge system are detectable at −160°C (*50*). In Table II.2.2 following, DTA data of phase reactions of some of the substances are compared with the results reported in the literature. The reliability of DTA for low temperature studies is evident.

Organic Chemistry

Metal Chelates. The mono- and di-substituted derivatives of 8-quinolinol are used as precipitants for metal ions in the gravimetric method of estimation. One of the important considerations that determines the applicability of this method is the thermal stability of these complexes.

DTA has been used to study the thermal stabilities of 8-quinolinol and substituted 8 quinolinol-metal chelates of 15 metal ions (*52*). Thermograms

TABLE II.2.2

Low Temperature Differential Thermal Analysis of Some Substances [Markowitz and Boryta (51)]

Sample	*Phase Reaction*	*DTA (°C)*	*Published (°C)*
Br_2	Fusion	− 6.8	− 7.3
Hg	Fusion	−40.0	−38.9
17.5% $NaNO_3$+82.5% H_2O (mole %)	Eutectic fusion	−18.0	−17.5
28.6% LiCl+71.4% H_2O (mole %)	Hydration transitions		
	(i) $LiCl \cdot 5H_2O \longrightarrow LiCl \cdot 3H_2O + 2H_2O$	−68.0	−68.0
	(ii) $LiCl \cdot 3H_2O \longrightarrow LiCl \cdot 2H_2O + H_2O$	−20.5	−20.0
	(iii) $LiCl \cdot 2H_2O \longrightarrow LiCl \cdot H_2O + H_2O$	+18.8	+12.5

show endothermic and exothermic peaks owing to fusion and decomposition. The stability is in the order

$$\begin{aligned}&Sc < Th < Sb < Pb < Co(III)\\ &= Cu < Bi < Zn < Fe < Al < UO_2 < Ni < Co(II) < Mn\\ &= Cd\end{aligned}$$

Wood Chemistry. Not much work has been done on the application of DTA in wood chemistry. The work of Arseneau (*53*) on balsam fir should indicate the potentiality of DTA. DTA of balsam fir shows a number of endothermic and exothermic peaks in the region 145-360°C, which could be attributed to reactions of compounds such as cellulose, acid lignin, and other compounds extractable with alcohol–water, benzene–alcohol, etc.

In the examination of birchwood, exothermic peaks at 389°, 414° and 452°C have been detected (*54*). Thermograms of cellulose, holocellulose, lignin, xylan, and xylose indicate that holocullulose has the combined thermal characteristics of birchwood and xylan. Decomposition of xylan does not seem to lead to the formation of xylose

The role of fire retardants on teak, poplar, birch, larch, bongossi, and beech has been discussed on the basis of thermal curves (*55*).

Organic Acids. Attempts to characterize substances within a group by DTA has met with some success. Wendlandt and Hoiberg (*56, 57*) obtained thermograms of 50 organic acids, which served as a 'finger print' for their characterization. Oxidation was prevented by carrying out DTA in an atmosphere of argon. The organic acids showed endothermal effects and occasionally exothermal effects caused by reactions such as dehydration, decarboxylation, sublimation, decomposition, and solid–liquid transition. The endothermic effects of the solid–liquid transition were generally 10-30°C higher than those reported for fusion temperatures. This can be due to the inherent differences in the dynamic and quasi-static methods. The organic acids studied by Wendlandt and Hoiberg included salicylic acid and its derivatives, nitro, triiodo, p-amino, p-fluoro, p-chloro, p-iodo, p-bromo and m-nitro benzoic acids, and oxalic, malonic, succinic, and glutaric acids.

Explosives. Many compounds have been synthesized for use as a rocket fuel. As the materials may be available in small amounts it is useful to have a quick method of determining the heats of explosion with a reasonable accuracy. This would give an index of its destructiveness, the potentiality of its use as a rocket fuel, and would also estimate the theoretical specific impulse (*58*).

DTA has been used to determine the heat of explosion using only 5-25 mg of the material (*59*). The explosives and propellants studied comprised nitroglycerine, tetranitromethane, 2,4,6-trinitrotoluene, cyclotrimethylene trinitroamine, 2,4,6-trinitro phenyl methyl nitramine, pentaerythritol tetra nitrate, double-base propellants N-5 and JPN, and aluminized double-base propellant ABL-2056.

Polyethylene. Recently, DTA has attracted increasing attention in polymer chemistry (*60*). DTA is a rapid and elegant technique for studying the thermal stability of polyethylene blends formed under fabrication conditions. A plot of differential temperature against the period of exposure to oxygen at 150°, 170°, and 190°C provides the induction period for rapid oxidative degradation. The induction periods determined by DTA are in close agreement with those obtained by other methods. The activation energy for the induction time for polyethylene labelled Marlex 50, DYNH-3, DYNH-Blend B, and DYNK ranges from 21.6 to 33.0 kcal/mole. These values show close agreement with published results.

In polyethylene industries using high-pressure polyethylene blends, estimation of their linear contents is of great value. Clampitt (*61*) investigated the possibility of the application of DTA for the estimation of the linear content of blends containing a linear polyethylene and a high-pressure polyethylene. Blends were prepared containing linear polyethylene from 5 to 40% and thermal peaks were used to estimate the linear content. The high pressure polyethylene showed a peak at 118°C and the linear polyethylene gave a peak at 136°C. The area of the endotherm at 118°C decreased in proportion to the linear polyethylene content. The area of the thermal peak represents crystallinity, and hence it is possible to calculate the crystallinity values of blends (*62*). The values increase with increase in the linear polyethylene content.

Epoxide Reactions. Insoluble or amorphous cross-linked epoxy systems do not yield sharp X-ray patterns and DTA seems to offer a better method for studying them. DTA of six epoxides was carried out alone or in the presence of polymerizing agents such as m-phenylene diamine, methane diamine, maleic anhydride, and nadic methyl anhydride (*63*). DTA exhibited characteristic endo and exothermic reactions for the identification of the epoxy systems. Most of the catalyzed epoxides showed exothermic peaks at low temperatures and at 300-400°C, which could be ascribed to the isomerization of the epoxy groups to carbonyl groups and to the thermal polymerization of the epoxy groups. The temperature at which peaks appeared clearly indicated the catalytic nature of the polymerization. Maleic anhydride is more reactive than nadic methyl anhydride in polymerizing Epon 828 as would be evident from the appearance of exothermal effects at relatively lower temperatures. The extent of polymerization and the heats of polymerization can also be determined from the low temperature peak (*7*).

Coals. In coal chemistry, attempts have been made to estimate the calorific values by thermal peak areas and to evaluate the rank of coal. Gamel and Smothers (*64*) investigated semibituminous coals and observed that the exothermal area corresponded to the calorific value of the coal determined by the peroxide bomb method.

The applicability of vacuum DTA to the determination of the rank of coal has been described by Whitehead (*65*), Breger and Whitehead (*66*), and King and Whitehead (*67*). The studies showed that with increasing rank from lignite to semi-anthracite the exothermic peaks were displaced towards higher temperatures. Further studies indicated that there is no correlation between the fixed carbon content and the peak temperature. However, the peak temperatures of the vitrains selected from a number of coals of different ranks showed a good correlation with the fixed carbon content.

Physical Chemistry

Catalysts. An extensive study by Ramachandran and co-workers (*68-75*) has shown that DTA can be an advantageous method for the study of catalysts.

In Fig. II.2.4. are shown the thermograms of chromic oxide gels prepared by precipitation of chromium nitrate, chromium chloride, or chromium sulfate with ammonium hydroxide. The first endothermic peak below 200 °C represents the loss of loosely held water and that between 200 and 300 °C represents the loss of rigidly-bound water. The gel prepared from chromium nitrate gives an intense exothermic peak at 395 °C (sample 1), but the gel pre-

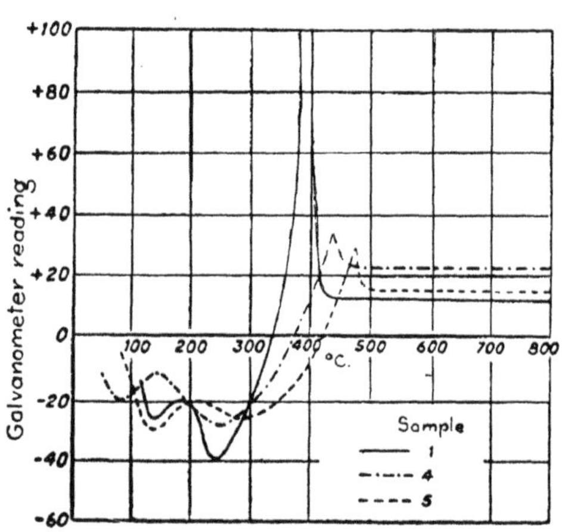

Fig. II.2.4: DTA of chromic oxide gel prepared from (1) chromium nitrate, (*4*) chromic sulfate, (5) chromic chloride (*72*).

pared from chloride (sample 5) or sulfate (sample 4) shows subdued exothermic effect at 480 or 430°C. The exothermic peaks at 395-480°C in all cases are due to the formation of crystalline chromic oxide from the amorphous form. The catalytic activity of Cr_2O_3 is suddenly destroyed when heated to temperatures above 380°C and this may be due to a decrease in the surface area consequent on crystallization. This proves that the crystalline variety is not active. Moreover, a considerable decrease in the value of the exothermic effects in the gels prepared from chromium chloride or chromium sulfate should indicate that the enormous heat liberated by nitrate-precipitated gels is associated with a large surface area. The superiority of nitrate-precipitated catalyst over others is hence apparent. Surface area data show that the nitrate-precipitated gel has values of 235.9, 301.0, and 315.3 m²/g at 100, 200, and 300°C respectively. At 350°C the value decreases to 123 m²/g and the surface area is only 19.2 m²/g at 400°C, which corresponds to the crystallization temperature in DTA.

Surface Area Determination. The DTA method is potentially a useful tool for the determination of surface area. If a clay is exposed to a constant humidity and then analyzed it would show a low-temperature endothermic effect, the area of which would be proportional to the adsorbed water. In clay minerals the peak area should be in the order kaolinite$<$illite$<$montmorillonite. Roy (*76*) made an attempt to calculate surface areas of clays by DTA but he has not indicated the conditions of equilibration and has also concluded that part of the constitutional water is lost at very low temperatures. In montmorillonites the low-temperature endothermic peak area is proportional to the surface area (*77*).

Surface area can be expressed in terms of the equation

$$(\bar{A}) = k(\Delta H_\theta)$$

where k is an unspecified proportionality factor and ΔH_θ is the integral heat of adsorption per gram of the adsorbent.

For surface area determinations, Schreiber and Waldman (*78*) used two cups, one containing carbon black and the other a reference material. After evacuation of the system water, n-butane or ethyl chloride was admitted into the sample cell. The adsorption was completed within 10 min and resulted in an exothermic peak. The area under the peak was proportional to the surface area. The values obtained by this method corresponded to those determined by the standard B.E.T. method.

A carbon black having a surface area in the range 43-327 m²/g had thermal peak areas varying in the range 5.5-88.0 cm²/g. Compared with B.E.T., electron microscopic, and permeability methods of surface area determination the DTA technique is simple and yields quicker results.

Order–Disorder Transitions. Order–disorder reactions are amenable to study by DTA. Jaffray (*79*) in the thermal analysis of barium oxyacetate found a weak transition at 40±0.5°C. This was ascribed to a kind of order-disorder transition manifested by the simultaneous rotation of carboxyl oxygen atoms around the carbon–carbon bonds (*80*).

An exothermal peak of small intensity was detected in the DTA of calcium and magnesium carbonates 40-80°C above the decarbonation peak (*81*). This was attributed to an order reaction of the metastable oxide formed immediately after decarbonation.

Impurity Effects. Small amounts of impurities of the order of 0.1-0.01% may alter the thermal peak to a great extent. This has both a theoretical and practical significance. Interest has been shown in the effect of additives on the temperature of decomposition of dolomite. Berg (*82*) reported that in the presence of 0.1% NaCl the $MgCO_3$ decomposition in dolomite is lowered. Suppression of the evolution of CO_2 was also brought about by an addition of 1.0% NaCl (*83, 84*). Sodium fluoride has been reported to bring down the characteristic temperature of $MgCO_3$ decomposition by as as much as 235°C (*85*).

Partially calcined dolomites may be used for making Sorel cement. But it should be ensured that only $MgCO_3$ is decomposed because otherwise the formation of CaO may have a deleterious effect. In practice it is difficult to calcine under conditions that would prevent the decomposition of $CaCO_3$. Ramachandran and Kacker (*86*) added 0.1-1.0% NaCl to dolomite and found that less stringent conditions are required to calcine such a mix for making an oxychloride cement.

It should be noted that NaCl, in addition to decreasing the characteristic temperature of $MgCO_3$, also decreases the peak area. Thus DTA should prove a useful tool for quality control purposes in the utilization of magnesite deposits contaminated with soluble salts (*87*).

Radioactivity. A metamict mineral is typified by the formula $A_mB_nO_{2mn}$, where A=U, Th, rare earths, Ca, Na, etc. and B=Nb, Ta, Ti, etc. Volchok and Holland (*88*) have discussed the possibility of using metamict minerals as age indicators. The metamict mineral is continuously exposed to α activity from uranium or thorium present in it. Due to this exposure the bonds in the lattice are borken, depending on the α-activity, the period of α-bombardment, and the strength of the crystal. When these minerals are heated recrystallization occurs with a liberation of heat in the form of exothermic peaks beyond 700°C. The peak area would be proportional to the degree of lattice disorganization. Hence the age of the metamict mineral can be estimated by knowing the α-activity and differential thermal peak area.

DTA is more sensitive than X-ray or infrared analysis in detecting the changes that occur when ammonium perchlorate is exposed to high energy radiations. DTA shows differences but X-ray and infrared analysis fail to indicate any difference (*89*).

Curie Temperature. The Curie temperature corresponds to a second-order transition and is determined by magnetic methods. Magnetite exhibits two exothermic peaks in DTA. The first exothermic peak starts at about 580°C. Grinding the magnetite increases the intensity of the first exothermic peak. This peak is attributed to the oxidation of Fe^{+2} formed as a protective film on the surface of magnetite. The second exothermic peak is due to the oxidation of the remaining magnetite, facilitated through the diffusion of oxygen after the crystallization of the protective film. It is an interesting fact that the beginning of the second exothermic peak corresponds to the Curie temperature, representing a ferromagnetic to paramagnetic transition. A small reversible endothermic peak at about 585°C represents in fact the change from ferromagnetism to paramagnetism (*90*).

Kinetic Constants. The temperature at which maximum deflection occurs in DTA varies with the heating rate. An expression can be derived relating this variation with kinetics of the reaction. Kinetic constants determined by DTA for kaolinites compare with those obtained by conventional isothermal techniques (*91*). In the equation

$$\frac{d(ln\ \varphi/Tm^2)}{d(1/Tm)} = \frac{-E}{R}$$

the heating rate $\varphi = dT/dt$ and Tm=peak temperature. A plot of $ln\ \varphi/Tm^2$ against $1/Tm$ yields a straight line with a slope equal to $-E/R$. If E is known, A can be calculated from the equation

$$Ae^{\frac{-E}{RT}} = \frac{E}{RT^2} \cdot \frac{dT}{dt}$$

The activation energy E for kaolins determined by DTA varies from 34.97 to 38.83 kcal/mole and the frequency factor ($\log_{10}$A), from 6.11 to 7.05. Applying isothermal technique, the corresponding values are 38.81-42.06 and 7.10- to 8.66 kcal/mole.

Vaughan (92), from the dehydration studies of kaolinite, found E to follow the first-order reaction. The activation energy values calculated from the Arrhenius equation decreased in the order dickite>kaolinite>halloysite >ball clay>fire clay.

Specific Heat. The base line shift in DTA may be due to differences in heat capacities of the sample and the inert material, poor heat transfer, lack of symmetry, and differences in packing, particle size, or concentration. By the use of a sensitive instrument and small amounts of samples of about 1-10 mg, reproducible results may be realized. In fact the base line shift can be turned to advantage for determining the specific heat of a sample. A mathematical expression has been derived by David (*93*) for determining the specific heat. The final equation takes the form

$$C_p = \frac{K'K(\Delta T dt)}{\varrho g dT} \times \text{area due to sample}$$

where g =sample size in gm

C_p =average specific heat of the sample (cal/gm/°C)

$K' = \left(\varrho + \frac{k}{k\Delta T} \cdot \frac{d^2T}{dX^2}\right) = 1.68$ (experimentally determined)

K =heat transfer coefficient of the system in cal/mm²

dT =temperature range of interest

$(T_1—T_2)\ dt = T \cdot dt$ over a time interval

T_1 and T_2 =temperatures of the reference thermocouple and the sample thermocouple respectively

ϱ =density of the sample

k =thermal conductivity of the sample

The specific heat values were determined for p-dinitrobenzene, benzoic acid, oxalic acid, nickelous nitrate, dicyandiamide, naphthalein, polyethylene, and polystyrene. The values by DTA very nearly corresponded to the known values.

Application of Clausius–Clayperon Equation. Heats of vaporization of water can be calculated from the well known Clausius–Clayperon equation. Sarasohn (*94*) found the boiling endotherms of water at pressures 766, 206, and 98 mm to be 100.0, 66.1, and 49.8°C respectively. The ΔHv value was calculated by applying the equation

$$\Delta Hv = 1.987[T_2T_1/(T_2 - T_1)] ln(p_2/p_1)$$

The heat of vaporization was 541 cal/gm. at 766-206 mm and 545 cal/gm. at 98-766 mm.

In the decarbonation of magnesite Stone (*95*) found that with increasing CO_2 pressures the decarbonation peak shifted to higher temperatures. The slope of the plot of log p against $1/T$ gave the value of $-\Delta H/R$. The heat of reaction obtained by computation from the peak area of magnesite decomposition agrees fairly well with that calculated from the Clausius-Clayperon equation (*96*).

Techniques Derived from Differential Thermal Analysis

Derived Differential Thermal Curves (DDTA). It is not always possible to determine a thermal peak area accurately due to the difficulty in locating the beginning and end of the thermal peak. Peak temperatures are also not easily discernible if the thermal effects show as broad humps or valleys. These difficulties can be largely eliminated by a plot of the first derivative of the differential temperature as proposed by Burgess (*97*) and Rosenhaim (*98*).

Frederickson (*99*) compared DTA and DDTA curves of kaolinite and demonstrated the sharper nature of the peaks in the DDTA plot. Further studies on fusion and crystallization of potassium nitrate, potassium perchlorate, calcium oxalate monohydrate, and calcium carbonate have shown the relative advantages of DDTA (*100, 101*).

Dynamic Differential Calorimetry (DDC). In the DTA method thermal effects are largely influenced by the sample in the immediate vicinity of the thermocouple and also by the thermal conductivity between the sample and the thermocouple. These factors could be eliminated by adopting the DDC method suggested by Boersma (*102*) and devised by Schwiete and Ziegler (*103*).

In the DDC technique the samples are placed in two metallic crucibles and the heat exchange occurs through the walls. Thermocouple beads are not in contact with the sample (refer to Fig. V.2.3, Chapter V). The application of DDC to a study of the hydration products of C_3S and C_2S is discussed in Chapter V. DDC in comparison with DTA may not have better sensitivity but shows better reproducibility (*104*).

Double Differential Technique. Many natural minerals containing two materials may show interference effect in DTA. Peak effects of one of the materials can be eliminated by mixing an estimated amount of it with the inert substance. The resultant DTA shows the peak for a single material and hence facilitates quantitative estimation. This method, known as Double Differential Technique, was developed by Roberts and Brindley (*105*). Small amounts of kaolinite in ganister and silica rocks can be estimated by including the required amounts of quartz with the inert material (*8*).

Bi-Differential Thermal Analysis. In DTA the inert material has different thermal characteristics from the sample. Gerard-Hirne and Levy suggested the use of an active material such as kaolinite for studying the kaolinitic minerals. This method was found to be more sensitive than DTA (*106*).

Differential Scanning Calorimeter (DSC). An instrument has been described by Watson *et al* (*107*) that directly estimates the thermal energy changes in a reaction. The method requires a few milligrams of the sample and operates at 80°C/min. The peak area gives directly the heat of reaction and is independent of specific heat, sample geometry, and scanning rate. In DSC technique the sample is maintained isothermal to an inert material by a supply of heat to the sample or inert material. The heat energy required to maintain isothermal conditions is recorded as a function of time or temperature.

Differential Enthalpic Analysis. In the course of DTA, if extra power is supplied to maintain the temperature of the sample and that of the inert sample at the same level, then the extra energy will be directly proportional to the heat of transition. This method was designated as Differential Enthalpic Analysis by Eyraud *et al* (*108*, *109*). Refinements in the technique by Speros and Woodhouse (*110*) enabled accurate determination of the heat of fusion of a number of metals.

REFERENCES

1. Grim, R.E., *Annals N.Y. Acad. Sci.*, *53*, 1031, 1951.
2. Ramachandran, V.S., and Bhattacharyya, S.K., *J. Sci. Ind. Res.* (*India*), *13A*, 365, 1954.
3. Kauffman, A.J., *World Oil*, July 1948.
4. Honeyborne, D.B., *Chem. and Ind.*, p. 662, 1955.
5. Ramachandran, V.S., and Patwardhan, N.K., *J. Nat. Bldg. Org.*, *11*, 1, 1957.
6. Webb, T.L., *S.A. Ind. Chem.*, *7*, 224, 1953.
7. Murphy, C.B., *Anal. Chem.*, *30*, 867, 1958; *32*, 168, 1960; *34*, 298R, 1962; *36*, 347R, 1964; *38*, 443R, 1966.
8. Foldvari-Vogl, M., *Acta Geol. Acad. Sci., Hungary*, *5*, 1, 1958.
9. Pask, J.A., and Warner, M.F., *Bull. Am. Ceram. Soc.*, *33*, 168, 1954.
10. Norton, F.H., *J. Am. Ceram. Soc.*, *22*, 54, 1939.
11. Lehmann, H., Das, S.S., and Paetsch, H.H., *Tonind.-Ztg. u Keram. Rundschau*, *No. 1*, 55, 1954.
12. Lukaszewski, G.M., and Redfern, J.P., *Lab. Pract.*, *10*, 630, 1961.
13. Garn, P.D., *Anal. Chem.*, *33*, 1247, 1961.
14. Redfern, J.P., *Thermal Analysis Review*, Stanton Instruments Ltd., London, No. 1, 1962.
15. Ramachandran, V.S., Feldman, R.F., and Sereda, P.J., *Highway Res. Rec. No. 62*, 1964.
16. Mackenzie, R.C., and Mitchell, B.D., *Analyst*, *87*, 420, 1962.
17. Webb, T.L., and Krüger, J.E., *Symp. Analysis Calcareous Materials*, London, 1963.
18. Bhattacharyya, S.K., and Ganguly, N.D., *Proc. Natl. Inst. Sci., India*, *A27*, 588, 1961.
19. Wendlandt, W.W., *Thermal Methods of Analysis*, Interscience Publishers, New York, 1964.
20. Smothers, W.J., and Yao Chiang, *Handbook of Differential Thermal Analysis*, Chemical Publishing Co., New York, 1966.
21. Mackenzie, R.C. (Editor), *The Differential Thermal Analysis of Clays*, Mineralogical Society, London, 1957.
22. Ramachandran, V.S., and Garg, S.P., *Differential Thermal Analysis—Its Application in Building Materials*, Central Building Research Institute, Roorkee, India, 1959.
23. Berg, L.G., '*Introduction to Thermography*,' Izd. Akad. Nauk SSSR, Moscow, 1961.
24. Berg, L.G., Nikolaev, A.V., and Rode, Y., '*Termografiya*,' Izd. Akad.

Nauk SSSR, Moscow, 1944.
25. Kissinger, H.E., and Newman, S.B., *DTA in Analytical Chemistry of Polymers*, Vol. XII, Part II, Interscience, New York, 1962.
26. Gordon, S., *Encylopedia of Science and Technology*, Vol. 13, McGraw-Hill, New York, 1960.
27. Meites, L. (Ed.), *Handbook of Analytical Chemistry*, McGraw Hill, New York, 1963.
28. Janassen, H.B., and Weissberger, A. (Eds.), *Technique of Inorganic Chemistry*, Vol. I, Interscience, New York, 1963.
29. West, R.R., and Others, *The Defect Solid State*, Interscience New York, 1957.
30. Clark, G.L. (Ed.), *Encylopaedia of Spectroscopy*, Reinhold, New York, 1960.
31. Grim, R.E., *Clay Mineralogy*, McGraw-Hill, New York, 1953.
32. Searle, B., and Grimshaw, R.W., *The Chemistry and Physics of Clays*, Ernest Benn Ltd., London, 1959.
33. Elias, M., Stovick, M., and Zahradnik, L., *Differencni Thermicka Analysa*, Nakladatelstvi Ceskoslovenske Akadmie Ved Prague, 1957.
34. Taylor, H.F.W. (Ed.), *The Chemistry of Cements*, Academic Press, New York, 1964.
35. *Proceedings of the First International Conference on Thermal Analysis*, Aberdeen, Scotland, Sept. 1965.
36. *Proceedings of the First Symp. Thermal Analysis*, Toronto, 1965.
37. *Symp. Thermal Analysis*, London, April 1965.
38. *Trans. 1st Conference Thermography*, Kazan, 1953.
39. Mackenzie, R.C., Macmillan Co. Ltd., London, 1962.
40. Sadtler Research Lab. Inc., Philadelphia, Pennsylvania, U.S.A.
41. King, E.J., *Medical Res. Council, Special Rept. Series No. 250*, London, H.M.S.O., 1945.
42. Dempster, P.B., and Ritchie, P.D., *J. Appl. Chem. (London)*, *3*, 182, 1953.
43. Ramachandran, V.S., Ahmad, F.U., and Jain, L.C., *Proc. Indian Acad. Sci.*, *50*, 314, 1959.
44. Langer, H.G., and Gohlke, R.S., *Anal. Chem.*, *35*, 1301, 1963.
45. Bondi, R., Straub, W.A., Buyok, E.G., and Melnick, L.M., *Anal. Chem.*, *38*, 1336, 1966.
46. Ramachandran, V.S., and Majumdar, N.C., *J. Appl. Chem. (London) 11*, 449, 1961.
47. Noble, W., *National Brick Advisory Council*, London, 1950.
48. Kato, C., *Yogyo Kyokai Shi.*, *67*, 243, 1956.
49. Everhart, J.O., and Van Der Beck, R.R., *Bull. Am. Ceram. Soc.*, *32*, 239, 1953.
50. Levina, S.S., *Trudy Inst. Fiz. Metal.*, Akad. Nauk SSSR, Ural Filial, *22*, 67, 1959.
51. Markowitz, M.M., and Boryta, D.A., *Anal. Chim. Acta*, *31*, 397, 1964.
52. Wendlandt, W.W., and Horton, G.R., *Anal. Chem.*, *34*, 1098, 1962.
53. Arseneau, D.F., *Can. J. Chem.*, *39*, 1915, 1961.
54. Sergreeva, V.N., and Vaivads, A., *Latvijas PSR Zinatnu Akad. Vestis, No*, *9*, 103, 1954.
55. Keylwerth, R., and Christoph, N., *Materialprufung*, *2*, 281, 1960.
56. Wendlandt, W.W., and Hoiberg, J.A., *Anal. Chim. Acta*, *28*, 506, 1963.

57. Wendlandt, W.W., and Hoiberg, J.A., *Anal. Chim. Acta*, *29*, 539, 1963.
58. Lewis, B., Pease, R.N., and Taylor, H.S., *Combustion Processes*, Princeton Univ. Press Princeton, N.J., 1956.
59. Bohon, R.L., *Anal. Chem.*, *35*, 1845, 1963.
60. Shreiber, H.P., and Waldman, M.H., *Ind. & Eng. Chem.*, *53*, 137, 1961.
61. Clampitt, B.H., *Anal. Chem.*, *35*, 577, 1963.
62. Ke, B., *J. Polymer, Sci.*, *42*, 15, 1960.
63. Anderson, H.C., *Anal. Chem.*, *32*, 1592, 1960.
64. Gamel, C.M., and Smothers, W.J., *Anal. Chim. Acta*, *6*, 442, 1952.
65. Whitehead, W.L., *Conf. Origin and Constitution of Coal*, Crystal Cliff., Nava Scotia, pp. 100-110, 1950.
66. Breger, I.A., and Whitehead, W.L., *Fuel*, *30*, 247, 1951.
67. King, L.H., and Whitehead, W.L., *Econ. Geol.*, *50*, 22, 1955.
68. Ramachandran, V.S., and Bhattacharyya, S.K., *Proc. Natl. Inst. Sci., India*, *No. 12*, p. 23, 1959.
69. Ramachandran, V.S., and Bhattacharyya, S.K., *J. Sci. & Eng.*, *1*, 73, 1957.
70. Ramachandran, V.S., and Bhattacharyya, S.K., *J. Sci. Ind. Res.* (*India*), *13B*, 882, 1954.
71. Bhattacharyya, S.K., and Ramachandran, V.S., *J. Sci. Ind. Res.* (*India*), *12B*, 279, 1953.
72. Ghosh, J.C., Bhattacharyya, S.K., Gopalaswamy, S.N., and Ramachandran, V.S., *J. Sci. Ind. Res.* (*India*), *11B*, 547, 1952.
73. Bhattacharyya, S.K., and Ramachandran, V.S., *J. Sci. Ind. Res.* (*India*), *11B*, 549, 1952.
74. Bhattacharyya, S.K., and Ramachandran, V.S., *J. Sci. Ind. Res.* (*India*), *11B*, 550, 1952.
75. Bhattacharyya, S.K., Ramachandran, V.S., and Ghosh, J.C., *Advances in Catalysis*, *Vol. IX*, Acad. Press, New York, 1956.
76. Roy, H., *Trans. Indian Ceram. Soc.*, *23*, 181, 1964.
77. Zul'fugarov, Z.G., Aliev, A.S., Rosulova, S.M., and Smirnova, V.E., *Kinetika i Kataliz.*, *3*, 565, 1962.
78. Schreiber, H.P., and Waldman, M.H., *Can. J. Chem.*, *37*, 1782, 1959.
79. Jaffray, J., *C.R. Acad. Sci., Paris*, *225*, 106, 1947.
80. Watanabe, T., and Saito, T., *Nature*, *163*, 226, 1949.
81. Rao, C.N.R., and Pitzer, K.S., *J. Phys. Chem.*, *64*, 282, 1960.
82. Berg, L.G., *Compt-Rend. Acad. Sci., USSR*, *38*, 24, 1943.
83. Schwob, Y., *Compt. Rend.*, *224*, 47, 1947.
84. Esin, O.A., Geld, P.V., and Popel, S.I., *Zh. Prikl. Khim.*, *22*, 345, 1949.
85. Murray, J.A., Fischer, H.C., and Shade, R.W., *M.I.T. Fellowship Rept., Proc. Natl. Lime Asscn.*, *49*, 95, 1951.
86. Ramachandran, V.S., and Kacker, K.P. (*Unpublished results*).
87. Webb, T.L., Lecture Series, *S.A. Chem. Inst., Johannesberg*, July 1966.
88. Volchok, H.L., and Holland, H.D., *Am. Mineralogist*, *37*, 709, 1952.
89. Freeman, E.S., and Anderson, D.A., *J. Phys. Chem.*, *63*, 1344, 1959.
90. Schmidt, E.R., and Vermaas, F.H.S., *Am. Mineralogist*, *40*, 422, 1955.
91. Kissinger, H.E., *J. Res. Nat. Bur. Std.*, *57*, 217, 1956.
92. Vaughan, F., *Clay Minerals Bull.*, *2*, 265, 1955.
93. David, D.J., *Anal. Chem.*, *36*, 2162, 1964.
94. Sarasohn, I.M., *Du Pont Thermogram*, *2*, January 1965.

95. Stone, R.L., *J. Am. Ceram. Soc.*, *37*, 46, 1954.
96. Ellis, B.G., and Mortland, M.M., *Am. Mineralogist*, *47*, 371, 1962.
97. Burgess, G.K., *Nat. Bur. Std. Bull.* *5*, 199, 1908-1909.
98. Foote, P.S., Fairchild, C.O., and Harrison, T.R., *U.S. Bur. Std. Tech. Paper No. 170*, 1921.
99. Frederickson, A.F., *Am. Mineralogist*, *39* 1023, 1954.
100. Freeman, E.S., *Anal. Chem.*, *31*, 624, 1959.
101. Campbell, C., Gordon, S., and Smith, C.L., *Anal. Chem.*, *31*, 1188, 1959.
102. Boersma, S.L., *J. Am. Ceram. Soc.*, *38*, 281, 1955.
103. Schwiete, H.E., and Ziegler, G., *Ber. Deut. Keram. Ges.*, *35*, 193, 1958.
104. Torklar, K., Lasser, K., and Fritzer, H.P., *Sprechsaal*, *95*, 239, 1962.
105. Roberts, A.L., and Brindley, G.W., *Clay Minerals Bull.*, *1*, 15, 1947.
106. Gerard-Hirne, J., and Levy, C., *Bull. Soc. Franc. Ceram.*, pp. 26-40, 1951.
107. Watson, E.S., O'Neill, M.J., Austin, J., and Brenner, N., *Anal. Chem.*, *36*, 1233, 1964.
108. Eyraud, C., *Compt. Rend.*, *238*, 1511, 1954.
109. Eyraud, C., Goton, R., Trambouze, Y., Tran Hun The, and Prettre, M., *Compt. Rend.*, *240*, 862, 1955.
110. Speros, D.M., and Woodhouse, R.L., *Nature*, *197*, 1261, 1963.
111. Garn, P.D., Thermo-Analytical Methods of Investigations, Academic Press, New York, 1965.

III. Formation of Portland Cement Clinker and Polymorphism of C_3S and C_2S

III.1 FORMATION OF PORTLAND CEMENT CLINKER

Portland cement clinker is formed by firing to a temperature of 1300-1500°C a calcareous material such as limestone, marine shells, chalk, or marl with a siliceous material such as clay, shale, or slate. The resulting clinker is mixed with 4-5% gypsum and ground to a fine powder of average diameter 10 μ. Anhydrite or a mixture of anhydrite and gypsum are also used in place of gypsum.

Principally, Portland cement comprises four phases, *viz.* tricalcium silicate (C_3S), dicalcium silicate (C_2S), tricalcium aluminate (C_3A), and a ferrite phase approximating to the composition C_4AF, known as tetracalcium alumino-ferrite. The existence of a glassy phase ranging from 2-20% has been reported in clinkers by earlier workers but data obtained recently are not in accordance with this finding.

Five types of Portland cement are recognized, containing differing proportions of the four principal phases. In Table III.1.1 the average percentage of each of the four phases contained in the five types of American ce-

TABLE III.1.1
Potential Phase Compositions of Five Types of Cement [Bogue (1)]

Cement		*Potential Phase Composition, %*			
Type	*Description*	C_3S	C_2S	C_3A	C_4AF
I	General use	45	27	11	8
II	Moderate heat of hardening	44	31	5	13
III	High early strength	53	19	11	9
IV	Low heat	28	49	4	12
V	Sulfate resisting	38	43	4	9

ments is shown.

The four Portland cement phases do not have the exact chemical compositions as described above but are modified by slight substitution of other ions. The C_3S phase may contain some Al_2O_3 and MgO in solid solution, the C_2S phase may contain K_2O, and the C_3A phase, some Na_2O.

The common impurities that may be present in the Portland cement are magnesia, fluorine compounds, phosphates, lead oxide, zinc oxide, alkalies, and sulfides. Magnesia if present as periclase causes expansion and deterioration of cements on long exposure, and its presence in excess of 5% is not desirable. Fluorine in excess of 0.1% and phosphate in excess of 0.5% are known to have deleterious effects (*2*, *3*). Excessive amounts of zinc oxide or lead oxide are not permitted in cements. Alkalies if present in larger amounts lead to the well known 'alkali-aggregate' expansion. The sulfides may oxidize to sulfates and form a deposit in the kiln and in cements may cause deterioration.

Chemistry of the Reactions. The chemical processes that occur in the formation of Portland cement are extremely complex. Some of the techniques applied to the study of these reactions include high-temperature microscopy, X-ray diffraction, electron microscopy and diffraction, and spectroscopic, polarographic, chemical, and thermal analysis. The rate of reactions is dependent on various factors such as the chemical and mineralogical composition of the raw mixes, the proportions of the raw mixes, the intimacy of mixing, the particle size, crystallinity, and defects in the crystals.

Lea (*4*) has summarized the approximate temperatures at which different reactions take place in the kiln, but there is always a possibility of overlapping of reactions. Below a temperature of 800°C calcium carbonate is decomposed and the clay is dehydroxylated with the formation of CA and probably of CF. Between 800 and 900°C C_2S starts forming, and at 900-1000°C C_5A_3 and probably C_2AS are also produced. Gehlenite (C_2AS) that may exist below 1000°C is decomposed at 1000-1100°C. Both C_3A and C_4AF are identified at about 1200°C. At 1260°C liquid appears but a substantial quantity of C_3S is formed only at 1450°C.

Tricalcium silicate in cements, originally designated as alite, is modified by substitution of MgO and Al_2O_3 to give a compound of composition $C_{54}S_{16}AM$ (*5*, *6*). Brunauer *et al* (*7*) gave the formula as $52CS_3.C_6AM$. Alite is unstable below 1275°C and decomposes to C_2S and CaO, but the rate of decomposition is so slow that it can exist in a metastable state at normal temperatures. The dicalcium silicate phase occurs as α, α', β, or γ-modification. In clinker, C_2S occurs in the β-form with the possibility of MgO, Al_2O_3, P_2O_5, Na_2O, CaO, and K_2O entering into solid solution.

The study of the phase equilibria of binary and ternary oxide systems is an important approach to the study of clinker chemistry, the most pertinent being the ternary system $CaO–Al_2O_3–SiO_2$. In the binary system $CaO–SiO_2$, four compounds of composition CS, C_3S_2, and C_3S are formed, the last two forming preferentially at higher lime concentrations. In the $CaO–Al_2O_3$ system, C_3A, C_5A_3 or $C_{12}A_7$, CA and C_3A_5 or CA_2 are detected. In Portland cement only C_3A and C_5A_3 may be present. In clinker formation the rate of reaction of clay–$CaCO_3$ mixtures is faster than that in pure oxide systems.

Differential Thermal Analysis

The DTA technique yields information on the mineralogy of the siliceous and calcareous raw materials used to make cement clinker. The solid state reactions in the binary and ternary systems relevant to clinker chemistry and the polymorphic transformations in C_3S and C_2S are also advantageously studied by DTA.

Siliceous Materials. In the formation of various phases in clinker the clay or shale contributes SiO_2, Al_2O_3, Fe_2O_3, alkalies, and alkaline earths. Clay minerals such as attapulgite, sepiolite, and chlorite containing MgO in larger proportions are not suitable as raw materials for the manufacture of Portland cement. Kaolinite, containing mainly alumina and silica, would be the most desirable clay mineral, especially for the production of white cement. The montmorillonite clay mineral may be used provided it satisfies the chemical requirements. Illitic clays are suitable if they do not possess excess alkalies. Nontronitic clay mineral would not be suitable because of a high Fe_2O_3 content. In the wet manufacturing process kaolinite is preferred to illite because of its superior burning and nondusting characteristics (*8*).

DTA can be used to identify the clay mineral making the siliceous source in the manufacture of clinker. In Fig. III.1.1 are given the typical thermograms of kaolinite, illite, montmorillonite, chlorite, sepiolite, and attapulgite (*56*). The endothermic reaction in kaolinite at 550-600°C is accompanied by dehydration and the exothermic reaction at 980°C is associated with the formation of γ-Al_2O_3 (*9*) or nucleation of mullite. The endothermic and exothermic peaks of large intensity are characteristic of kaolinite and are detectable even in a concentration of 10%. Poorly crystallized kaolinites may indicate a small endothermic peak in the vicinity of 100°C.

Illite exhibits endothermic peaks at 100-200°C, 500-650°C, and at about 900°C, and an exothermic peak immediately following the third endothermic effect. The first endothermic effect corresponds to the loss of interlayer water, the second and the third to the expulsion of water from the lattice and the destruction of illite structure respectively. The exothermic peak is probably associated with the formation of a spinel.

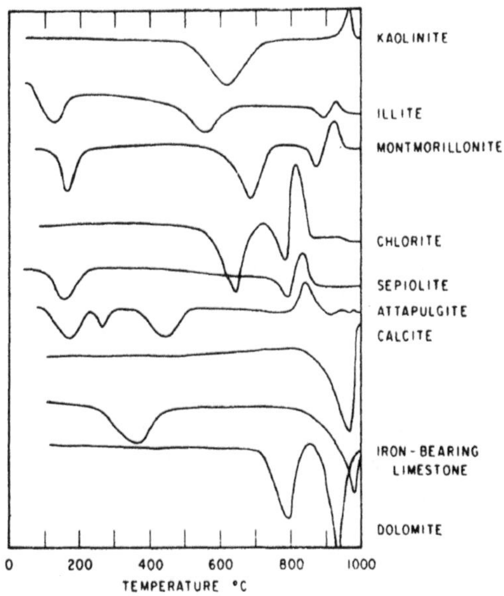

Fig. III.1.1: DTA of some raw materials for the manufacture of Portland cement (*56*).

Intensity of the peaks in illites is of lower magnitude than that in kaolinite. Illite differs from kaolinite in showing additional endothermal effects at 100-200°C and 900°C. The montmorillonitic clay shows an initial endothermal peak at 100-250°C owing to the loss of interlayer water and another at 600-700°C due to loss of lattice water. The endothermic peak at about 900°C corresponds to the final breakdown of the montmorillonitic lattice. Immediately after the third endothermic peak, an exothermal peak appears and is associated with the formation of a spinel. Chlorite indicates an endothermic reaction at 500-700°C and another at 800°C. The second endothermal effect is followed by an exothermic peak. It is generally believed that the (OH) water in chlorite is driven off in two stages (*10*, *11*).

The dehydration of brucite layer corresponds to the first endothermic peak and the dehydration and loss of mica layer structure result in an endothermic effect at 750-800°C. The exothermal effect at 800-900°C is adduced to the formation of olivine. Sepiolite exhibits a large endothermic reaction starting below 100°C followed by a series of endothermal effects between 200 and 700°C. In addition there is a sharp endothermal peak at 700°-800°C and a sharp exothermal effect.

Hygroscopic and zeolitic water are expelled below 350°C (*12*, *13*). The high temperature endothermic effect is caused by the final dehydration and loss of structure. The formation of enstatite may be responsible for the exothermal reaction at 850°C. Attapulgite shows three endothermal reactions due to loss of water at about 200°C, 225-350°C and 400-530°C. It is believed that the high temperature exothermic peak is due to the formation of enstatite.

In Table III.1.2 are presented the differential thermal data for some of the typical clay minerals. The peak temperature of any mineral is shown to occur over a range because of the possibility of variations due to impurities and operative factors.

Though DTA is extensively applied to identify clay minerals it has certain limitations. The method cannot distinguish illite from halloysite, nontronite, or poorly crystallized kaolinite as all these minerals exhibit the characteristic endothermic peaks between 500 and 600°C due to the removal of (OH) water. The exothermic peak at 900-1000°C is shown by most of the clay minerals and is not recommended as a guide for identification purposes. Ramachandran *et al* (*14-19*) have discovered that the DTA of clay mineral-dye complexes is able to readily distinguish the above clay minerals by characteristic exothermal peaks. In this method the clay is treated with 0.2% aqueous solution of methylene blue, methyl violet, or malachite green. After 24 hr the unadsorbed dye is removed by filltration and washing.

Fig. III.1.2 shows the thermograms of MB-complexes of four mixtures containing illite and nontronite. In all the complexes the first characteristic exothermic peak occurs at 470-490°C and the second at 580-600°C due respectively to the presence of illite and nontronite. A mixture of pure illite and nontronite would exhibit only one endothermic peak between 500 and 600°C. In the dye-clay mineral complexes the exothermal peak areas could be used to determine the proportions of component clay minerals in mixtures. The principal exothermal peaks for the identification of kaolinite, halloysite, illite, nontronite, and montmorillonite are 430, 450, 470, 590, and 670°C respectively.

The application of this method for identification of clay minerals in binary and ternary mixtures such as illite–nontronite, illite–montmorillonite, kaolinite–montmorillonite, kaolinite–nontronite, kaolinite–illite, montmorillonite–nontronite, kaolinite-illite–montmorillonite, kaolinite–illite–nontronite, kaolinite–nontronite–montmorillonite, and illite–nontronite–montmorillonite has also been discussed by Ramachandran and co-workers (*15*, *16*).

TABLE III.1.2
DTA Data of Some Clay Minerals

		Endothermic effects			Exothermic effects		
Clay mineral	*Ionic composition*	*peak temperature, °C*	*intensiy*	*interpretation*	*peak temperature °C*	*intensity*	*interpretation*
Kaolinite	$Al_4Si_4O_{10}(OH)_8$	105-115	Small	Adsorbed water (poorly crystallized kaolinite)	980-990	Very intense	γ-Al_2O_3 and/or nucleation of mullite
		550-600	Intense	Dehydroxylation			
Halloysite	$Al_4Si_4O_{10}(OH)_84H_2O$	110-125 540-560	Moderate Intense	Interlayer water Dehydroxylation	980-990	Intense	〃
Illite	$(K\cdot H_3O)Al_2(Si\cdot Al)_4O_{10}(OH)_2$	110-125 540-580 890-910	V. small Small Moderate	Adsorbed water Dehydroxylation Lattice destruction	940-950	Medium	Spinel
Mont-morillonite	$(Al_{3.33}\cdot Mg_{0.67})(Si_8)O_{20}(OH)_4$ ↓ $M^+_{0.67}$	130-160	Moderate	Interlayer water	900-920	Small	Spinel
		680-710	Moderate	Dehydroxylation			
Nontronite	$Fe^{+3}_{4.00}(Al_{0.66}Si_{7.34})O_{20}(OH)_4$ ↓ $M^+_{0.66}$	140-150	Moderate	Interlayer water	900-910	Small	Spinel
		480-500	Moderate	Dehydroxylation			
Attapulgite	$Mg_5Si_8O_{20}(OH)_2\cdot 8H_2O$	90-110 225-250	Small Moderate	Zeolitic water Water coordinated with magnesium	850-870	Moderate	Enstatite, sillimanite, and cristobalite
		400-430	Moderate	Water from silicate unit			
Chlorite	Alternate mica-like $(Mg\cdot Fe)_6(Si\cdot Al)_8O_{20}(OH)_4$ and brucite-like $[(Mg\cdot Al)_6(OH)_{12}]$ layers	180-200 500-650	Small Moderate	Adsorbed water Decomposition of brucite layer	900-930	Moderate	Olivine and Spinel
		800-850	Small	Dehydration of mica layer			

Note: M+ refers to a monovalent cation.

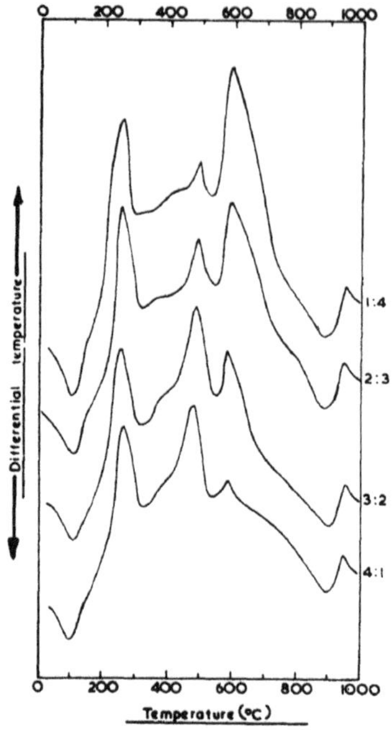

Fig. III.1.2: DTA of methylene blue complexes of illite–nontronite mixtures (*57*).

An example of the potentiality of DTA for studying the economics of cement manufacture is provided by Mehta (*20*). The siliceous source for clinker formation may be kaolinite or quartz from clays and limestones. If quartz is used as a siliceous source it would involve expensive grinding and burning. The presence of quartz in mixes is identifiable by a reversible endothermic effect at about 575°C. Also, if a clay contains a bituminous residue, an exothermal effect should occur over a wide range (*250-500*°C) and this makes possible an assessment of the heat effects in the kiln below 500°C.

Calcareous Materials. The calcareous source will not be suitable for cement production if it is associated with a high MgO content. The presence of pyrites is also not desirable. If pyrites, iron oxide, magnesite, brucite, or dolomite is present as a contaminant DTA can identify its presence. Limestone is identified by a single endothermal peak at a temperature of 850-950°C due to decarbonation. An iron-bearing limestone may show an endothermic peak at about 350-400°C (*21*) (Fig. III.1.1), whereas a brucite-bearing limestone may show an endothermic effect at about 475°C caused by the dehydration of $Mg(OH)_2$.

Magnesite is identified by an endothermic effect at about 700°C. Pure dolomite exhibits two endothermic peaks at about 800 and 950°C. The first endothermic effect is caused by the decomposition of $MgCO_3$, and the second, by the decomposition of $CaCO_3$. If dolomite is present as a contaminant the peak temperatures are slightly shifted. In carbonate rocks dolomite could be estimated even if present to an extent of only 0.3% (*22*).

Lime–Clay Reactions. A rapid heating of the binary and ternary mixtures in a thermal analyzer may provide indication of the course of reactions through the appearance of thermal inflections. High-temperature DTA units have been developed to study solid state reactions up to 1600°C (*23, 24*). A thermal analysis of a mixture of calcareous substance and clay shows up as several endothermal and exothermal effects (*25-30*).

A raw mix used for the production of cement clinker may exhibit four thermal peaks at 1200, 1280, 1380, and 1420°C (*30*). The exothermic peaks at 1200°C corresponds to the formation of β-C_2S, and that at 1280°C, to the formation of C_3S and liquid. The peak at 1380°C represents the melting of the ferrite phase and the endothermic effect at 1420°C denotes the transformation from β to α-C_2S.

De Keyser (*25*) heated a mixture of kaolin and lime and detected gehlenite (C_2AS) and a small quantity of β-C_2S at 900°C. At 1050°C the amounts of C_2AS decreased and C_5A_3 formed. At 1100°C β-C_2S increased and C_3A appeared. At 1400°C C_3S was formed by the combination of CaO and C_2S.

DTA has also been applied to study the reactions in the production of white cement (*27*) and to examine the nature of the dolomite–kaolinite reaction (*29*). A mixture of calcined clay and lime may show an exothermic peak at 890°C for the lime–clay reaction.

Lime–Silica Reactions. Thermal reactions in the system CaO–SiO_2 and $Ca(OH)_2$–SiO_2 have been studied by DTA. In the thermal investigations of mixtures of $3CaO+SiO_2$, $2CaO+SiO_2$, and $CaO+SiO_2$ an exothermal effect was indicated in all the mixes and this was ascribed to a change in the lattice structure of silica (*32*). Barta and co-workers (*34-37*) carried out extensive work on the thermal reactions of binary and ternary oxide systems. In Fig. III.1.3 is shown the formation of C_3S from a mixture of $3CaO+SiO_2$ and also the effect of Al_2O_3, Fe_2O_3, and $Al_2O_3+Fe_2O_3$ on the temperature of formation of C_3S. The exothermic peak at 1450°C indicates the formation of C_3S. The peak temperature decreases to 1380, 1350, and 1320°C with additions of 5% Al_2O_3, 2% Al_2O_3+2% Fe_2O_3, and 5% Al_2O_3+5% Fe_2O_3 respectively.

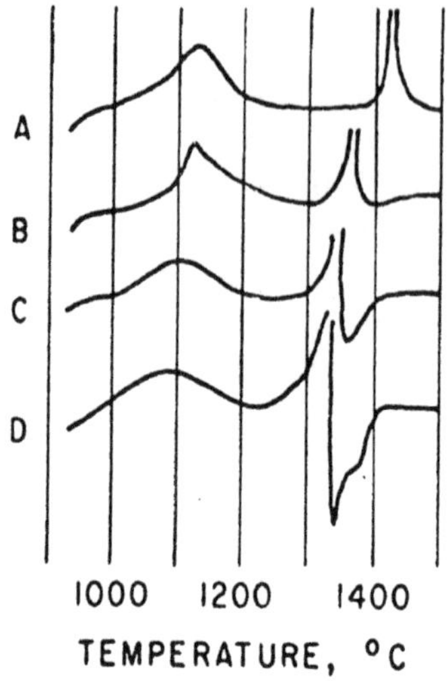

Fig. III.1.3: Thermograms of $3CaO \cdot SiO_2$ (A) $3CaO+SiO_2$, (B) with 5% Al_2O_3, (C) with 2% Al_2O_3+2% Fe_2O_3, (D) with 5% Al_2O_3+5% Fe_2O_3 (*37*).

DTA along with X-ray technique is used to study the effect of Al_2O_3, Fe_2O_3, MgO, Na_2O, and K_2O on the temperature of formation of alite (*38*). DTA shows that the decomposition of C_3S with B_2O_3 results in the formation of CaO and a saturated solution of B_2O_3 in C_5BS (*39*).

The rate of formation of a silicate is enhanced if the starting materials have high surface areas. This has been shown in the reaction of SiO_2 of differing reactivities with calcium carbonate (*31*).

De Keyser (*33*) heated a mixture of 2 parts $Ca(OH)_2$ and one part SiO_2 in a DTA furnace up to 1000°C. After obtaining a thermogram a second run was made and samples were removed for X-ray examination at the cha-

Thermal treatment	*Phases identified*
470°C	$Ca(OH)_2$, little γ-C_2S
480°C	Quantity of γ-C_2S becomes considerable
552°C	γ-C_2S increases and β-C_2S appears
806°C	Decrease of γ-C_2S and formation of β-C_2S in increasing quantities through inversions
837°C	The above tendencies accentuated
912°C	Disappearance of γ-C_2S and increase of β-C_2S
960°C	β-C_2S the only silicate detected

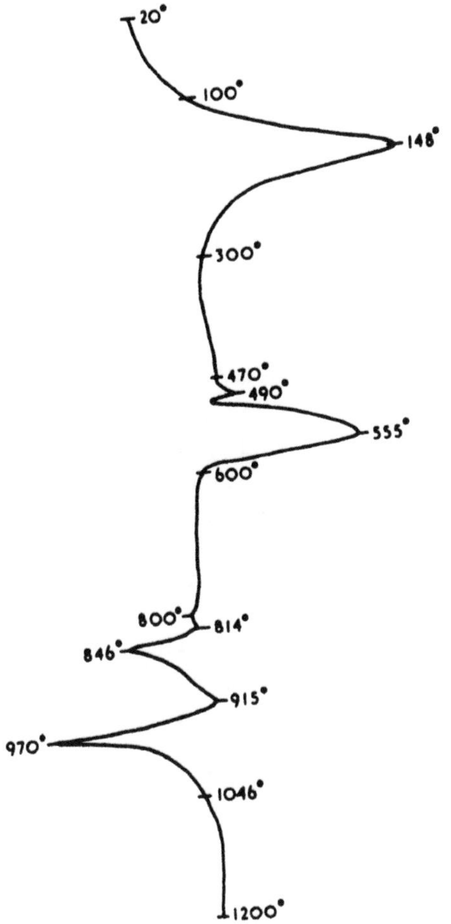

Fig. III.1.4: Thermogram of $Ca(OH)_2$–SiO_2 mixture (*33*).

racteristic temperatures of thermal inflections. Fig. III.1.4 shows the thermogram of a $Ca(OH)_2$–SiO_2 mixture (endothermal effects are shown as upward peaks). The identification of the thermal effects is facilitated through the X-ray study as described on p. 62.

$CaCO_3$–SiO_2 Reactions. It has been observed that the nature of the reactions in the system CaO–SiO_2 may not reflect that occurring in a cement kiln, because of the prevalence of CO_2-atmosphere. Hence some attention has been directed to the study of the system $CaCO_3$–SiO_2. Pure mixtures of $CaCO_3$ and SiO_2 fired in a jet mill show two endothermal effects at 900 and 976°C corresponding to triclinic→monoclinic→trigonal transformations in C_3S (*40*). Against this De Keyser (*41*) reported that a mixture of $CaCO_3$

and SiO_2 behaved similarly to $Ca(OH)_2$–SiO_2 in indicating the formation of C_2S at 808 °C, transformation of γ-C_2S to α'-C_2S at 912°C, and the transformation of α'-C_2S to β-C_2S at 965°C.

Evidence of formation of C_3S at as low a temperature as 840°C has been advanced in studies on $CaCO_3 + SiO_2$ (*42*). Wilburn and Thomasson (*43*) in their studies of silica flour and precipitated $CaCO_3$ found an exothermic peak at 1400°C and an endothermic peak at 1430°C.

A mineralizer such as calcium fluoride or fluorspar is added to the raw cement mix to accelerate the formation of cement minerals. The effect of fluorides on the formation of various silicate and aluminate phases has been studied extensively by Courtault (*44*), Mukerji (*45*), Budnikov and Sologubova (*26*, *27*), Sychev and Barkan (*46*), and Bereczky (*47*).

The effect of additions of alkali and alkaline earth fluorides on the $CaCO_3$ –SiO_2 reaction is conveniently studied by DTA (*44*). In Fig. III.1.5 is presented the thermogram of $CaCO_3 + SiO_2$ mixture and that containing additions of LiF, NaF, or KF (fluorine content added in each case *0.39*%). Without the addition of any fluoride the curve shows a slope at about 920 °C, while that treated with alkali fluórides shows a slope starting at a temperature of about 700 °C.

It is believed that spurrite [$(C_2S)_2 \cdot CaCO_3$] formed in an atmosphere of CO_2 is a transition phase that converts to C_3S. The amount of spurrite formed is determined.by DTA during the cooling cycle, as can be seen in Fig. III.1.6. The height of the thermal peak indicates that the effectiveness of the alkali fluorides is in the order LiF>NaF>KF. Amongst alkaline

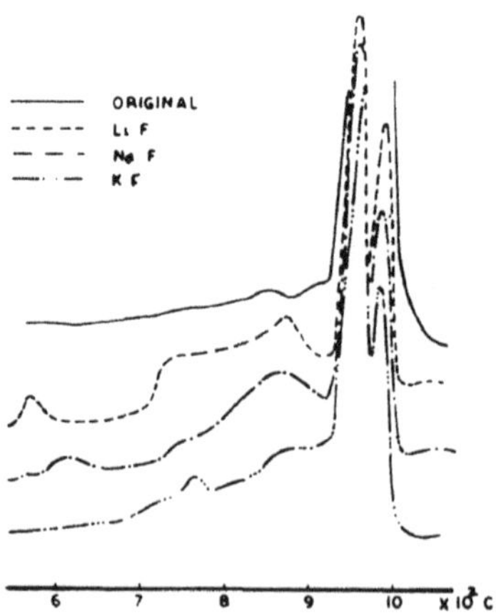

Fig. III.1.5: DTA of $CaCO_3$–SiO_2 with additions of fluorides (*44*).

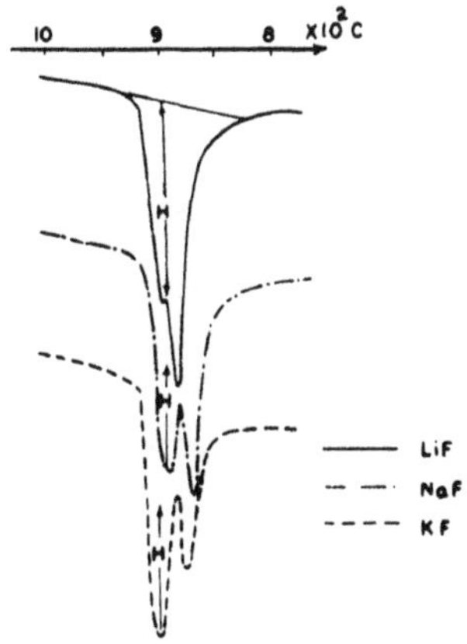

Fig. III.1.6: Cooling curves of $CaCO_3$–SiO_2 containing fluorides (*44*).

earth fluorides the mineralizing effect is in the order $CaF_2 > SrF_2 > BaF_2$. With MgF_2 as a mineralizer there is a possibility of formation of isotropic compounds at about 400°C (*46*).

$CaCO_3$–Al_2O_3 System. Thermal reactions in the synthesis of calcium aluminates from a mixture of $CaCO_3$ and Al_2O_3 in molar ratios of 1.5 to 5.1 show an endothermal effect in the eutectic temperature range 1375-1388°C (*48*). The products identified are C_5A_3 and CA. In Fig. III.1.7 are shown the DTA curves of different proportions of CaO in Al_2O_3 (*34-37*). At all concentrations an exothermic effect is observed in the range 950-1000°C, which can be ascribed to a simultaneous formation of CA and $C_{12}A_7$. The endothermal effect at 1170°C represents the change of γ-Al_2O_3 to the α-form and the other at 1200°C denotes the onset of formation of C_3A. Melting in the system is indicated by an endothermic effect at about 1300°C.

The mineralizing action of CaF_2 on the system $CaCO_3 + Al_2O_3$ has been studied by Courtault (*44*). The reaction is accelerated in presence of CaF_2. In this case the formation of a compound of the formula $C_{12}A_7.2.5CaCO_3$ is postulated to explain the catalyzing action of CaF_2.

CaO–Fe_2O_3 System. Reactions in the CaO–Fe_2O_3 system containing different proportions of CaO are shown in Fig. III.1.8. In all mixtures CF forms at 950°C and is independent of the relative proportions of CaO and Fe_2O_3. Melting occurs at about 1150°C; this is reflected as an endothermal effect and should correspond to the eutectic temperature of 1200°C in the phase diagram. The compound C_2F is formed subsequently, the endothermic valley at 1300°C corresponding to an incongruent melting of the mix-

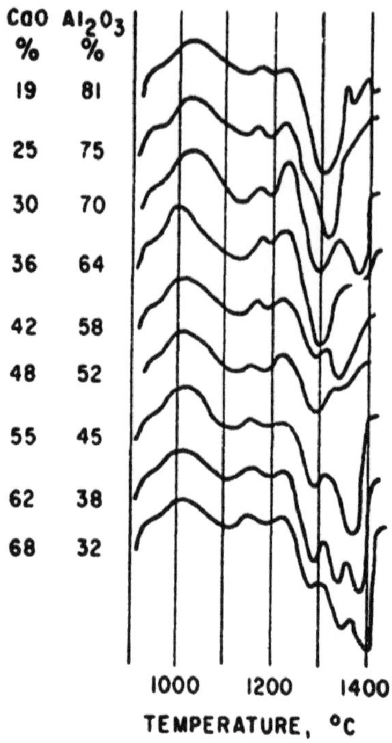

Fig. III.1.7: Thermograms of mixtures of CaO and Al_2O_3 (*37*).

ture.

CaO–Al₂O₃–Fe₂O₃ System. Interesting results have been obtained in the study of the ternary system containing CaO, Al_2O_3, and Fe_2O_3 (*34-37*, *49*). The DTA of three mixes containing different proportions of CaO, Al_2O_3, and Fe_2O_3 is shown in Fig. III.1.9. The exothermal formation of C_4AF takes place at about 980°C. Dicalcium ferrite (C_2F) is formed at lower contents of Al_2O_3, as indicated by an endothermal dip at about 1160°C. De Keyser (*49*) examined the thermal reactions in three ternary mixtures, *viz.* $6CaO + Al_2O_3 + 2Fe_2O_3$, $4CaO + Al_2O_3 + Fe_2O_3$, and $6CaO + 2Al_2O_3 + Fe_2O_3$, and compared them with a binary mixture $2CaO + Fe_2O_3$. In all the mixtures C_2F was identified at about 800°C with CF forming subsequently. At higher temperatures CaO combined with CF to give C_2F. The compound C_5A_3 was also detected in the temperature range 1000-1250°C. De Keyser's results

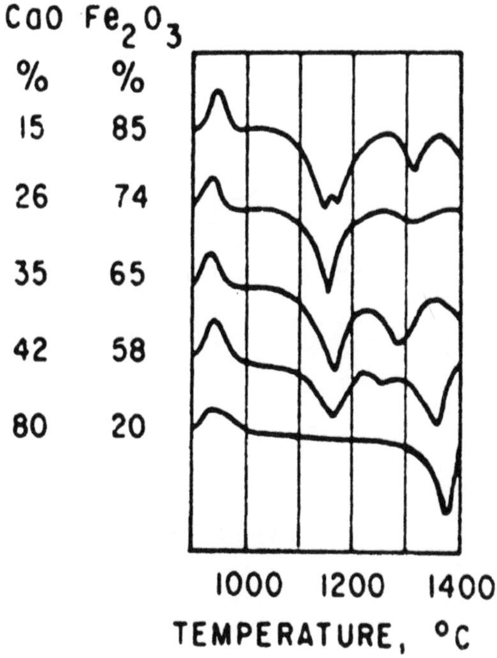

Fig. III.1.8: Thermograms of mixtures of CaO and Fe_2O_3 (*37*).

agree with those of Barta in detecting the maximum formation of C_2F at about 1200°C.

Other Systems. The other ternary system studied by De Keyser (*25, 50*) is $CaO–Al_2O_3–SiO_2$ in the temperature range 1000-1400°C. Warburton and Wilburn (*55*) have obtained thermograms in the system $CaCO_3–SiO_2–Al_2O_3$.

In the quaternary system $CaO–Al_2O_3–Fe_2O_3–SiO_2$ van Bemst (*51*) detected β-C_2S, which was followed by β-CS and C_2AS. At temperatures above 1000°C C_3A and $C_6A_XF_Y$ were indicated and above 1100°C β-C_2S formed in addition. In mixtures containing higher proportions of Fe_2O_3 and alkalies C_3S was preferably formed at 1300°C. The mechanism of formation of alite in the quaternary system has been studied by Ueda (*52*). In the system C_3A-C_5A_3-C_2F Ueda did not detect thermal reactions corresponding to the formation of calcium ferrite and calcium aluminoferrite. There was, however, an endothermal effect at about 1380°C, which was ascribed to the melting on the celite phase.

By using quench and DTA methods Stutterheim (*53*) studied the quaternary system $CaO–Al_2O_3–SiO_2–MgO$. The phase diagram of the binary system gehlenite–spinel could be elucidated by these methods.

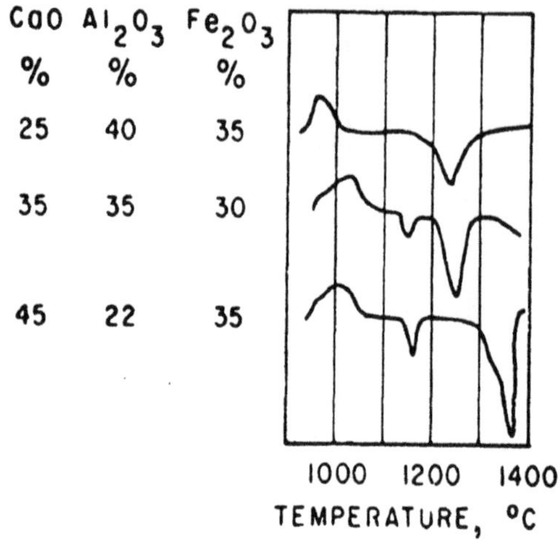

Fig. III.1.9: Thermograms of ternary mixture containing CaO, Al_2O_3 and Fe_2O_3 (*37*).

REFERENCES

1. Bogue, R.H., *Chemistry of Portland Cement*, Reinhold, New York, 1955.
2. Taylor, H.F.W., *The Chemistry of Cements*, Vol. I, Academic Press, London, 1964.
3. Ershov, L.D., *Tsement*, *4*, 19, 1955.
4. Lea, F.M., *The Chemistry of Cement and Concrete*, Edward Arnold Ltd., London, 1956.
5. Jeffery, J.W., *Acta. Cryst.*, *5*, 26, 1954.
6. Jeffery, J.W., *Proc. 3rd Intern. Symp. Chem. Cements*, London, p. 30, 1952.
7. Brunauer, S., Copeland, L.E., Kantro, D.L., Weise, C.H., and Schulz, E.G., *Am. Soc. Testing Mater. Bull.*, *59*, 1091, 1959.
8. Schwiete, H.E., *Zement-Kalk-Gips*, *9*, 351, 1956.
9. Insley, H., and Ewell, R.H., *J. Res. Nat. Bur. Std.*, *14*, 615, 1935.
10. Barshad, I., *Am. Mineralogist*, *33*, 655, 1948.
11. Nutting, P.G., *U.S. Geol. Surv. Profess. Paper 197E*, 197, 1943.
12. Longchambon, H., *Bull. Soc. Franc. Mineral.*, *60*, 232, 1937.
13. Migeon, G., *Bull. Soc. Franc. Mineral.*, *59*, 6, 1936.
14. Ramachandran, V.S., Kacker, K.P., and Patwardhan, N.K., *Nature*, *191*, 696, 1961.
15. Ramachandran, V.S., Kacker, K.P., and Patwardhan, N.K., *Trans. 8th Intern. Ceramic Congr.*, Copenhagen, 293, 1962.

16. Kacker, K.P., and Ramachandran, V.S., *Trans. 9th Intern. Ceramic Congr.*, Brussels, 483, 1964.
17. Ramachandran, V.S., Garg, S.P., and Kacker, K.P., *Chem. & Ind. (London)*, *23*, 790, 1961.
18. Ramachandran, V.S., and Kacker, K.P., *J. Appl. Chem. (London)*, *14*, 455, 1964.
19. Ramachandran, V.S., Kacker, K.P., and Patwardhan, N.K., *Am. Mineral.*, *47*, 165, 1962.
20. Mehta, P.K., *Pit & Quarry*, p. 141, 1964.
21. Verma, M.P.C., *Indian Concrete J.*, *30*, 175, 1956.
22. Rowland, R.A., and Beck, C.W., *Am. Mineral.*, *37*, 76, 1952.
23. Newirk, T.F., *J. Am. Ceram. Soc.*, *41*, 409, 1958.
24. Courtault, B., *Rev. Mater. Construct.*, *No. 569*, 37, 1963.
25. DeKeyser, W.L., *Bull. Soc. Chim. Belges*, *63*, 40, 1954.
26. Budnikov, P.P., and Sologubova, O.M., *Ukr. Khim. Zh.*, *19*, 92, 1953.
27. Budnikov, P.P., and Sologubova, O.M., *Silikat Tech.*, *4*, 503, 1953.
28. Surovkin, V.M., and Krylov, G.M., *Uzbeksk Khim. Zh.*, *6*, 68, 1962.
29. Ivanova, V.I., and Tatarskii, V.B., *Dokl. Akad. Nauk. SSSR*, *73*, 341, 1950.
30. Ueda, M., *Semento Gijutsu Nenpo*, *6*, 116, 1952.
31. Schramli, W. and Becker, F., *Zement-Kalk-Gips*, *13*, 265, 1960.
32. Vasenin, F.I., *Zh. Prikl. Khim.*, *21*, 389, 1948.
33. De Keyser, W.L., *Proc. 3rd Intern. Symp. Chem. Cements*, London, p. 84, 1952.
34. Barta, R., and Satava, V., *Stavivo*, *31*, 15, 1953.
35. Barta, R., and Satava, V., *Proc. Glass and Ceramics*, *SNTL 2*, Praha, 1954.
36. Barta, R., *Chemie*, p. 257, 1958.
37. Barta, R., *Chemistry and Technology of Cement*, Ceskoslovenske Akad. Ved., Praha, 1961.
38. Ueda, M., *Semento Gijutsu Nenpo*, *9*, 13, 1955.
39. Mircea, S., *Silikaty*, *9*, 34, 1965.
40. Sychev, M.M., and Korneev, V.I., *Zh. Prikl. Khim.*, *37*, 2295, 1964.
41. De Keyser, W.L., *Bull. Soc. Franc. Ceram.*, p. 2, 1953.
42. Subba Rao, B.V.S., Mehdi, S., and Datar, D.S., *J. Sci. Ind. Res. (India)*, *21D*, 249, 1962.
43. Wilburn, F.W., and Thomasson, C.V., *Phy. Chem. Glasses*, *2*, 126, 1961.
44. Courtault, B., *Rev. Mater. Construct. No. 570*, 67; *No. 571*, 110; *No. 572*, 143; *No. 573*, 190, 1963.
45. Mukerji, J., *J. Am. Ceram. Soc.*, *48*, 210, 1965.
46. Sychev, M.M., and Barkan, A.S., *Trudy Leningrads kogo Tekhnologi Cheskogo Instituta imeni Lensoveta, Vypusk 56*, 81, 1960.
47. Bereczky, A., *Silikat Tech.*, *11*, 474, 1960.
48. Ao, Tokeo, *Assn. Japan. Portland Cement Engrs. Rev.*, *23*rd Gen. Meeting, Tokyo, p. 22, 1936.
49. De Keyser, W.L., *Bull. Soc. Chim. Belges*, *64*, 395, 1955.
50. De Keyser, W.L., *Rev. Gen. Sci. Appl.*, *1*, 84, 1952.
51. Van Bemst, *Silicates Ind.*, *26*, 290, 1961.
52. Ueda, M., *Semento Gijutsu Nenpo*, *8*, 16, 1954.
53. Stutterheim, N., *Ind. Chem. (S. Africa)*, *6*, 136, 1952.
54. Longuet, P., and Courtault, B., *Silicates Ind.*, *28*, 28, 1963.

55. Warburton, R.S., and Wilburn, F.W., *Phys. Chem. Glasses*, *4*, 91, 1963.
56. Ramachandran, V.S., Feldman, R.F., and Sereda, P.J., *Highway Res. Rec. No. 62*, 1964.
57. Kacker, K.P., Ph.D. thesis, Banaras Hindu University, 1963.

III.2. POLYMORPHISM OF C_3S

In 1897, Törnebohm (*1*) reported that Portland cement comprises four compounds of which the dominant phase is alite. The formula of alite was given as $3CaO \cdot SiO_2$ by Guttmann and Gille (*2*) in 1931. It was shown by Jander and Wuhrer (*3*) and Koyangi and Sudoh (*4*) that alite is not a pure C_3S but contains small amounts of MgO and C_3A. Jeffery (*5-7*) has conclusively shown that the alite phase can be regarded as C_3S modified by slight 'solid solution,' i.e. by the substitution of two Al atoms and one Mg atom for two Si atoms, giving the formula $C_{54}S_{16}AM$.

C_3S occurs in three modifications at room temperature. The synthetic preparation is triclinic, that formed in open hearths is trigonal, while in Portland cement clinker it is monoclinic (alite).

The cementing properties of C_3S and those of alite are not much different. Nurse (*8*) compared the compressive strengths of cubes containing C_3S or alite with 3 parts sand and found the strength to be slightly higher in C_3S cubes, especially at earlier periods of curing.

Interesting results have been reported on the thermal transformations in C_3S and alite using X-ray and DTA.

Differential Thermal Analysis

Nurse and Welch [quoted by Jeffery (*7*)] studied the thermograms of pure C_3S and alite (Fig. III.2.1). The C_3S preparation exhibits as many as six endotherms at 464, 622, 750, 923, 980, and 1465°C. The peak at 464°C

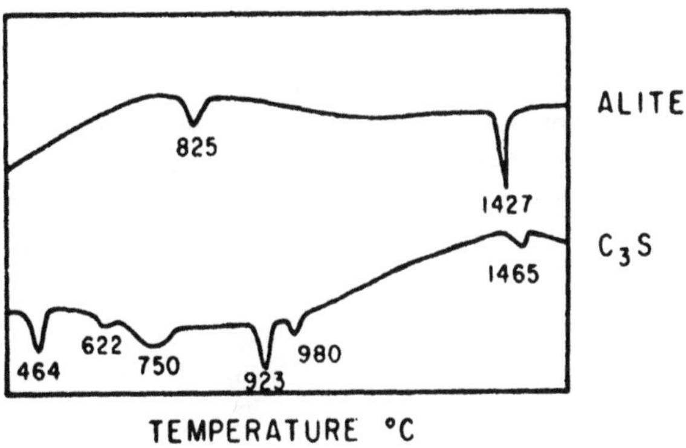

Fig. III.2.1: Thermograms of alite and pure tricalcium silicate (*7*).

is attributed to the dehydration of $Ca(OH)_2$ formed as a consequence of the hydration of free lime present in the C_3S preparation. C_2S is present as a contaminant in C_3S as is evidenced by three peaks at 622, 750 and 1465°C due respectively to $\beta \rightarrow \alpha'$, $\gamma \rightarrow \alpha'$, and $\alpha' \rightarrow \alpha$ transitions. The peaks at 923° and 980°C are characteristic of C_3S and may be due to triclinic→monoclinic and monoclinic→trigonal transitions or to triclinic→trigonal and trigonal →trigonal transition + rotation of anions respectively.

The thermogram of synthetic alite differs from pure C_3S by exhibiting only two endotherms at 825 and 1427°C. The effect at 825°C is due to the monoclinic→trigonal transition and corresponds to a similar effect in C_3S at 980°C. This transition occurs at a lower temperature in alite owing to the solid solution effect. The peak at 825°C should not be ascribed to $\gamma \rightarrow \alpha'$ C_2S transition because C_2S is present as the β or α'-form in the alite preparation. Yamaguchi and Miyabe (*9*) also obtained similar results.

X-ray and DTA studies of Yannaquis and co-workers (*10*) showed transitions for C_3S at 615, 750, 917, 973, and 1050°C. They inferred that before the triclinic form converts to monoclinic form three transitions occur at 615, 750, and 917°C. The triclinic→monoclinic transition occurs at 973°C and the final monoclinic→trigonal transition, at 1050°C.

Yamaguchi *et al* (*11*) have compared the thermograms of pure C_3S with those of two types of solid solutions of composition $3[(1-x)CaO \cdot xMgO]$ SiO_2 and $3[(1-y)CaO \cdot CaF_2]$ SiO_2. Pure C_3S exhibits two characteristic endothermal effects at 925 and 985°C. In solid solutions these peaks occur at lower temperatures, *viz.* at 895 and 940°C.

A systematic investigation on the role of alumina and magnesia in the polymorphism of C_3S has been carried out by Midgley and Fletcher (*12*).

In Table III.*2.1* is presented the variation of peak temperatures and X-ray diffraction lines by substituting CaO with varying amounts of MgO.

TABLE III.2.1
DTA and Composition Data on Magnesium Substituted C_3S
[Midgley and Fletcher (12)]

Sample	*DTA temperature at which endothermic reactions are observed, °C*			
C_3S	595	747	916	969
$C_{155}MS_{52}$	—	721	911	954
$C_{154}M_2S_{52}$	—	663	907	923
$C_{153}M_3S_{52}$	—	—	901	—
$C_{152}M_4S_{52}$	—	—	886	—
$C_{151}M_5S_{52}$	—	—	871	—
$C_{150}M_6S_{52}$	—	—	864	—
$C_{146}M_{10}S_{52}$	—	—	860	—

Four transitions are evident in pure C_3S at 595, 747, 916, and 969°C. The peak at 750°C for the transition in the triclinic form was also reported by Yannaquis *et al* (*10*). In the solid solutions the temperatures corresponding to triclinic→monoclinic and monoclinic→trigonal transition are lowered. The monoclinic form is stabilized with the composition in the range $C_{152}M_4S_{52}$ and $C_{151}M_5S_{52}$.

In preparations containing Al^{3+}, three Al^{3+} replace $3Si^{4+}$ with the excess charge being balanced by Al^{3+} in the interstitial spaces. Substitution of Al^{3+} does not influence polymorphism. It was also observed that the C_3S preparation containing saturated amounts of Al_2O_3 and MgO did not show any inflection in the thermograms. Midgley and Fletcher also presented DTA results for the systems C_3S–$C_{4.5}A$ and C_3S–C_6A and for the composition $C_{156-x}M_xS_{50.5}A$.

A re-examination of the polymorphism in C_3S was carried out by Miyabe and Roy (*13*) using a precise X-ray diffractometer and DTA. In Fig. III.2.2 thermograms of C_3S are presented for three cycles of heating and cooling. In the heating cycle two sharp endothermic effects with characteristic temperatures at 910±5°C and 970±5°C can be seen, the former being of greater intensity. The transition of triclinic→monoclinic form takes place at 910°C and monoclinic→trigonal form, at 970°C. In the cooling cycle these endotherms show up as exothermal effects without much shift in the peak temperatures. Additional endothermal humps were also detected at 500°-600°C and 650°-690°C. The first endothermic effect may indicate $\alpha \rightarrow \beta$ transition in triclinic C_3S and the second, its completion. It is quite probable that the two endothermal effects correspond to two triclinic→triclinic transitions.

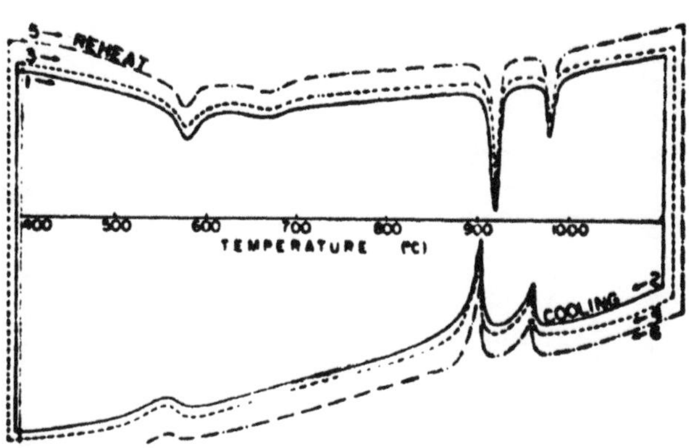

Fig. III.2.2: Differential thermal analysis of $3CaO \cdot SiO_2$ showing successive heating and cooling cycles, with a heating rate of 10°C/min (*13*).

REFERENCES

1. Törnebohm, A.E., *Tonind.-Z. Keram. Rundschau*, *21*, 1148, 1897.
2. Guttann, A., and Gille, F., *Zement*, *20*, 144, 1931.
3. Jander, W., and Wuhrer, J., *Zement*, *27*, 73, 86, 337, 1938.
4. Koyanagi, K. and Sudoh, T., *Zement*, *28*, 563, 1939.
5. Jeffery, J.W., Ph.D. Thesis, London University, 1950.
6. Jeffery, J.W., *Acta Cryst.*, *5*, 26, 1952.
7. Jeffery, J.W., *Proc. 3rd Intern. Symp. Chem. Cements*, London, pp. 30-48, 1952.
8. Nurse, R.W., *J. Appl. Chem. (London)*, *2*, 244, 1952.
9. Yamaguchi, G., and Miyabe, H., *J. Am. Ceram. Soc.*, *43*, 219, 1960.
10. Yannaquis, N., Rigourd, M., Maziers, Ch., and Guinter, A., *Bull. Soc. Franc. Mineral. Crist.*, *85*, 271, 1962.
11. Yamaguchi, G., Ikegami, H., Shirosuga, K., and Amano, K., *Semento Gijutsu Nenpo*, *11*, 24, 1957.
12. Midgley, H.G., and Fletcher, K.E., *Trans. Brit. Ceram. Soc.*, *62*, 917, 1963.
13. Miyabe, H., and Roy, D.M., *J. Am. Ceram. Soc.*, *47*, 318, 1964.

III.3. POLYMORPHISM OF C_2S

The polymorphism of C_2S has aroused much interest in clinker chemistry. The phase equilibrium studies reveal the existence of three stable forms α, β, and γ-C_2S (*1-3*). The α-form is shown to be stable above 1420°C. This form transforms to the β-modification, which in turn converts to the γ-form at 675°C. The fourth form of C_2S, denoted by α', was predicted by Bredig (*4*, *5*). Tromel (*6*) confirmed the presence of α'-C_2S above 850°C by X-ray diffraction studies. Bredig (*7*) provided a detailed sequence of transformations in C_2S.

According to Bredig the low temperature γ-C_2S transforms to α'-C_2S at about 850°C, the $\alpha' \rightarrow \alpha$ transition occurring at about 1450°C. During the cooling cycle α'-C_2S inverts to β-C_2S below 675°C instead of changing to the γ-form. The β-form ultimately converts to the γ-form, producing the so called 'dusting.' Nurse (*8*), at the Third International Symposium on Cement Chemistry, presented data on the temperatures of transformation of various forms of C_2S. The sequence of transformations during heating is $\gamma \rightarrow \alpha' \rightarrow \alpha$ and that during cooling is $\alpha \rightarrow \alpha' \rightarrow \beta \rightarrow \gamma$.

The data of Toropov *et al* (*10*) are not in agreement with those of Bredig (*7*), Nurse (*8*), and Roy (*9*). Toropov and co-workers suggested that in the heating cycle, transformations occur in the sequence $\gamma \rightarrow \beta \rightarrow \beta' \rightarrow \alpha' \rightarrow \alpha$ and in the cooling cycle the transformations are reversible and are shown by $\alpha \rightarrow \alpha' \rightarrow \beta' \rightarrow \beta \rightarrow \gamma$. The free energy-temperature diagrams of both Nurse and Toropov are compared in Fig. III.3.1.

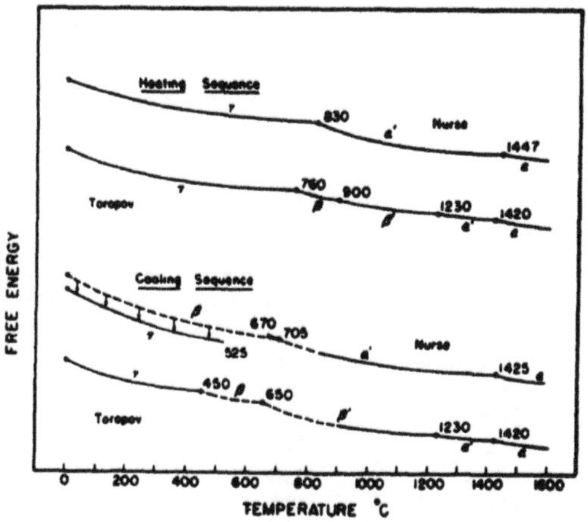

Fig. III.3.1: Free energy–temperature diagram of C_2S (*11*).

TABLE III.3.1
Compressive Strength Data for 1:3 Mortar Specimens of Various Modifications of C_2S [Toropov and Fedorov (17)]

Sample (1:3 mortar)	Compressive Strength, kg/cm²				
	3 days	7 days	28 days	Autoclave treatment	Neat paste 28 days
γ-C_2S	0	0	0	42.5	0
β-C_2S	9.0	13.5	22.5	125.0	22
α'-C_2S	3.0	4.0	8.8	13.0	21
α-C_2S	0	0	0	6.5	0

At ordinary temperatures α, α', and β-C_2S are stable only in the presence stabilizers. The β-form in the Portland cement clinker is stabilized by the addition of some impurity ions. The additives such as P_2O_5, As_2O_3, V_2O_5, Cr_2O_3, Mn_2O_3, BaO, and $2BaO \cdot SiO_2$ prevent the $\beta \rightarrow \gamma$ inversion. Yannaquis (*12*) believes that β-C_2S may exist in a metastable condition even in absence of a stabilizer provided the γ modification is heated to about 1100°C for longer periods and then cooled slowly. The main product however, is γ-C_2S if cooling starts from 1500°C (*13*, *14*).

In Portland cement clinker, C_2S is present mainly in the β or γ-form. Metzer (*15*) detected α'-C_2S in Portland cement clinker but this must have formed under abnormal conditions of high K_2O or P_2O_5 content in the raw mix.

Of the four forms of C_2S, β-C_2S develops maximum strength, α'-C_2S shows lower strength, and γ and α-C_2S are practically inert under normal conditions of curing (Table III.3.1). Budnikov and Azelitskaya (*16*) have reported that the γ -form has hydraulic properties. The strength development in β-C_2S depends on the type of stabilizer used. The P_2O_5-stabilized

TABLE III.3.2
Strength of 1:3 Mortars made from C_2S [Nurse (8)]

Stabilizer	Gypsum addition	Compressive strength, lb/in.²		
		7 days	28 days	6 months
$Fe_2O_3 + Na_2O$	None	295	436	—
	3%	212	350	1544
½% B_2O_3	None	350	763	1869
	3%	401	726	2417
5% $Ca_3(PO_4)_2$	None	485	1837	4380
	3%	553	1402	3607

C_2S yields better strengths than that stabilized with B_2O_3 or $Fe_2O_3 + Na_2O$ (Table III.*3.2*).

Differential Thermal Analysis

The quenching technique is not of much applicability in studying the polymorphism of C_2S because neither the α nor the α'-form is stable at room temperature. Newman and Wells (*18*) recognized only three polymorphs of C_2S, *viz.* α, β, and γ. They studied by DTA the inversions in the presence of additives such as CaO, SiO_2, Al_2O_3, MgO, Fe_2O_3, CaF_2, TiO_2, BaO, Cr_2O_3, B_2O_3, Mn_2O_3, V_2O_5, P_2O_5, Na_2O, and K_2O. All the additives except CaO promoted the $\alpha \rightleftharpoons \beta$ transformation at lower temperatures. The compounds Cr_2O_3, V_2O_5, P_2O_5, B_2O_3, Mn_2O_3, BaO, and $2BaO \cdot SiO_2$ prevented the $\beta \rightarrow \gamma$ inversion.

Welch, quoted by Nurse (*8*), obtained DTA curves of C_2S on heating and cooling cycles and the results are shown in Fig. III.3.2. Curve A shows the thermogram of γ-C_2S. The endothermal hump between 780 and 830°C indicates a sluggish transition of γ to α'-C_2S. At 1447°C the sharp endothermal peak can be correlated with the conversion of $\alpha' \rightarrow \alpha$-$C_2S$. The thermal effects are all reversible, as is evident from the cooling curve.

The $\alpha \rightarrow \alpha'$ conversion is exhibited at 1425°C and an exothermal effect at 670°C corresponds to the $\alpha' \rightarrow \beta$ transition. There is also an irregular exothermal dent starting at 525°C for the $\beta \rightarrow \gamma$ inversion and this effect is accompanied by a sudden increase in the volume of the mass. Curve C shows the behaviour of β-C_2S stabilized by CaO. The endotherms at 705 and

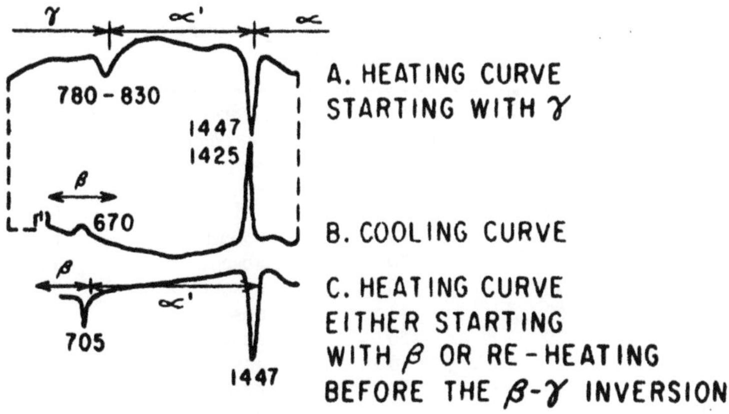

Fig. III.3.2: Inversion in C_2S (*8*).

1447°C are due respectively to $\beta \rightarrow \alpha'$ and $\alpha' \rightarrow \alpha$ transitions.

Roy (*9*) also applied DTA to detect the possible thermal changes in γ-C_2S but failed to confirm the results of Welch (*8*). He also thought that higher pressures would promote the $\gamma \rightarrow \alpha'$-$C_2S$ transition but DTA carried out in a nitrogen pressure of 3000-12,000 lb/in.^{2} failed to show any thermal effect.

Majumdar and Smith (*11*) re-examined the polymorphism of C_2S by DTA and X-ray and obtained inflections at 500 and 400°C in cooling curves. The peak at 400°C was ascribed to the $\beta \rightarrow \gamma$ transformation and it was also observed that cooling from about 1420±10°C resulted in the complete conversion of $\beta \rightarrow \gamma$ transformation, while cooling from lower temperatures gave a mixture of γ and β-forms at room temperature.

Toropov *et al* (*10*) expressed the enantiotropic transformation in C_2S as $\gamma \rightleftharpoons \beta \rightleftharpoons \beta' \rightleftharpoons \alpha' \rightleftharpoons \alpha$. Vasenin (*19*) had also suggested the existence of the β'-form, based on cooling curves. De Keyser (*20*) found in his thermal studies an unmistakable endothermic peak at 925°C and this may be ascribed to the transition $\beta \rightleftharpoons \beta'$.

Addition of certain chemical agents in small quantities shifts the peak temperatures in DTA. Toropov (*21*) and Toropov and Borisenko (*22*) studied a series of solid solutions by treating C_2S with barium orthosilicate. The results are shown in Table III.3.3. Pure C_2S shows two thermal effects at *810* and *1440*°C, representing $\gamma \rightarrow \alpha'$ and $\alpha' \rightarrow \alpha$ transformations. Addition of $2BaO \cdot SiO_2$ decreases the temperature of formation of α-C_2S and stabilizes the α' form at lower temperatures, as can be seen from the absence of any thermal effect for $\gamma \rightarrow \alpha'$ transition. With $2BaO \cdot SiO_2$ up to *70*% the

TABLE III.3.3
Polymorphous Transformations of C_2S–B_2S Found by Thermal Analysis [Toropov (21)]

No.	*Composition* $2CaO \cdot SiO_2$ *mole %*	$2BaO \cdot SiO_2$ *mole %*	*Temperatures of the polymorphous transformations* $\gamma \rightarrow \alpha'$, °C	$\alpha' \rightarrow \alpha$, °C
1	100	0	810	1440
2	90	10	absent	1340
3	80	20	absent	1230
4	70	30	absent	1130
5	60	40	absent	1040
6	50	50	absent	950
7	40	60	absent	900
8	30	70	absent	thermal effects not observed
9	20	80	absent	
10	10	90	absent	
11	0	100	absent	

TABLE III.3.4

Thermal Curves of C_2S in Presence of Mineralizers [Kurdowski (23)]

Specimen	Reaction temperature, °C heating	cooling	reheating
β-C_2S (stabilized by SiO_2)	605	560	605
β-C_2S+1% B_2O_3	705	650	705
γ-C_2S	750	680	750
γ-C_2S+5% CaF_2	770	720	770
γ-C_2S+2% Na_2SiF_6	750	700	750
γ-C_2S+1% BaO	720	650	720
γ-C_2S [according to Nurse (*8*)]	780	670	705

thermal effect corresponding to $\alpha' \rightarrow \alpha$ transformation is maintained. Above *70*% the solid solution α-C_2S–B_2S is stabilized.

Kurdowski (*23*) also observed the effect of stabilizing agents on the thermal peaks of γ-C_2S and β-C_2S and his data are summarized in Table III.3.4. β-C_2S shows a peak at *605*°C, which decreases to *560*°C on cooling. Nurse explained this as due to $\beta \rightleftharpoons \alpha'$ transformation. γ-C_2S shows a peak at *750*°C on heating and is ascribed to $\gamma \rightarrow \beta$ transformation as reported by Toropov. Kurdowski seems to imply that γ transforms to β' through β, and this is not in agreement with the results of Nurse.

Recently Feng Shu-gee and co-workers (*24*) reported the stability of β-C_2S with respect to burning temperature by using DTA, X-ray, and optical methods of analyses.

REFERENCES

1. Day, A.L., and Shepherd, E.S.S., *J. Am. Chem. Soc.*, *28*, 1089, 1906.
2. Shepherd, E.S., Rankin, G.A., and Wright, F.E., *J. Ind. Eng. Chem.*, *3*, 211, 1911.
3. Rankin, G.A., and Wright, F.E., *Am. J. Sci.*, 4th Series, *39*, 1, 1915.
4. Bredig, M.A., *Am. Mineralogist*, *28*, 594, 1943.
5. Bredig, M.A., *J. Phys. Chem.*, *49*, 537, 1945.
6. Tromel, G., *Naturwissenschaften*, *36*, 88, 1949.
7. Bredig, M.A., *J. Am. Ceram. Soc.*, *33*, 188, 1950.
8. Nurse, R.W., *Proc. 3rd Intern. Symp. Chem. Cements*, London, pp. 56-76, 1952.
9. Roy, H.M., *J. Am. Ceram. Soc.*, *41*, 293, 1958.
10. Toropov, N.A., Volkonskii, B.V., and Sadkov, V.I., *Dokl. Akad. Nauk SSSR*, *112*, 467, 1957.
11. Majumdar, A.K., and Smith, D.K., *Proc. 4th Intern. Symp. Chem. Cements*, Washington, p. 23, 1960.
12. Yannaquis, N., *Rev. Mater. Construct.*, *480*, 213, 1955.

13. Guinier, A., and Yannaquis, N., *Compt. Rend.*, *244*, 2623, 1957.
14. Yannaquis, N., and Guinier, A., *Bull. Soc. Franc. Mineral Crist.*, *82*, 126, 1959.
15. Metzer, A., *Zement-Kalk-Gips*, *6*, 269, 1953.
16. Budnikov, P.P., and Azelitskaya, R.D., *Dokl. Akad. Nauk SSSR*, *108*, 57, 1956.
17. Toropov, N.A., and Fedorov, N.F., *Zh. Prikl. Khim.*, *35*, 2585, 1962.
18. Newman, E.S., and Wells, L.S., *J. Res. Nat. Bur. Stad.*, *36*, 137, 1946.
19. Vasenin, F.I., *Zh. Prikl. Khim.*, *21*, 10, 1948.
20. De Keyser, W.L., *Bull. Soc. Chim. Belges*, *62*, 235, 1953.
21. Toropov, N.A., *Proc. 4th Intern. Symp. Chem. Cements*, Washington, p. 113, 1960.
22. Toropov, N.A., and Borisenko, A.I., *Proc. 4th Conf. Exptl. Mineral & Petr., Edn. Acad. Sci. USSR*, *11*, 214, 1953.
23. Kurdowski, W., *Proc. 6th Conf. Silicate Ind.*, Budapest, p. 269, 1963.
24. Feng Shu-Gee, Hwang Jinyang, Tao Ging-Yang and Lee Ya Chin, *Kuei Suan Yen Hsueh Pao*, *3*, 105, 1964.

IV. Hydration of Portland Cement

IV.1. SYSTEMS CONTAINING CaO, Al_2O_3, SiO_2, and H_2O AT ORDINARY TEMPERATURES

The ternary and quaternary systems comprising CaO, Al_2O_3, SiO_2, and H_2O are of importance in studying the chemistry and morphology of hydrous calcium silicates formed in hydrating cements. These systems could either be studied under autoclaving conditions or at ordinary temperatures. At low temperatures equilibrium is attained only slowly. Chapter VII deals with the high temperature behaviour of these systems. In this chapter the products formed at room temperature are discussed.

In the literature there has been some variance in the nomenclature of calcium silicate hydrates adopted by different workers. Table IV.1.1 presents some of the designations adopted.

There are various calcium silicate hydrate compounds that are classified under the 'tobermorite group.' Because of their significance in the chemistry of cement hydration a reasonable classification of these hydrates becomes necessary.

Tobermorite ($C_6S_6H_5$) is a natural mineral that can also be synthesized

TABLE IV.1.1
Various Designations of some of the Calcium Silicate Hydrates [Bogue (1)]

Composition	*Thorvaldson* (2)	*Bessey* (3)	*Taylor* (4), *Taylor* & *Bessey* (5)	*Bogue* (6)
$C_2SH_{0.9-1.25}$	C_2S hydrate (I)	C_2S α-hydrate	C_2S α-hydrate	$C_2SH(A)$
$C_2SH_{1.1-1.5}$	C_2S hydrate (II)	C_2S β-hydrate	C_2S β-hydrate	$C_2SH(B)$
$C_2SH_{0.3-1.0}$	C_2S hydrate (III)	C_2S γ-hydrate	C_2S γ-hydrate	$C_2SH(C)$
$C_2SH_{0.67}$	—	—	C_2S δ-hydrate	$C_2SH(D)$
C_2SH_{2-4}	—	—	C_2SH (II)	C_2SH_2
$CSH_{1.1}$	—	—	Flint's CSH	CSH(A)
$C_{1.0-1.5}SH_{1.0-2.5}$	—	—	CSH (I)	CSH(B)

TABLE IV.1.2
The Tobermorite Group of Compounds [Taylor (7)]

CaO:SiO₂ ratio	*Degree of Crystallinity*		
	High	*Low*	*Very low*
Below 1.5 (near 0.83 when crystalline)	14.0 Å tobermorite, 11.3 Å tobermorite, 9.35 Å tobermorite	CSH (I)	Plombierite
1.5 and over (1.5-1.75 when crystalline)	12.6 Å tobermorite, 10 Å tobermorite	CSH (II)	CSH gel

from lime and silica in an autoclave. In fact most synthesized calcium silicate hydrates resemble tobermorite. Tobermorite has a layer structure, is very similar to vermiculite clay mineral, and has a basal spacing of 14.0, 11.3, or 9.35 Å. Heat treatment of 14.0 or 11.3 Å tobermorite at 300°C causes its conversion to the 9.35 Å type. At 800°C tobermorite converts to β-CS. The $CaO:SiO_2$ ratio for the three types of tobermorite varies in the range 0.8-1.33 (Table IV.1.2). A synthetic tobermorite with 4 or 5% Al_2O_3 has been prepared by Kalousek. The aluminium-substituted tobermorite exhibits characteristic thermogram for its identification (*8*).

A poorly crystallized calcium silicate hydrate structurally related to tobermorite is formed by mixing calcium hydroxide solution with silica gel. Taylor (*4*) has designated it as CSH(I), Bogue (*6*) as CSH(B), and Kalousek and Prebus (*9*) as C/S *0.8-1.33* hydrate. CSH(I) exhibits an intense exothermic peak at *835-900*°C. On thermal treatment CSH(I) first forms β-C_2S and then transforms to β-CS at higher temperatures (*10*). Plombierite is a very poorly crystallized tobermorite. It transforms to β-CS at about 600-700°C (*10*).

Two more types of tobermorites are known having a CaO/SiO_2 ratio of 1.5-2.0 and with a basal spacing of 12.6 or 10 Å. A lime-rich phase is designated as CSH(II) by Taylor (*7*), as C_2SH_2 by Bogue (*1*), and as C_2SH by Bernal (*11*). CSH(II) may be prepared under hydrothermal treatment, by treating silica gel with calcium glycerate, or by bottle hydrating C_3S or β-C_2S at low temperatures. A very poorly crystallized and lime-rich calcium silicate hydrate gel is formed in the paste hydration of C_3S, β-C_2S, or Portland cement at room temperature and is known as CSH gel or tobermorite (G) (*12*).

Differential Thermal Analysis

The DTA technique has proved very useful in the investigation of the system $CaO–SiO_2–H_2O$ at room temperatures. The products formed in this

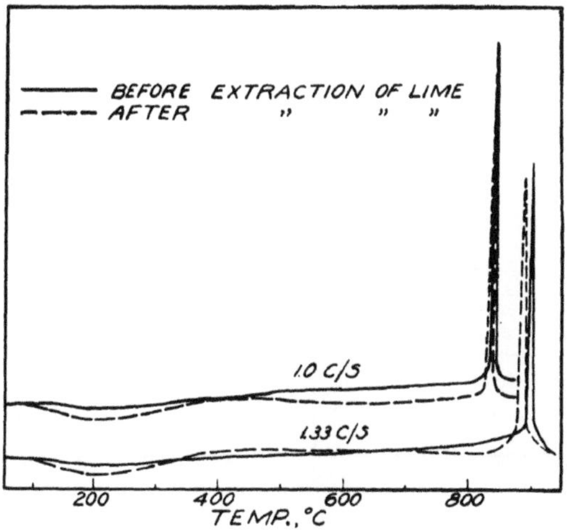

Fig. IV.1.1: DTA curves of 1.0 and 1.33 C/S solids before and after extraction of lime with aceto-acetic ester (*13*).

system at higher temperatures are different and are dealt with in Chapter VIII. By applying DTA Kalousek (*13*) could differentiate two types of calcium silicate hydrates in this system, one containing a CaO/SiO_2 molar ratio of 0.8, 1.0, 1.25, and 1.33 and the other containing a CaO/SiO_2 molar ratio of 1.5 and above. The differential thermal behaviour of the hydrate containing a CaO/SiO_2 ratio of 1.0 or 1.33 is shown in Fig. IV.1.1. The specimens with a CaO/SiO_2 ratio of 0.8, 1.0, 1.25, and 1.33 exhibit an exothermic peak at 830±5°C, 830±5°C, 865±5°C and 900±5°C respectively, showing thereby that the exothermic peak is shifted to a higher temperature as the CaO/SiO_2 ratio in the hydrate increases.

Some experiments were also done to investigate the extractability of lime from samples with CaO/SiO_2 ratio of 1.0 and 1.33 by aceto-acetic ester. Thermograms of the extracted specimens remained substantially the same, as denoted by the dashed-line curve. These results demonstrate that lime in these two samples is strongly bound.

In another series of experiments Kalousek examined the solids with a CaO/SiO_2 ratio of 0.8, 1.0, or 1.2 after treating them with excess lime for about 18 days. The results are shown in Fig. IV.1.2. The solid with CaO/SiO_2 =1.2 does not exhibit any change, whereas in that with CaO/SiO_2=1.0 the exothermic peak is shifted to a higher temperature. The solid with CaO/

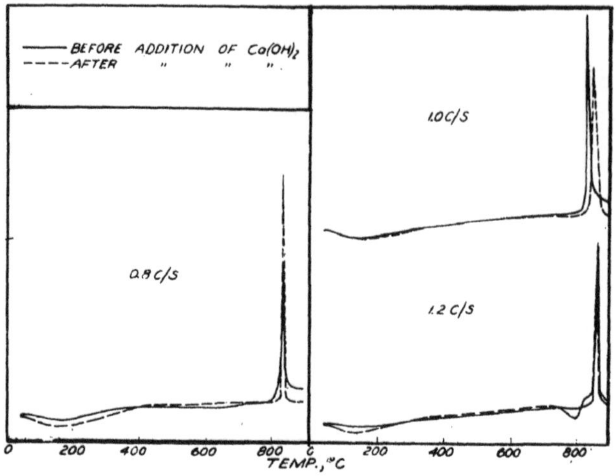

Fig. IV.1.2: DTA curves of samples before and after treatment in solution of $Ca(OH)_2$ (*13*).

SiO_2=0.8, however, shows a pronounced enhancement of the exothermal effect.

A chemical examination reveals that the 0.8 CaO/SiO_2 sample converts to a solid with a CaO/SiO_2 ratio of 0.98. The solid with a CaO/SiO_2 ratio of 1.2 adsorbs 4.1% CaO that could be extracted easily. The endothermic effect at 780-820°C for the 1.2 CaO/SiO_2 specimen is presumably due to adsorbed lime. It appears from these results that in the system CaO–SiO_2–H_2O a continuous series of solids exist between the end members $C_4S_5H_n$ and $C_4S_3H_n$. These could be differentiated by DTA and not by the X-ray diffraction method.

The investigation of the system $Ca(OH)_2$–silica gel–water by Newman (*14*), in addition to showing some agreement with the results of Kalousek, has yielded some more interesting data. Newman studied 15 mixes containing a CaO/SiO_2 ratio of 0.3 to 3.3 and cured at periods ranging from 1 day to 2½ years. The thermal curves generally exhibited a broad endothermal valley in the range 80-350°C owing to the loss of free and adsorbed water. The endothermic effect registered in the range 350-500°C could be ascribed to the dehydration of calcium silicate hydrate. In some solids the presence of $Ca(OH)_2$ was indicated by an endothermal effect between 525 and 570°C.

All the samples showed two exothermic peaks for the reaction between

CaO and SiO_2 at a temperature of 830 and 900°C. At 16 weeks the samples with a CaO/SiO_2 ratio of 0.3, 0.4, 0.5, 1.5, 1.6, 2.0, or 3.3 showed two exothermic peaks, while those with a CaO/SiO_2 ratio of 0.7 and 1.2 showed a single sharp exothermic peak.

The sequence of thermal reactions could be better understood with respect to a solid formed at a CaO/SiO_2 ratio of 1.5 (Fig. IV.1.3.). The solid at 1 day exhibits a peak at 500-600°C, representing the presence of a considerable amount of $Ca(OH)_2$. With an increase in the curing period, calcium hydroxide reduces in amount and at about 16 weeks is completely absent. An interesting feature of this sample is that it gives two exothermic peaks at earlier periods. Of the two exothermic effects that at higher temperature is relatively smaller at 1 day but grows larger at the expense of the other and at 1-2 years there is only a single exothermic effect. The reactions causing the exothermic peaks have not been resolved.

Yoshii and Sudo (*15*) also studied the $CaO–SiO_2–H_2O$ system by DTA. The solids formed had a CaO/SiO_2 ratio of 0.89-1.51 and showed similar thermal behavior as reported by Kalousek (*13*).

The DTA method can be used to differentiate tobermorite gel from CSH(I)

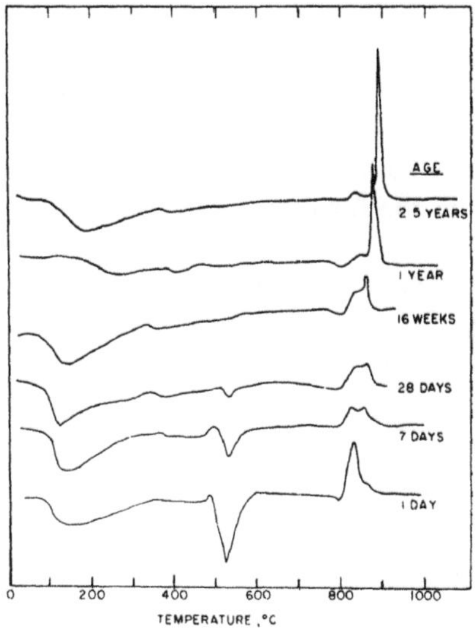

Fig. IV.1.3: DTA of solids formed with CaO/SiO_2=1.5 (*14*).

(*16*). CSH(I) sample synthesized at room temperature from lime and silica shows a strong exothermic reaction at 825°C, indicating conversion to β-wollastonite (*16*, *17*). CSH(I) also indicates a low temperature endotherm at 160-165°C and two endotherms at 665 and 790°C. The tobermorite gel prepared by the hydration of C_3S, C_2S, or alite shows similar thermal characteristics. An endothermal effect at 140-150°C for the desorption of water, another at 570°C representing decomposition of $Ca(OH)_2$, and a small exothermic effect at about 860°C are typical of a tobermorite gel.

The exothermal reaction in tobermorite gel is of lesser intensity than that of CSH(I) and this could form a basis for differentiating them. CSH(II) is different from CSH(I) in showing an exothermal bulge at 400-450°C and a small exothermal dent at 850°C (*13*). Both CSH(I) and CSH(II) may be detectable in CaO–silica gel–H_2O (*18*).

In soil stabilization, lime reacts with aluminosilicate hydrates to form a strong bond. Eades and Grim (*19*) investigated the possible formation of calcium silicate hydrates in $Ca(OH)_2$–clay mineral mixtures containing kaolinite, halloysite, bentonite, or illite. The kaolinite–$Ca(OH)_2$ mixture showed a peak at 300-400°C, suggesting the presence of calcium silicate hydrate, but montmorillonite did not seem to form a calcium silicate hydrate.

Against this, Glenn and Handy (*20*) obtained evidence by DTA that bentonite reacts with lime. They reported that in a mixture containing bentonite–$Ca(OH)_2$–MgO and cured for 1 year no peak could be detected at 680°C, corresponding to the dehydroxylation of bentonite. An endothermic effect observed at 790°C could be attributed to the presence of calcium silicate hydrate.

Lime also reacts with kieselguhr and gives an exothermic peak at 850-900°C, corresponding to the formation of wollastonite from calcium monosilicate (*21*).

In the system CaO–Al_2O_3–$CaSO_4$–H_2O studied by Turriziani and Schippa (*22*) the DTA and X-ray results showed the formation of compounds similar in behavior to C_4AH_{13} and $C_3A \cdot CaSO_4 \cdot 12H_2O$.

REFERENCES

1. Bogue, R.H., *The Chemistry of Portland Cement*, Reinhold Publishing Corporation, New York, p. 509, 1955.
2. Thorvaldson, T., *Proc. Symp. Chem. Cements*, Stockholm, p. 246, 1938.
3. Bessey, G.E., *Proc. Symp. Chem. Cements*, Stockholm, p. 178, 1938.
4. Taylor, H.F.W., *J. Chem. Soc.*, p. 3682, 1950.
5. Taylor, H.F.W., and Bessey, G.E., *Mag. Concrete Res.*, *2*, 15, 1950.
6. Bogue, R.H., *Mag. Concrete Res.*, *3*, 87, 1953.
7. Taylor, H.F.W., *Proc. 4th Intern. Symp. Chem. Cements*, Washington, p. 167,

1960.
8. Kalousek, G.L., *J. Am. Ceram. Soc.*, *40*, 74, 1957.
9. Kalousek, G.L., and Prebus, A.F., *J. Am. Ceram. Soc.*, *41*, 124, 1958.
10. Gard,J.A., Howison, J.W., and Taylor, H.F.W., *Mag. Concrete Res.*, *11*, 151, 1959.
11. Bernal, J.D., *Proc. 3rd Intern. Symp. Chem. Cements*, London, p. 216, 1952.
12. Brunauer, S., and Greenberg, S.A., *Proc. 4th Intern. Symp. Chem. Cements*, Washington, p. 135, 1960.
13. Kalousek, G.L., *Proc. 3rd Intern. Symp. Chem. Cements*, London, p. 296, 1952.
14. Newman, E.S., *J. Res. Nat. Bur. Stad.*, *59*, 187, 1957.
15. Yoshii, T., and Sudo, G., *Semento-Gijitsu-Nenpo*, *11*, 48, 1957.
16. Diamond, S., Dolch, W.L., and White, J.L., *Highway Res. Rec. No. 62*, p. 62, 1964.
17. Taylor, H.F.W., *J. Chem. Soc.*, p. 163, 1953.
18. Jambor, J., *Silikaty*, *6*, 162, 1962.
19. Eades, J.L., and Grim, R.E., *Nat. Acad. Sci.—Nat. Res. Council Publ. No. 771*, 51, 1960.
20. Glenn, G.R., and Handy, R.L., *Iowa Acad. Sci. Proc.*, *69*, 390, 1962.
21. Lapoujade, P., and Vogein, M.P., *Bull. Soc. Franc. Ceram.*, *No. 28*, 3, 1955.
22. Turriziani, R., and Schippa, G., *Ric. Sci.*, *24*, 2356, 1954.

IV.2. HYDRATION OF PORTLAND CEMENT

It is indubitable that the exploration of the physical and chemical processes in the hardening of cements is the key to an understanding of the setting phenomenon, dimensional changes, strength development, and durability in concretes. An understanding of the basic chemistry of the hydration of Portland cement is greatly facilitated through a study of the hydration of the individual cement compounds such as C_3S, C_2S, C_3A, and the ferrite phase, and also of the ternary systems $CaO–SiO_2–H_2O$ and $CaO–Al_2O_3–H_2O$.

The hydration of Portland cement is a subject of many investigations, and at the Fourth International Symposium on Cement Chemistry held in 1960 at Washington a large proportion of the papers dealt with cement hydration.

It is a remarkable fact that though the hydration of cements has been investigated by techniques such as petrographic microscopy, X-ray diffraction, electron microscopy and diffraction, infrared absorption, nuclear magnetic resonance spectroscopy, thermogravimetric and differential thermal analysis, there are still some unresolved problems.

The ground Portland cement clinker reacts with water with avidity and evolves considerable amount of heat before flash setting. Gypsum in amounts of 4-5% is always added to regulate the initial reactions. A ground clinker reacts with water to form hexagonal calcium aluminate hydrates C_2AH_x and C_4AH_x, which ultimately convert to the stable cubic hydrate C_3AH_6. Tricalcium silicate and dicalcium silicate react with water more slowly, giving calcium silicate hydrate. The ferrite phase reacts faster than the silicates.

The initial reaction of C_3A is modified in the presence of gypsum. In Portland cement pastes gypsum reacts with C_3A to form calcium aluminate trisulfate of formula $3CaO{\cdot}Al_2O_3{\cdot}3CaSO_4{\cdot}31H_2O$ (corresponding to the natural mineral ettringite) and calcium aluminate monosulfate of formula $3CaO{\cdot}Al_2O_3{\cdot}CaSO_4{\cdot}12H_2O$. The trisulfate or the high sulfoaluminate crystallizes as needles and the low sulfoaluminate, as hexagonal plates. There is evidence of a solid solution of the low sulfoaluminate in $3CaO{\cdot}Al_2O_3{\cdot}Ca(OH)_2{\cdot}12H_2O$.

The retarding action in cements is generally believed to be due to a coating of the sulfoaluminate on the surface of C_3A. A coating of the hexagonal compounds typified by C_4AH_{13} cannot be ruled out, especially under conditions of low SO_4^{2-} ion concentration.

There has been some dispute on the type of sulfoaluminate that forms first during the hydration of Portland cement. The compound stable under equilibrium conditions is the high sulfoaluminate (*1-3*). It has also been suggested that the low sulfoaluminate forms as a metastable product before conversion to the high sulfate form (*4*). But the general opinion is more in favour of considering the high sulfoaluminate as the initial product of hydration in Portland cement (*5-9*).

Though the cubic hydrate C_3AH_6 may be detected in the hydrated ground clinkers its presence has not been reported in Portland cement pastes. Very small amounts of sulfate ions seem to inhibit the formation of C_3AH_6.

The ferrite phase in cements is reported to hydrate to cubic C_3FH_6 and hexagonal C_4FH_{13}. But it is more general to treat both cubic and hexagonal hydrates in hardened pastes as ferrite–aluminate solid solutions (*10*).

The silicate phases C_3S and C_2S together are present to an extent of 70-80% in Portland cement and hence it is obvious that the hardened cement paste should predominate in calcium silicate hydrate. As with the C_3S and C_2S pastes, Portland cement paste consists of tobermorite gel in the form of distorted plates and large fibers.

The identification of calcium silicate hydrate by X-ray diffraction is not easy. The strong reflection for CSH(I) or CSH(II) at 3.06 Å is subject to interference by alite and calcite. The 1.82 Å line for tobermorite is distinguishable in pastes but this does not provide a clue to the composition of the gel. Other techniques such as DTA and TGA also confirm that tobermorite gel is the major silicate hydrate in cement pastes (*11*, *12*).

Tobermorite gel is always associated with impurities such as Al_2O_3 and alkalies. Taylor (*13*) gives the approximate composition of the CSH gel or tobermorite gel as $C_{1.7}SH_x$. The gel has a low crystallinity, is related to the tobermorite mineral, and is more nearly equivalent to CSH(II) than CSH(I).

In hydrated Portland cement, crystalline calcium hydroxide forms in a substantial quantity, the main source being C_3S. Probably calcium hydroxide also exists in a colloidal form. The hydrating cement paste invariably absorbs CO_2 from the atmosphere to give calcium carbonate.

It can be assumed that to a good approximation the components of the Portland cement should react at the same rate as they do when hydrated independently. Bogue and Lerch (*17*) investigated the hydration rate of individual Portland cement compounds and in combination with gypsum and inferred that the compounds in cements hydrated at their characteristic rates without mutual interference. Other workers, notably Woods, Steinour, and Starke (*14,15*) and Verbeck and Foster (*16*), using calorimetric measurements, came to the same conclusion. Against this the research workers (*18-20*) at the Portland Cement Association, U.S.A., have put forward the hypothesis of 'fractional rates,' according to which the fractional rate of hydration of the individual phases in a cement paste is the same. This

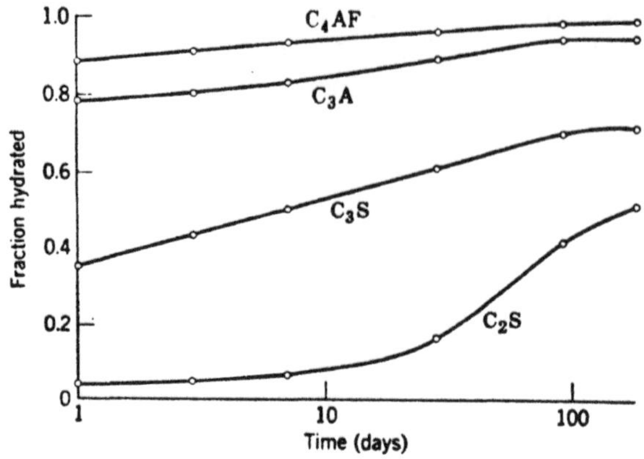

Fig. IV.2.1: Rate of hydration of pure compounds (*18*).

would imply that if the cement is hydrated to an extent of *50*%, each of the components, *viz* C_3S, C_2S, C_3A, or the ferrite phase, is also hydrated to an extent of *50*%. This was found to be valid only for type I and type III cements. It is doubtful if this hypothesis is valid for the first few days of hydration.

An examination of the hydration products of cement from very early times by X-ray diffraction shows that during the first five hours the reactivity is in the order $C_3A > C_4AF > C_3S > C_2S$, while after a day when sulfate has reacted with C_3A to a considerable extent, the order of reactivity is $C_3S > C_3A > C_4AF > C_2S$ (*21*). Copeland and Bragg (*18*) studied the fractions of the four major phases hydrated at different periods and their results are shown in Fig. IV.2.1.

Extensive data on the physical properties of Portland cement paste are to be found in the publications of Powers (*22*, *23*). The cement gel has a surface area of 220 m^2/g and occupies 2.2 times the volume of the cement. The porosity of the gel is 28% and the average diameter of the gel pore is 18 Å.

The mechanism of hydration and hardening has been one of the most debated topics in the field of cement chemistry. The first attempt to explain the cementing action was made by Le Chatelier (*24*) in 1887. He proposed that cement dissolves in water and becomes supersaturated with respect to the hydration product. Hence, precipitation of the hydrated product occurs and the crystals so produced interlock each other. Michaelis (*25*) in 1893 put forth the gel hypothesis according to which the gel is mainly responsible for the development of strength though some crystalline product may be present in cement pastes.

For several years now there have been supporters for both hypotheses. The history of the controversy is described by Bogue (*26*). Recently, Hansen (*27*, *28*), Jeffery (*29*), and Chatterji and Phatak (*30*, *31*) have suggested a topochemical or a solid state mechanism for the hydrating cements. The general tendency amongst many cement chemists is to treat the early reaction as one of solution-precipitation and that at later stages mainly as a topochemical type of reaction.

Differential Thermal Analysis

A study of the hydration reactions in Portland cement is complex because it involves several hydration products of the individual phases, *viz.* C_3S, C_2S, C_3A, and the ferrite phase in presence of gypsum. Some of the products that may be detected in the early hydration of Portland cement are calcium silicate hydrate, low sulfoaluminate hydrate, hexagonal hydroaluminates, solid solution of low sulfoaluminate with C_4AH_{13}, high sulfoaluminate, $Ca(OH)_2$, $Mg(OH)_2$, $CaCO_3$, calcium carboaluminates, and compounds analogous to low and high sulfoaluminates containing Fe_2O_3 and Al_2O_3.

Lepingle (*32*) in 1941 suggested the use of the differential thermal analysis method for the study of hydraulic binders such as Portland cement, alumina cement, and slag cement. Kalousek, Davis, and Schmertz (*8*) in 1949 showed for the first time that DTA is a powerful tool in following the hydration of Portland cement.

Thermograms have yielded useful data on the qualitative and quantitative aspects of the hydration products formed in Portaland cement under different conditions. DTA studies have also been applied for investigating false set; the reaction of carbon dioxide, sulfate, and chloride solutions on Portland Cement; the role of alkalies in the hydration; and the changes occurring in cements under different conditions of storage. A large number of publications are available describing the DTA of hydrated Portland cements. Fairly comprehensive data have been obtained by Petzold and Göhlert (*33*), Petzold and Talke (*34*), Bozhenov and Suvorova (*35*), Barta and Satava (*36*, *37*), Rey (*38*), Esenwein (*39*), Levcuk (*40*), Govorov (*41*, *42*), Kurdowski (*43*), Krawtschenko (*44*), Lommatzsch (*45*, *46*), Mtschedlov-Petrosian *et al* (*47*, *51*), Greenberg (*48*), Odler (*49*), Greene (*9*), Turriziani (*12*), Tamas (*50*), Kalousek and Adams (*52*), Heller and Ben-Yair (*53*), Midgley and Rosaman (*54*), Nemecek (*55*), Diamond, Dolch and White (*56*), Budnikov, Gorshkov and Pototskaya (*57*), Tamas and Liptay (*58*), and Midgley (*59*).

Drying of Specimens. In the DTA examination of hydrated products of cements the drying procedure adopted for preparing the samples is very critical. Insufficient drying may not effectively remove the physically-bound water that might interfere with the low temperature endothermal effects. On the other hand too severe a drying might dehydrate the chemically-held water from the hydrated products.

Non-specification of the drying procedure often makes comparison of the DTA curves reported by different investigators difficult. Greene (*9*) suggested vacuum drying of the samples at a pressure of 1 or 2 mm of Hg for 2 to 3 hr. Kalousek *et al* (*8*) washed the samples 4 times with acetone, twice with ethyl ether, and then dried them at a pressure of 0.01 mm for 1 hr. In his recent studies Kalousek crushed the samples to pass through 30 mesh, inundated them in pure acetone and left for 3 min, and then suction-filtered to remove the liquid acetone. Finally, the samples were placed in a desiccator and evacuated at 4-5 $\times$ 10^{-2} torr for 24 hr.

Rey (*38*) found that washing twice with acetone and ether and then evacuation in vacuum for 6 hr was not sufficient to remove all the physically bound water. But treatment with acetone followed by 5 washings with ether and drying in 0.1 mm of Hg for 1 hr was more efficient. In their determination of the non-evaporable water in cement pastes Copeland and Hayes (*60*) suggested placing the sample in a vacuum desiccator connected to a trap maintained at the temperature of dry ice at —*79*°C.

Organic solvents should be used with caution because it is possible that treatment with some of them may be equivalent to drying the samples to about 100-150°C.

Hydrated Cement Pastes. Petzold and Göhlert (*33*) compared the thermograms of hydrated cement pastes reported by different investigators and found a broad agreement. In Fig. IV.2.2 the first eight thermograms of the cement pastes have been constructed from those of Barta and Satava (*36*, *37*), Bozhenov and Suvorova (*35*), Rey (*38*), Esenwein (*39*), Levcuk (*40*), Govorov (*41*, *42*), Petzold and Talke (*34*), and Kurdowski (*43*), and the ninth was computed from the above as the most probable standard curve. All the curves show at least one endothermal effect below 200°C due to dehydration, an endotherm between 500° and 600°C corresponding to the dehydration of $Ca(OH)_2$, and a high temperature endothermal effect beyond 800°C representing the decarbonation of $CaCO_3$.

The standard curve was compared with that of the hydrated product of a synthetic mixture containing C_3S, C_2S, C_3A, C_4AF, and gypsum, and a great similarity was observed. These results show that clinker minerals hydrate without any interference with each other.

For a complete interpretation of the DTA of cement pastes a study of the hydrated products of individual cement minerals as well as those of ground clinker becomes necessary. The thermal behaviour of several Portland cement clinkers without the addition of gypsum has been studied by Kalousek

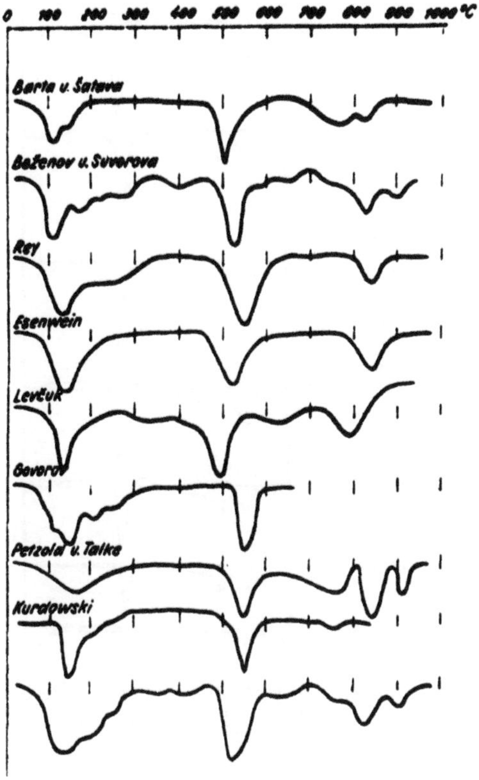

Fig. IV.2.2: Comparison of thermograms of cement pastes reported by different investigators (*33*).

et al (*8*).

In Fig. IV.2.3 thermograms of hydrated clinkers formed at different curing periods are shown. An endothermic effect at 130-140°C increases in size with curing time and is attributed partly to the desorption of capillary water from the lime-silica gel. The clinker in the presence of gypsum also showed an endotherm at this temperature and this was ascribed to the phase X, which denotes a single phase containing all the cement components. The endothermic effect at 195-200°C represents the dehydration of SO_3—poor member of the solid solution approaching C_4AH_{13}. Another endothermic effect at 315-330°C exhibited by clinker 1 is due to C_3AH_6, and that at 415°-430°C, to the dehydration of $Mg(OH)_2$. The presence of $Ca(OH)_2$ is indicated by the endotherm at 515-590°C. An exothermal hump at 450-500°C and an endothermal valley at 800-900°C were also evident in some clinkers.

A systematic DTA investigation was carried out by Greene (*9*) on the

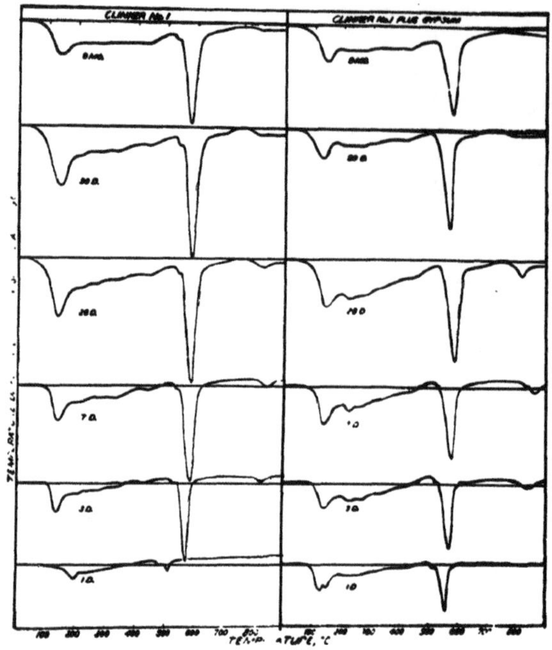

Fig. IV.2.3: Thermal analysis of a cement clinker at different ages of test with and without addition of 1.8% SO_3 as gypsum (*8*).

hydrated products of cement formed at periods ranging from 5 min to 7 days. Fig. IV.2.4 shows the thermal behavior of unhydrated cement and that hydrated for 5 min, 1 hr, 4 hr, 24 hr, and 7 days. The endothermic peaks at

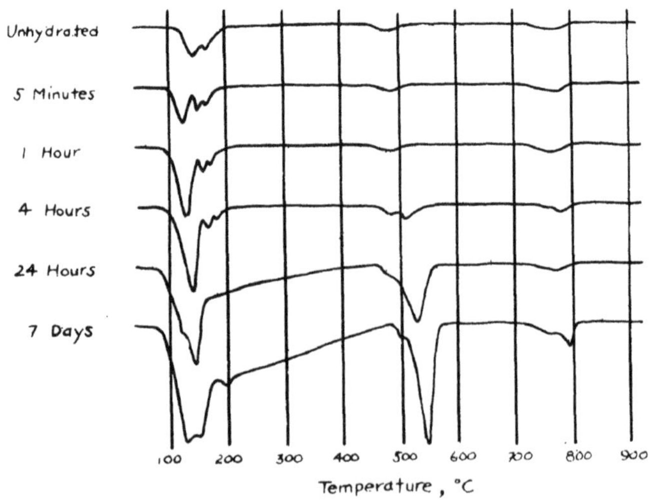

Fig. IV.2.4: DTA of cement hydrated at different periods (*9*).

140 and 170°C characterize gypsum, and the endothermal humps below 500 and 800°C are due respectively to the decomposition of $Ca(OH)_2$ and $CaCO_3$ formed during exposure to air.

Within 5 min, an endothermic effect appears at about 130°C caused by the dehydration of the high sulfoaluminate. Gypsum is partly consumed at this period, as is evident from the reduced intensities of the 140 and 170°C peaks. At 1 hr there is an increase in the sulfoaluminate peak at the expense of gypsum peaks. Further increase in the amount of sulfoaluminate is evident at 4 hr.

The onset of an endothermic effect at about 500°C after 4 hr of hydration is attributed to the dehydration of $Ca(OH)_2$ formed by the hydration of the C_3S phase. A small endothermic effect registered just below 500°C may be due to the desorption of the chemisorbed water on the free lime of the cement. The free lime estimation in the cement corresponds to the intensity of the endothermic effect below 500°C. The hydrated cement cantains $Ca(OH)_2$ partly in a very finely divided state, which might be decomposed at a slightly lower temperature than crystalline $Ca(OH)_2$.

At 5 hr of hydration Mtschedlov-Petrosian *et al* (*47*) observed a peak at 280°C, which they ascribe to the presence of C_2AH_7. The $Ca(OH)_2$ decomposition peak increases in intensity considerably between 4 and 24 hr and also between 24 hr and 7 days. Gypsum is almost absent at 24 hr. At 24 hr. endothermic effect appears just below the temperature of the sulfoaluminate peak and can be clearly seen at 7 days. This peak denotes the presence of calcium silicate hydrate.

Kalousek *et al* (*8*) also detected an endothermic peak at about this temperature, which was attributed to the phase X. Another endothermic peak at 200°C after 7 days may be attributed to low sulfoaluminate, a solid solution of this compound with C_4AH_{13}, or to C_4AH_{13} itself. An upward slope starting from about 200°C and continuing up to a temperature of about 500°C in the samples cured for 24 hr and 7 days may indicate a gradual loss of water from calcium silicate hydrate. TGA also shows a loss of weight in this temperature range (*61-63*).

After 7 days the intensity of the peak at 800°C for $CaCO_3$ decomposition is greater because the sample at this period is exposed to air for longer periods. DTA does not show the presence of C_3AH_6, though it is a stable phase in the system $CaO–Al_2O_3–CaSO_4–H_2O$

There is controversy as to whether the high or the low sulfoaluminate is the initial product of hydration of Portland cement. Under equilibrium conditions the high sulfoaluminate is the stable form, the low sulfate form being metastable with respect to the high sulfoaluminate. The low sulfate form is preferably formed in a saturated solution of $Ca(OH)_2$ and $CaSO_4$.

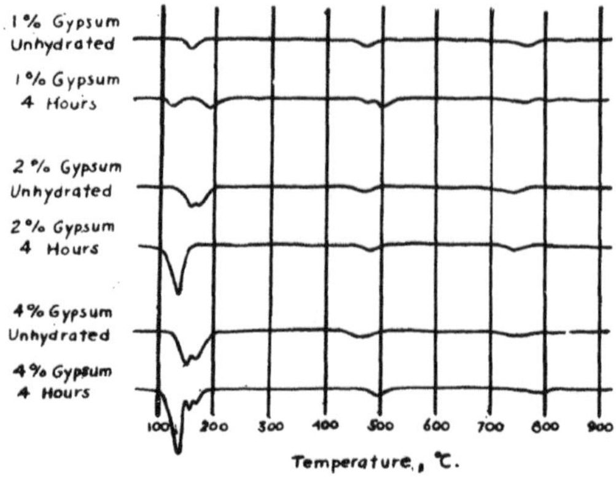

Fig. IV.2.5: DTA curves of Portland cement clinker hydrated with various amounts of added gypsum (*9*).

But there is no evidence in DTA for the formation of low sulfoaluminate at early periods in the hydration of cements in $CaSO_4$ and $Ca(OH)_2$. Govorov (*41*) and Gracheva and Vikulin (*64*) detected the crystallization of monosulfate together with high sulfoaluminate in concentrated aqueous suspensions of Portland cement containing insufficient gypsum.

In hydrating cements the amount and formation of sulfoaluminate and C_4AH_{13} are dependent on the gypsum content. This can be studied by DTA as shown in Fig. IV.2.5 in which thermograms represent the hydrated cements containing 1, 2, or 4% gypsum.

At 4 hr the peak at 200°C could be correlated with the presence of C_4AH_{13}, the formation of which could not be prevented with 1% gypsum. The high sulfoaluminate is also formed with 1% gypsum. With 2% gypsum the formation of C_4AH_{13} is prevented as is evident from the absence of an endothermic effect at 200°C. Further increase in the high sulfoaluminate content can be seen with 4% gypsum.

The DTA method was applied by Mtschedlov-Petrosian and co-workers (*47*) to examine the effect of surface area and of vibration of 1:3 cement/quartz mortars on the reaction velocity of hydration. The endothermal peak area corresponding to $Ca(OH)_2$ decomposition was used to calculate the degree of hydration of C_3S.

No detectable difference was observed in the reaction velocity over a surface area of 2500–5400m²/g. This indicated that the increase in surface area was contributed mainly by the softer grains of gypsum. For vibration periods ranging from 1 min to 3 days, the intensity of the $Ca(OH)_2$ decomposition peak did not vary much. The vibration seems to be effective only on about 3% by weight of gypsum and aluminate phase present in the mortar.

The possibility of a solid solution between $C_3A.3CaSO_4 \cdot aq$ and $C_3A \cdot 3Ca(OH)_2.aq$ was suggested by Kalousek (*65*) in 1941. Flints and Wells (*66*) reported the formation of the compound $C_3A \cdot 3Ca(OH)_2 \cdot aq$ in 1944. Midgley and Rosaman (*54*) sought an explanation for the difference observed in *d* values and DTA peak temperatures between the pure sulfoaluminate and the ettringite phase in the hydrated Portland cement.

DTA was particularly useful in resolving this problem. Midgley and Rosaman found that the endothermic peak temperature *T* and the area *A* of the endothermic peak (of the ettringite phase) followed the relationship

$$T=a+b \ lnA$$

where *a* and *b* are arbitrary constants. From the straight line plot the temperature corresponding to an arbitrary area of 1000 mm^2 was designated as the normalized peak temperature. The normalized peak temperatures for the cement paste hydrated for 1 day, 7 days, 14 days, 28 days, 3 months, and 6 months were respectively 136, 144, 149, 152, 156, and 156°C. The *d* values varied from 9.75 Å at 1 day to 9.63 Å at 6 months.

The differences in the normalized peak temperatures could not be explained by the presence of $C_3F \cdot 3CaSO_4 \cdot 32H_2O$, $C_3A \cdot CaSO_4 \cdot 12H_2O$ or β-$C_3A \cdot Ca(OH)_2 \cdot 12H_2O$. Members of the solid solution $C_3A \cdot 3CaSO_4 \cdot aq$ –$C_3A \cdot 3Ca(OH)_2.aq$ were prepared and a straight line was obtained in a plot of the relationship between their normalized peak temperature and the composition. From the endothermal peak temperature of the ettringite phase the probable composition of the solid solution could be computed. The results showed that the ettringite phase corresponding to the pure sulfate is formed initially, while the hydroxy-end member is formed at later periods.

The types of water associated with a Portland cement paste may be examined by DTA. These are useful in studying the effect of pore structure on sorption behavior in cements (*91*).

Hydration Under Different Environments

Carbonation. The lime liberated during the hydration of cement takes up CO_2 from the atmosphere and forms $CaCO_3$. An endothermic effect beyond 750°C indicates the presence of $CaCO_3$, the intensity of which generally increases with increasing period of hydration. Cuband, Murat, and Eyraud (*67*) estimated by DTA the carbonation of hydrated cement cured in air or in water and found a considerable amount of carbonation in both cases. Esenwein (*39*) found that after 7 days of hydration at 70% R.H. the cement showed endothermal effects corresponding to hydrated calcium silicate hydrate, calcium hydroxide, and calcium carbonate.

There is a possibility of the formation of carboaluminates in cements ex-

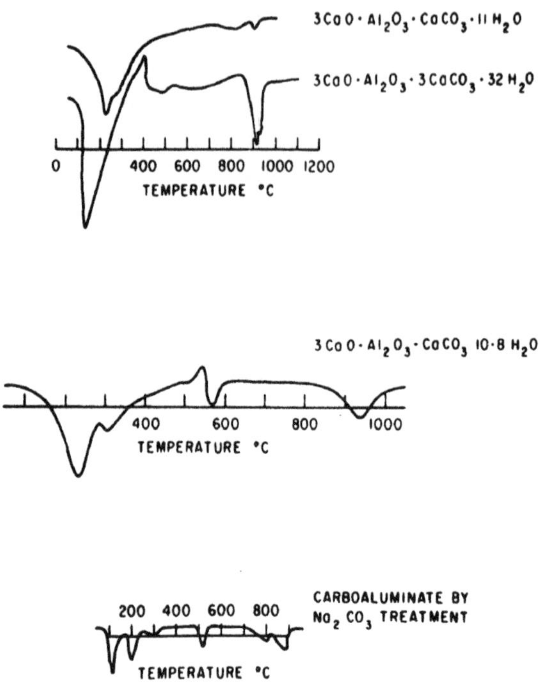

Fig. IV.2.6: DTA of carboaluminates reported by different workers (*9*, *68*, *69*).

posed to moist air or to alkali carbonates. Greene (*9*) examined the DTA of cement, with or without the presence of sodium carbonate. The cement containing 2% Na_2CO_3 showed a thermogram very similar to that of the carboaluminate reported by Turriziani and Schippa (*68*).

In Fig. IV.2.6 the thermograms of the carboaluminates reported by Carlson and Berman (*69*) and Turriziani and Schippa (*68*) are compared with the thermogram obtained by Greene (*9*). Not all the peaks in carboaluminates have been explained properly, and further work is needed in this direction.

Most workers have held that CO_2 is present as calcite in Portland cement pastes. Cole and Kroone (*70*, *71*), by applying the methods of DTA, TGA, and X-ray diffraction, have shown that CO_2 is chemically bonded largely as poorly crystallized vaterite, aragonite, and calcite and not as a silicate mineral. The samples studied by them covered eight Portland cement mortars cured in air or in CO_2 at 65 or 100% humidity for different periods, and two samples of calcium silicate hydrate either autoclaved or cured in a damp CO_2 atmosphere.

The DTA showed a small endothermal effect at a low temperature caused by the dehydration of interlayer water from hydrated cement minerals, two small endotherms at 500 and 550°C due to the dehydration of $Ca(OH)_2$, and a small endothermic effect at 600-750°C, followed by a more intense effect at about 800°C. The peak at 800°C could be attributed to the decomposition of well-crystallized calcite.

The endothermal area at 800°C in the DTA gave an indication of the amount of calcite formed. The $CaCO_3$ content obtained by chemical analysis (by assigning all CO_2 to $CaCO_3$) and that found by DTA vary, the DTA results showing lower values (*72*). These results (Table IV.2.1) demonstrate that CO_2 is only partly bound as well-crystallized $CaCO_3$ in mortars. The unaccounted CO_2 is partly expelled at 600°-750°C.

In the last column of the table the calculated endothermal areas of different samples at this temperature are shown. This figure, along with that obtained at 800°C for crystallized $CaCO_3$, did not fully account for all the CO_2 estimated by chemical analysis. It could be that part of the CO_2 is loosely held in mortars and is gradually desorbed without exhibiting any thermal effect. The thermogram of calcium silicate hydrate exposed to CO_2 shows an endothermic effect between 500 and 700°C.

The sequence of reactions during carbonation seems to be: cement minerals → siliceous residue + calcium hydroxide → vaterite + aragonite + poorly-crystallized calcite → well-crystallized calcite.

Chloride Solutions. Concretes exposed to the action of sea water are prone to attack by chloride solutions containing NaCl, KCl, $CaCl_2$, and $MgCl_2$. The effect of these chloride solutions on the formation of calcium chloroaluminates in hydrating cements may be followed by DTA. Heller and Ben-Yair (*73*, *74*) immersed the Portland cement paste containing 3.5% sodium, potassium, calcium, or magnesium chloride in sea water or in a solution containing sodium sulfate.

In distilled water the paste showed an expansion of 0.098% after 14 months and in 3.5% NaCl solution the value was 0.120%, while in the presence of 3.5% NaCl + 0.4% Na_2SO_4 the expansion was 0.129%. The increased expansion in presence of sodium sulfate may be explained by assuming that calcium sulfoaluminate occupies some of the pore space leaving the rest of the pores for the formation of calcium chloroaluminate.

The thermogram of the pure calcium chloroaluminate can be compared with the thermograms of the cement pastes exposed to distilled water or 3.5% NaCl solution (Fig. IV.2.7). The composite endothermic effect at 130°C represents CSH gel and the high calcium sulfoaluminate and the effects at 190 and 320°C may be ascribed to the chloroaluminate phase. The intensity of the chloroaluminate peak is greater for samples exhibiting

TABLE IV.2.1

$CaCO_3$ and Calcite Contents of Samples of Mortar and of Calcium Silicate Hydrate [Cole and Kroone (70)]

		Chemical analysis	*Differential thermal analysis (Calcite)*		
Material	*Sample*	*$CaCO_3$ total possible, %*	*$CaCO_3$ from $Ca(OH)_2$,† %*	*Well—crystallized,†† %*	*Poorly crystallized, ††† area in arbitrary units*
Portland cement mortars	Noncarbonated, sample 1	3.7	—	2.5	4.0
	Carbonated, sample 2	12.8	5.1	3.8	6.0
	Noncarbonated, sample 3	3.7	—	2.4	3.0
	Carbonated, sample 4	17.0	4.3	3.3	6.0
	Noncarbonated, sample 5	3.2	2.6	1.2	2.0
	Carbonated, sample 6	6.9	4.0	1.5	5.0
	Carbonated 3 months, sample 7	10.1	—	1.0	5.5
	Carbonated 30 months, sample 7	13.7	—	1.9	3.0
Calcium silicate hydrate	Noncarbonated, sample 8	2.4	—	4.7	26.0
	Carbonated, sample 9	33.3	—	5.3	115.0

†Based on the area of the endothermic peak at about 500°C for the non-carbonated sample of the pair.
††Based on the area of the endothermic peak at about 820°C.
†††Area of endothermic peak between 600-750°C.

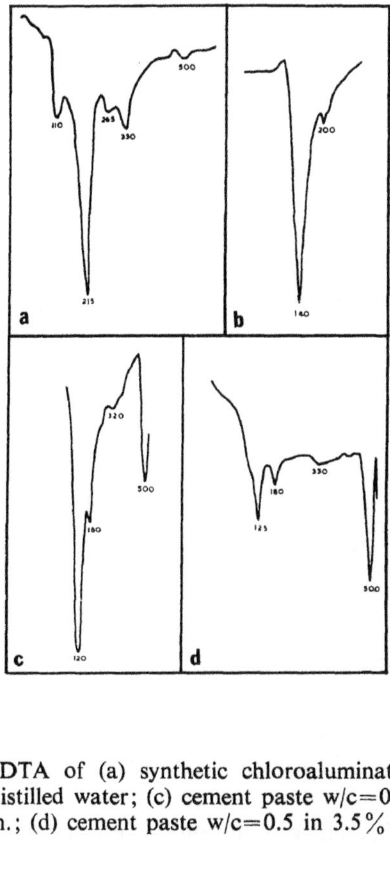

Fig. IV.2.7: DTA of (a) synthetic chloroaluminate; (b) cement paste w/c=0.5, in distilled water; (c) cement paste w/c=0.3 immersed in 3.5% NaCl soln.; (d) cement paste w/c=0.5 in 3.5% NaCl soln. (*74*).

more expansion.

Sulfate Solutions. The high sulfate resistance of Portland cement type V is attributed to the low amounts of sulfoaluminates formed in these cements. Portland cement I on exposure to 0.4% Na_2SO_4 solution for 1400 days shows an expansion of 0.128%, whereas under corresponding conditions Portland cement V shows an expansion of only 0.043%. On exposure to 10% Na_2SO_4 solution type V cement shows an expansion of 0.130% at 300 days, while type I cement disintegrates after 127 days with an expansion of 0.940% (*53*).

In Fig. IV.2.8 the thermogram of synthetic calcium sulfoaluminate is compared with the thermograms of Portland cement type I and V after exposing them to 0.4 or 10% Na_2SO_4 solution for 300 or 1400 days (*53*). The peak areas of sulfoaluminate are larger for both the concretes exposed to 10% Na_2SO_4 solution. Thermal effects also show that the concentration of sulfoaluminate in both type I and type V cements is of the same order of

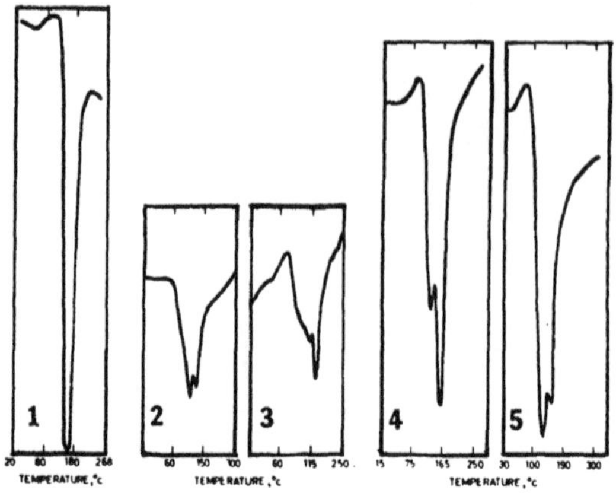

Fig. IV.2.8: Low temperature sulfoaluminate curves (*53*). (1) Synthetic calcium sulfoaluminate; (2) Portland cement I; (3) Portland cement V after exposure for 1400 days in 0.4% Na_2SO_4 solution; (4) Portland cement I; (5) Portland cement V after exposure for 300 days in 10% Na_2SO_4 solution.

magnitude. This indicates that the expansion in the Portland cements I or V is not related to the amount of crystalline calcium sulfoaluminate.

The first endotherm in Fig. IV.2.8c, d, e, and f may represent the transformation product of the sulfoaluminate or a part of a solid solution series with the hydrated calcium silicate similar to phase X described by Kalousek (*8*).

DTA results have confirmed that an autoclaved Portland cement, after immersion in a solution of Na_2SO_4, does not form $CaSO_4$ crystals (*90*).

Alkalies. The nature of hydration products of clinker minerals formed in the presence of alkalies is not well understood. In the clinker formation the alkalies may form solid solutions with the aluminate and silicate phases. Kryzhanovskaya, Mirak'yan, and co-workers (*75*) studied by DTA the hydration of $KC_{23}S_{12}$ and NC_8A_3, representing a solid solution of K_2O in C_2S and Na_2O in C_3A respectively. Both β-C_2S and $KC_{23}S_{12}$ do not show any difference in their thermograms.

The hydrated NC_8A_3 exhibits four endothermic effects, the peaks at 100 and 180°C representing the loss of water from the hexagonal hydroaluminates (*76*). Under similar conditions C_3A is converted to C_3AH_6 with the characteristic peak at 320-340°C. These results show that the alkali calcium aluminate is not easily converted to the cubic C_3AH_6.

The influence of alkalies on the hydration behaviour of C_3S and C_2S has also been studied by DTA (*77*). The samples C_3S–C_2S, C_3S–C_2S–Na_2O, and C_3S–C_2S–K_2O show similar thermograms with endothermal effects in the ranges 50-300°C, 500-600°C, and 780°-880°C.

The first endothermal effect is due to the removal of water from calcium silicate hydrate, the second, to the decomposition of $Ca(OH)_2$, and the third, to the dehydration of silicates. It is possible that alkalies do not form new compounds. In samples C_3S–C_2S + $CaCl_2$, C_3S–C_2S–Na_2O + $CaCl_2$, and C_3S–C_2S–K_2O + $CaCl_2$, an additional endothermic effect is obtained in the range 750-850°C, suggestive of a new compound in the presence of $CaCl_2$. The samples with SiO_2, *viz.* C_3S–C_2S + SiO_2, C_3S–C_2S–Na_2O + SiO_2, and C_3S–C_2S–K_2O + SiO_2, gave an exothermic peak at 780-860°C for the presence of CSH(I).

False Set. In some instances the cement exhibits an early set, commonly known as false set. The main cause of this is the production of hemihydrate and soluble anhydrite from gypsum during the final grinding process of the clinker in which the temperature may vary in the range 40-150°C. False set has also been observed in cements containing about 0.5% of alkali carbonate (*78*).

The dehydrated state of gypsum in a ground cement can be studied by DTA. Gilliland (*89*) compared the thermograms of $CaSO_4 \cdot 2H_2O$, $CaSO_4 \cdot \frac{1}{2}H_2O$ and $CaSO_4$ (insoluble) with the thermogram of a commercially ground cement sample (Fig. IV.2.9).

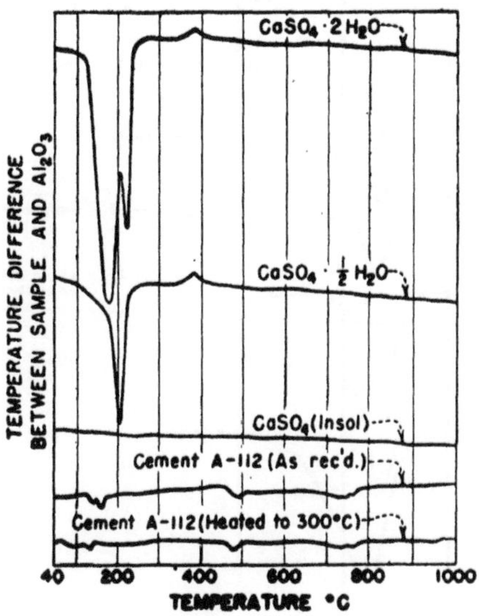

Fig. IV.2.9: Differential thermal curves of gypsum and its products (*89*).

Gypsum shows two endothermic peaks at 190 and 225°C for the stepwise loss of 2 molecules of water. The exothermic hump at about 380°C denotes the conversion of soluble to insoluble anhydrite. The hemihydrate has a single endothermic effect at about 200°C for the loss of ½ molecule of water. The insoluble anhydrite prepared by heating gypsum to 400°C does not show any thermal change.

A commercial ground cement designated by A-112 exhibits two endothermic peaks at 140° and 160°C, the latter of stronger intensity. In pure gypsum the second endothermic peak is more intense than the first. Hence it follows that in sample A-112 most of the calcium sulfate is present as the hemihydrate. Heating to 300°C converts the sample A-112 to soluble anhydrite, and this partially reconverts to hemihydrate.

Takemoto, Ito, and Hirayama (*79*) studied by DTA the extent to which dehydration in gypsum occurs in an experimental mill. Thermograms in-

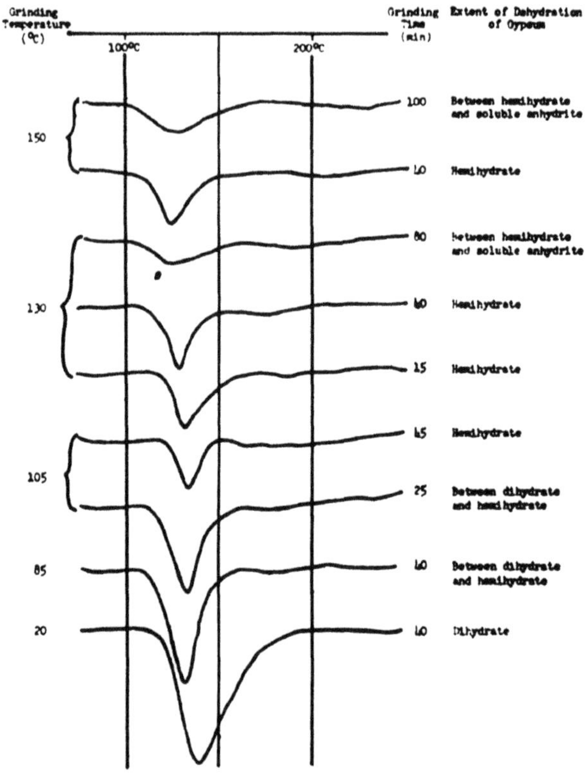

Fig. IV.2.10: Thermograms of gypsum at different grinding temperatures (*79*).

dicate the state of gypsum at different grinding temperatures (Fig. IV.2.10). The hemihydrate is formed at 85°C after a grinding period of 40 min and the soluble anhydrite is detected at 130°C at a grinding period of 80 min.

Murakami, Izuka, and Tanaka (*80*), Sulikowski *et al* (*81*, *82*), Fischer (*83*), Mtschedlov-Petrosian *et al* (*47*), Sato and Kanaya (*84*), and St. John (*85*) have used the DTA technique to estimate the various forms of calcium sulfate and their hydrates. Sato and Kanaya and Sulikowski used standard mixtures of $CaSO_4 \cdot 2H_2O$ and $CaSO_4 \cdot \frac{1}{2}H_2O$ for calibration. Since some gypsum may be present in the dehydrated state, St. John suggested the use of dehydrated gypsum for calibration. He compared the calibration curves by plotting A_1/A_2 (where A_1 is the endothermal area due to the loss of $1\frac{1}{2}$ molecules of water from gypsum and A_2 is the endothermal area of the hemihydrate) against the ignition loss in gypsum or dihydrate/hemihydrate mixtures, and observed some differences.

Portland cement exposed to aeration may sometimes show an abnormal setting behaviour distinct from the well known flash set. This phenomenon has not been explained satisfactorily. Kalousek *et al* (*92*) examined a ground cement clinker containing 76% alite, 7% belite, and 16% ferrite phase and also a ground clinker containing 2.0% SO_3. The samples were exposed to CO_2-free air at 60% R.H. or to laboratory air at 50% R.H. for periods ranging from 0.25 min to 15 min, and then subjected to quantitative DTA. Gypsum was indicated by two endotherms at 170 and 195°C. Ettringite or the Fe_2O_3-bearing solid solution could be identified by an endotherm at 140-160°C and $Ca(OH)_2$ by an endotherm at about 500-580°C. The hydrated ferrite phase formed in samples containing no gypsum gave an endotherm at 190°C.

These studies indicated that the abnormal setting was not induced by the rapid reaction of alite, belite or the ferrite phase, CO_2, or gypsum hemihydrate. It was suggested that the presence of both positively and negatively charged particles in the hydrated clinker caused the abnormal set.

Estimation of Hydrated Phases in Cements. One of the most important applications of DTA is the quantitative estimation of constituents in mixtures. Some success has been achieved in the study of the hydrated products of Portland cement.

Midgley (*59*) prepared calibration curves of pure constituents such as tobermorite gel, ettringite, and $Ca(OH)_2$. He was beset with the problem of determining the areas of the individual effects of triple peak in the cement below 200°C. He therefore constructed graphically the two peaks of free water and ettringite with reference to the standard peaks and the remainder was assigned to tobermorite gel. The estimation of $Ca(OH)_2$ was relatively easier. The percentage of tobermorite, ettringite, and $Ca(OH)_2$ determined

TABLE IV.2.2
Weight % of Cement Hydrate Minerals in a Hydrating 1:3 Mortar, W/C 0.52 [Midgley (59)]

Age	*Tobermorite*	*Weight % of Ettringite*	*Ca(OH)$_2$*
1 day	1	10	9
3 days	2	35	10
7 days	14	27	10
28 days	18	28	8

by DTA is shown in Table IV.2.2.

Recently, Kalousek (*86*) estimated by DTA gypsum, ettringite, monosulfate, and $Ca(OH)_2$ in cement pastes containing 0.4-4.0% SO_3 and cured at 24, 54, and 82°C for 19, 15, and 14 hr respectively.

Gypsum was estimated in cement pastes by carrying out DTA of prepared blends of gypsum with clinker. The ratio of the gypsum content to the average area of the endothermal effects gave an analytical factor of 0.04% gypsum per vernier unit. In cement pastes containing residual gypsum, the peaks were superimposed on the ettringite endotherm. Resolution of the peaks could be done by superimposing the endotherm of ettringite containing no gypsum.

Ettringite was estimated by blending known amounts of ettringite with cement paste. The analytical factor was 0.125% ettringite per vernier unit. Ettringite could be determined in amounts down to ±5%.

The pure monosulfate exhibits complex endothermal effects extending from about 150 to 350°C. A straight line was drawn to close the endother mal effect to determine the area. The analytical factor was found to be 0.120% monosulfate per vernier unit.

Calcium hydroxide exhibits an endothermal effect in the range 500-600°C and the endothermal peak area can be determined with good accuracy.

There is an overlapping of the endothermal effects of ettringite and tobermorite gel. Tobermorite gel was estimated by determining the area of $Ca(OH)_2$ peak as this was proportional to the amount of tobermorite gel according to the equation

$$2C_3S + nH_2O \longrightarrow C_3S_2H_{n-3} + 3Ca(OH)_2$$

By deducting this value from the total area in the range 100-200°C, ettringite could be calculated.

The total SO_3 obtained by the DTA method was compared with that determined on the basis of ignition. The DTA values were generally less than those determined from ignition loss and the missing SO_3 was to be accounted

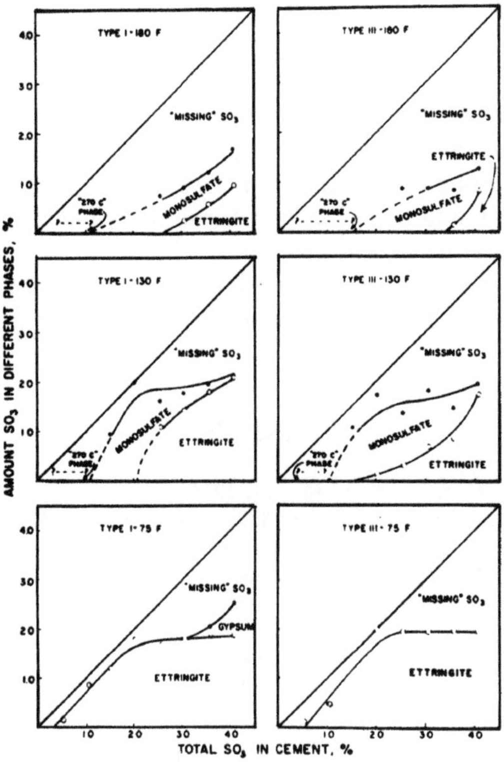

Fig. IV.2.11: Composite results on Type I and Type III cements cured at 75, 130, and 180°F for 19 hr (*86*).

for. In type I cement containing 4.08% SO_3, the missing SO_3 was 1.55% and the corresponding value in type III cement was 2.12%. The maximum values of missing SO_3 for type I and type II cements cured at 82°C were respectively 2.41 and 2.83%. The missing SO_3 was considered to be a lattice-substituted tobermorite gel designated as phase X by Kalousek.

In Fig. IV.2.11 the SO_3 phases as determined by DTA are shown at different curing temperatures. The amounts of SO_3-bearing phase are governed by the rate of the reactions:

ettringite + tricalcium aluminate→monosulfate

monosulfate + tricalcium aluminate→solid solution

solid solution + tobermorite gel→phase X

Proks and Siske (*87*) have demonstrated the use of low temperature DTA for estimating the chemically-combined water in cements at given time intervals. The freezing or the melting of free water results in the absorption or evolution of heat, and these can be determined by the areas of the thermal effects.

An instrument has been described by Budnikov *et al* (*88*) for determination of the heats of hydration of cements by graphic integration of the $\Delta T - t$ curve.

REFERENCES

1. Jones, F.E., *J. Phys. Chem.*, *48*, 311, 1944.
2. D'Ans, J., and Eick, H., *Zement-Kalk-Gips*, *6*, 302, 1953.
3. Eitel, W., *J. Am. Concrete Inst.*, *28*, 679, 1957.
4. Steinour, H.H., *Portland Cement Assoc. Res. Dep. Bull. 98*, November, 1958.
5. Taylor, H.F.W., *Ind. Chim. Belges*, *20*, 63, 1955.
6. Manable, T., and Kawada, N., *Semento Gijutsu Nenpo*, *12*, 46, 1958.
7. Manabe, T., and Kawada, N., *J. Am. Concrete Inst.*, *31*, 639, 1960.
8. Kalousek, G.L., Davis, C.W., and Schmertz, W.E., *J. Am. Concrete Inst.*, *20*, 693, 1949.
9. Greene, K.T., *Proc. 4th Intern. Symp. Chem. Cements*, Washington, p. 359, 1960.
10. Copeland, L.E., Kantro, D.L., and Verbeck, G., *Proc. 4th Intern. Symp. Chem. Cements*, Washington, p. 429, 1960.
11. Taylor, H.F.W., *27th Congr. of Industrial Chem.*, Brussels, p. 4, 1954.
12. Turriziani, R., *Ind. ital. Cemento*, *29*, 185; 219; 244; 276, 1959.
13. Taylor, H.F.W., *The Chemistry of Cements*, Vol. I., Academic Press, London, pp. 1-24, 1964.
14. Woods, H., Steinour, H.H., and Starke, H.R., *Ind. & Eng. Chem.*, *24*, 1207, 1932.
15. Woods, H., Steinour, H.H., and Starke, H.R., *Eng. News Record*, *110*, 431, 1933.
16. Verbeck, G.J., and Foster, C.W., *Proc. Am. Soc. Testing Mater.*, *50*, 1235, 1950.
17. Bogue, R.H., and Lerch, W., *Ind. & Eng. Chem.*, *26*, 837, 1934.
18. Brunauer, S., *The Science of Engineering Materials*, Edited by J.E. Goldman. John Wiley & Sons, Inc., New York, N.Y., 1957.
19. Powers, T.C., *Portland Cement Assoc. Res. Dep. Bull. 90*, July, 1958.
20. *Portland Cement Assoc., Yearly Rept. Res. Development Div.*, pp. 23-25, 1957.
21. Yamaguchi, G., Takemoto, K., Uchikawa, H., and Takagi, S., *Proc. 4th Intern. Symp. Chem. Cements*, Washington, p. 495, 1960.
22. Powers, T.C., The Physical Structure of Portland Cement Paste in The

Chemistry of Cements, Academic Press, London, p. 392, 1964.
23. Powers, T.C., *Proc. 4th Intern. Symp. Chem. Cements*, Washington, p. 577, 1960.
24. Le Chatelier, H., *Thesis 1887*, '*Recherches experimentales sur la constitution des Mortiers hydrauliques*', 2nd Ed., Dunold, Paris, 1904.
25. Michaelis, W., *Chem. Ztg.*, 982, 1893.
26. Bogue, R.H., *Chemistry of Portland Cement*, 2nd Ed., Reinhold, New York, 1955.
27. Hansen, W.C., *Proc. 4th Intern. Symp. Chem. Cements*, London, pp. 318-321, 1952.
28. Hansen, W.C., *Am. Soc. Testing Materials, Special Techn. Publn. No.* **266**, pp. 3-25, 1960.
29. Jeffery, J.W., *Chem. & Ind.* (*London*), p. 1756, 1955.
30. Chatterji, A.K., and Phatak, T.C., *Nature*, *203*, 138, 1964.
31. Chatterji, A.K., and Phatak, T.C., *Nature*, *207*, 711, 1965.
32. Lepingle, M., *Inst. Tech. Batiment Bull. 7*, 1941, and *Bull. 11*, 1943, Paris.
33. Petzold, A., and Göhlert, I., *Tonind.-Z. Keram Rundschau*, *86*, 228, 1962.
34. Petzold, A., and Talke, I., *Silicat. Tech.*, *11*, 122, 1960.
35. Bozhenov, P.I., and Suvorova, G.E., *Tsement*, *23*, 8, 1957.
36. Barta, R., and Satava, V., *Stavivo*, *31*, 15, 1953.
37. Barta, R., and Satava, V., *Chem. Prumysl*, *3*, 113, 1953.
38. Rey, M., *Silicates Ind.*, *22*, 533, 1957.
39. Esenwein, P., *Schweiz. Arch. Angew. Wiss. u Techn.*, *20*, 365, 1954.
40. Levcuk, N.A., *Trudy Khazkov Inst. Inzhanerov Zheleznodorozh. Transporta im. S.M. Kirova*, *39*, 19, 1960.
41. Govorov, A.A., *Trudy Kharkov Inst. Inzhenerov Zheleznodorozh. Transporta im. S.M. Kirova*, *39*, 11, 1960.
42. Govorov, A.A., *Visnik Akad. Constr. & Arch. Ukrain*, USSR, *1*, 30, 1960.
43. Kurdowski, W., *Cement-Wapno-Gips*, *1*, 4, 1960.
44. Krawtschenko, I.W., *Tsement*, *No. 1*, 11, 1957.
45. Lommatzsch, A., *Silikat Tech.*, *7*, 188, 1956.
46. Lommatzsch, A., *Silikat. Tech.*, *7*, 468, 1956.
47. Mtschedlov-Petrosian, O.P., Bunakov, A.G., Govorov, A.A., Latischew, F.A., Lewtschuk, N.A., and Sstrelkova, I.S., *Silikat. Tech.*, *9*, 556, 1958.
48. Greenberg, S.A., *J. Phys. Chem.*, *58*, 362, 1954.
49. Odler, J.G.I., *Zement-Kalk-Gips*, *55*, 303, 1966.
50. Tamas, F., *Proc. 6th Conf. Silicates Ind.*, Budapest, p. 425, 1961.
51. Mtschedlov-Petrosian, O.P., *Silikat. Tech.*, *11*, 189, 1960.
52. Kalousek, G.L., and Adams, M., *J. Am. Concrete Inst.*, *23*, 77, 1951.
53. Heller, L., and Ben-Yair, M., *J. Appl. Chem.* (*London*), *14*, 20, 1964.
54. Midgley, H.G., and Rosaman, D., *Proc. 4th Intern. Symp. Chem. Cements*, Washington, p. 259, 1960.
55. Nemecek, K., *Study Inst. Silicates*, Prague, 1954.
56. Diamond, S., Dolch, W.L., and White, J.L., *Highway Res. Rec.*, *62*, 62, 1965.
57. Budnikov, P.P., Gorshkov, V.S., and Pototskaya, T.A., *Khim i Prakt. Primerie Silikatov, Inst. Khim. Silikatov, Akad. Nauk SSSR*, p. 101, 1960.
58. Tamas, F., and Liptay, G., *Epitoanyag*, *14*, 281, 1962.
59. Midgley, H.G., *Proc. 4th Intern. Symp. Chem. Cements*, Washington, p. 479, 1960.

60. Copeland, L.E., and Hayes, J.C., *Am. Ceram. Soc. Bull.*, *194*, 70, 1953.
61. Nicol, A., *Rev. Mater. Constr.*, *No. 477*, 153; *No. 478-479*, 181, 1955.
62. Van Bemst, A., *Bull. Soc. Chim. Belges.*, *64*, 333, 1955.
63. Buckle, E.R., *J. Phys. Chem.*, *63*, 1231, 1959.
64. Gracheva, O.I., and Vikulin, A.Y., *Tr. Nil Asbesttsementa*, *14*, 24, 1962.
65. Kalousek, G.L., *Thesis*, Univ. of Maryland, 1941.
66. Flint, E.P., and Wells, L.S., *J. Res. Nat. Bur. Stad.*, *33*, 471, 1944.
67. Cuband, J.C., Murat, M., and Eyraud, C., *Rev. Mater. Constr. Rev.*, *Publ. No. 609*, 239, 1966.
68. Turriziani, R., and Schippa, G., *Ric. Sci.*, *26*, 2792, 1956.
69. Carlson, E.T., and Berman, H.A., *J. Res. Nat. Bur. Std.*, *64 A*, 333, 1960.
70. Cole, W.F., and Kroone, B., *J. Am. Concrete Inst.*, *31*, 1275, 1960.
71. Cole, W.F., and Kroone, B., *Nature*, *184*, 57, 1959.
72. Gaze, R., and Robertson, R.H.S., *Mag. Concrete Res.*, *8*, *7*, 1956.
73. Heller, L., and Ben-Yair, M., *Nature*, *191*, 488, 1961.
74. Heller, L., and Ben-Yair, M., *J. Appl. Chem.* (*London*), *16*, 223, 1966.
75. Kryzhanovskaya, I.A., Mirak'yan, V.M., Shokotova, B.G., and Kholodnyi, A.G., *Tsement*, *31*, 10, 1965.
76. Tulovskaya, Z.D., and Segalova, E.E., *Zh. Prikl. Khim.*, *37*, 2, 1964.
77. Anon, *Cement, Lime and Gravel*, *39*, 27, 1964.
78. Bogue, R.H., and Lerch, W., *Portland Cement Assoc.* Unpublished Fellowship Report, 1933. (*cf.* Copeland, L.E., and Kantro, D.L., *The Chemistry of Cements*, Academic Press, London, 1964.)
79. Takemoto, K., Ito, I., and Hirayama, K., *Rock Products*, *62*, 140, 1959.
80. Murakami, K., Izuka, S., and Tanaka, H., *Sekko To Sekkai*, *26*, 8, 1957.
81. Dyczek, J., Sawkow, J. and Sulikowski, J., *Cement-Wapno-Gips*, *18*, 99, 1963.
82. Sulikowski, J., *Proc. 7th Conf. Silicate Ind.*, Budapest, p. 137, 1965.
83. Fischer, W.K., *Silikat. Tech.*, *14*, 26, 1963.
84. Sato, N., and Kanaya, M., *J. Res. Onada Cement Co.*, *14*, 143, 1962.
85. St. John, D.A., *New Zealand J. Sci.*, *7*, 353, 1964.
86. Kalousek, G.L., *Mat. Res. & Stad.*, *5*, 292, 1965.
87. Proks, I., and Siske, V., *Silikaty*, *5*, 229, 1961.
88. Budnikov, P.P., Shotenberg, S.M., and Azelitskaya, R.D., *Tsement*, *24*, 15, 1958.
89. Gilliland, J.L., *J. Am. Concrete Inst.*, *47*, 809, 1951.
90. Rakhimbaev, S.M., and Kantsepolski, I.S., *Uzbesk Khim. Zh.*, *6*, 73, 1962.
91. Chekhovskoi, Y.V., Kazansii, V.M., and Leirikh, V.E., *Inzh.-Fiz. Zh.*, *Akad. Nauk Belorussk. SSR*, *6*, 50, 1963.
92. Kalousek, G.L., Jugovich, Z.T., and Gillam, J.L., *Am. Ceram. Soc. Bull.*, *46*, 270, 1967.

V. Hydration of Individual Phases in Portland Cement

V.1. HYDRATION OF C_3A

The tricalcium aluminate phase of composition $3CaO \cdot Al_2O_3$, although constituting a low proportion in Portland cement, exerts a significant influence on the heat development and setting property. The average C_3A content in type I, II, III, IV, and V cement is 11.0, 5.0, 11.0, 4.0, and 4.0% respectively (*1*). The phenomenon of 'flash set,' the formation of various calcium aluminate hydrates, and calcium carbo- and sulfoaluminate hydrates involve the reactions of C_3A.

Calcium Aluminate Hydrates. In the system $CaO–Al_2O_3–H_2O$ several calcium aluminate hydrates have been reported. At temperatures less than 50°C the hydrates of composition CAH_{10}, C_2AH_8, and C_4AH_{19} are formed readily. However, the thermodynamically stable compounds at room temperature are gibbsite, calcium hydroxide, and C_3AH_6. At temperatures in the range 50-200°C the compounds $Al(OH)_3$, C_3AH_6, and $Ca(OH)_2$ are detected at earlier periods. At temperatures exceeding 200°C, boehmite, diaspore, and $C_4A_3H_3$ are detected.

The study of the ternary system shows that in Portland cement, in the absence of sulfates and other interfering compounds, the metastable products of hydration of C_3A are expected to be C_2AH_x and C_4AH_x.

Tricalcium aluminate reacts with excess of water to form C_2AH_8 and either or both C_3AH_6 and $C_3A \cdot Ca(OH)_2 \cdot 12H_2O(C_4AH_{13})$. In saturated $Ca(OH)_2$ solution, C_2AH_8 reacts with $Ca(OH)_2$ to form C_4AH_{13} or C_3AH_6, depending on the conditions of hydration.

Of the various calcium aluminate hydrates only the structure of C_3AH_6 is established. The C_3AH_6 crystallizes in the cubic system and is related in structure to garnet. Garnet has a composition $Ca_3Al_2(SiO_4)_3$, and by replacement of each SiO_4^{-4} ion with $4(OH)^-$ ions the cubic hexahydrate of composition $Ca_3Al_2(OH)_{12}$ is obtained (*2*). The naturally occurring hydrogarnets have a composition intermediate between garnet and C_3AH_6. Both C_2AH_8 and C_4AH_x crystallize in thin hexagonal plates.

Roberts (*3*) recognized six crystalline modifications with the general composition C_4AH_x. The compound C_4AH_{19} exists in α_1 and α_2-modifications and both the forms convert to C_4AH_{13} at a relative humidity of 12-81%. The existence of α and β-C_4AH_{13} was also reported by Roberts but recent investigations of Seligman and Greening (*4*) have shown that the α-form is a CO_2-contaminated product of the formula $C_3A \cdot CaCO_3 \cdot 11H_2O$. The hexagonal C_2AH_8 compound is also reported to exist in three forms (*3*), *viz.* α_1, α_2, and β. The α_1-form has been designated simply as C_2AH_8 in the literature.

The tetracalcium trialuminate trihydrate $C_4A_3H_3$ is obtained (*5*) by autoclaving C_3A or $Ca(OH)_2$–Al_2O_3 mixtures at 350°C. The $C_4A_3H_3$ crystallizes in rectangular orthorhombic plates.

Thermal changes accompanying the heating of various calcium aluminate hydrates have been studied by TGA, DTA, and X-ray methods. Tricalcium aluminate hexahydrate shows little change in weight below 275°C. At 275°C a rapid loss in weight corresponds to the loss of 4.5 molecules from the C_3AH_6 (*6*). There is a gradual loss of water in the range 300-550°C, which denotes the loss of the remaining 1.5 molecules of water and formation of $C_{12}A_7$ and CaO. Steam curing of C_3AH_6 at 300-400°C is known to result in the conversion of the hexahydrate to $Ca(OH)_2$ and $C_4A_3H_3$, the latter decomposing to $C_{12}A_7$, CA_2, and H_2O at about 700°C.

The dehydration behaviour of β-C_4AH_{13} and C_2AH_8 has been systematically investigated by Buttler and others (*7, 8*). At least four temperatures are registered at which the curve shows a steep change in the weight-loss curves of C_4AH_{13}. At 150°C, C_4AH_{13} loses 6 molecules of H_2O and at about 300°C the compunds $C_4A_3H_3$ and $Ca(OH)_2$ are detected. A slight loss at 400°C corresponds to the dehydration of $Ca(OH)_2$. The $C_4A_3H_3$ loses all its water at about 700°C and forms $C_{12}A_7$. The C_2AH_8 decomposes to $C_2AH_{7.5}$, C_2AH_5 and C_2AH_4 successively at 34% R.H., 12% R.H. (*102*°C), and 120°C (*3, 9-11*).

The weight loss curve of C_2AH_8 shows a continuous drop in weight from room temperature to about 600°C. The compound $C_4A_3H_3$, which forms at higher temperatures, decomposes to $C_{12}A_7$ at about 750°C.

Differential Thermal Analysis

One of the advantages of the DTA technique in the investigation of products of hydration of C_3A is that it makes possible the detection of minute quantities of hydrates that may remain unidentified by the X-ray method. Hence this method has been applied by many workers to study the rate of conversion of the metastable hexagonal hydrates to the stable cubic hexahydrate under different conditions.

Differential thermal analyses of the hydration products of C_3A in pastes or suspensions have been investigated by Kalousek *et al* (*12*), Segalova *et al* (*13*, *24*), Lefol (*14*), Young (*15*), Majumdar and Roy (*16*), Jones (*17*), Budnikov *et al* (*19*), Sauman (*20*), Govorov (*21*, *30*), Brutskus and Segalova (*22*), Tulovskaya and Segalova (*23*), Okorokov and Golynkovolfson (*25*), Petzold and Göhlert (*26*), Rey (*27*), Feldman and Ramachandran (*18*, *28*), Kind *et al* (*29*), Gorshkov and Khmelevskaya (*31*), Turriziani (*32*), Turriziani and Schippa (*33*), and Midgley and Rosaman (*34*).

In the hydration of C_3A the development of the cubic C_3AH_6 is preceded by the formation of the hexagonal compounds C_2AH_{7-9} and C_4AH_{12-14} (*35-37*). The thermograms of C_3AH_6 reported by several workers seem to differ. A prominent endothermic bulge is always observed at about 315-330°C.

Jones (*17*), Majumdar and Roy (*16*), Young (*15*), and Feldman and Ramachandran (*28*) obtained two endothermal effects for C_3AH_6 representing dehydration in two stages, whereas Lefol (*14*) found a single endothermal effect at about 300°C. Budnikov *et al* (*19*), Sauman (*20*), and Petzold and Göhlert (*26*) reported three peaks. Govorov's (*21*) curves for C_3AH_6 contained as many as four endothermal peaks.

The difference in behavior of the hydrated C_3A reported by these investigators could have been caused by such variables as the method of preparation of C_3AH_6, the purity of the sample, the period of hydration, the temperature and the consistency of the paste during hydration, and the sensitivity of the DTA apparatus.

Kalousek *et al* (*12*) showed that the thermogram for C_4AH_{13} consists mainly of a large endothermic effect at 240°C, whereas Young (*15*) assigned three endothermic peaks at 120, 215, and 310°C to a mixture of C_4AH_{13} and C_2AH_8. Segalova and co-workers (*13*) showed large endothermic peaks at 190, 260, and 290°, with small peaks at 410 and 490°C representing the hexagonal aluminates. Rey (*27*) suggested that C_2AH_8 is identifiable mainly by peaks at 140 and 310°C.

Feldman and Ramachandran (*28*) made an investigation of the hydration of C_3A compacts at temperatures of 2, 12, 23, 52, and 80°C and correlated the DTA results with those of X-ray diffraction, porosimetry, and volume change. Based on their results they suggested a mechanism for the hydration of C_3A.

The hydration procedure and length change measurements were carried out by Feldman and Ramachandran as follows. The C_3A sample was hydrated at five temperatures, *viz.* 2, 12, 23, 52, and 80°C. The sample was in the form of a compact, approximately 0.20 cm thick, and compacted at a pressure of 48,000 lb/in. The porosity and pore size distribution of the

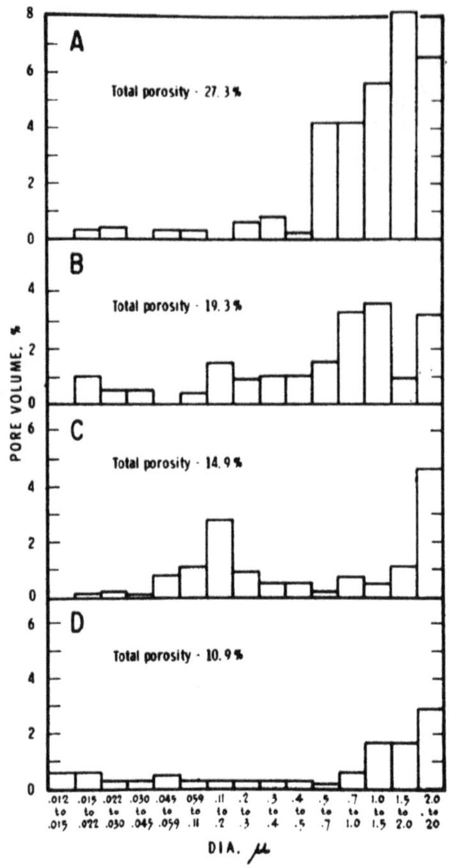

Fig. V.1.1: Pore size distribution of C_3A and hydrated C_3A compacts (*28*).

compact are shown in Fig. V.1.1. The procedures of compaction and other details are discussed in (*38-40*).

From the porosity value the water/solid ratio within the compact was computed to be 0.12. Length change measurements were made by introducing water to an inch-long sample mounted on a measuring gauge, placed in a vacuum tight chamber. Measurements were made with an accuracy of 2 $\times 10^{-6}$ in./in.

The reaction was stopped at predetermined periods by rapidly dipping the compact in liquid air and then grinding in an agate mortar with absolute alcohol cooled to a temperature of about −78 °C in solid carbon dioxide. The ground sample along with the alcohol was centrifuged and subsequently placed in the dessicator containing magnesium perchlorate and degassed by a vacuum pump for 6 hr. All the samples were ground to pass 100 mesh sieve in a gloved dry-box.

The DTA analysis was done in flowing nitrogen. A constant weight of the sample was used for each analysis. The heating rate was maintained at 10°C/min.

The pore size distribution and porosity were determined by a mercury porosimeter. The pore size distribution and porosity were calculated for four samples, *viz*. unhydrated C_3A compact, compact of C_3A hydrated for 7 days at 2°C, compact of C_3A hydrated for 25 hr at 23°C, and a compact of C_3A hydrated for 4 days at 52°C. The minimum pore size in which mercury could be injected was 0.012μ in diameter.

Fig. V.1.2 shows DTA thermograms of C_3A hydrated at 2°C for different periods. The unhydrated sample exhibits endothermic effects at 440 and

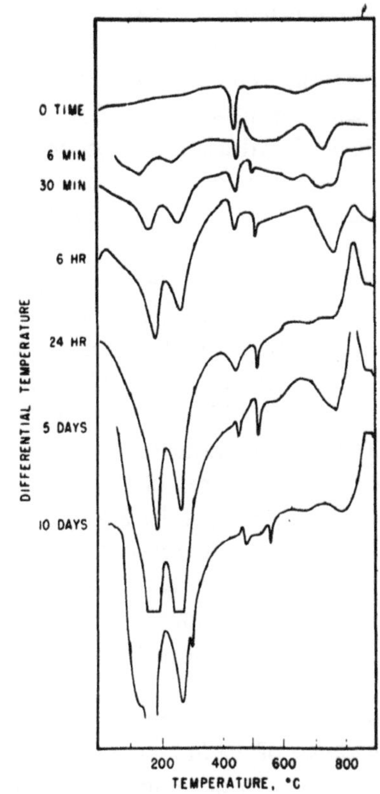

Fig. V.1.2: Hydration of C_3A at 2°C (*28*).

650°C due to $Ca(OH)_2$ and $CaCO_3$ respectively. At 6 min of hydration the endothermic effects at 130 and 235°C appear, these effects representing hexagonal aluminates C_2AH_{7-9} and C_4AH_{13} respectively. A small quantity of carbonate is also evident by an endothermic peak beyond 700°C.

At 30 min the effects at 130 and 235°C have increased in magnitude and shifted to 150 and 240°C respectively. Another endothermic effect is also observed at 500°C. At 6 hr together with further development of the peaks already reported (now at 160, 250, 450, and 510°C), an exothermic peak is observed at 840°C.

At 1 and 5 days the endothermic effect at 160, 250, and 510°C shift to 185, 260, and 530°C respectively and continue to increase in intensity. The exothermic peak at 840°C also shows a progressive increase in magnitude. At 10 days a new endothermic effect appears at 290°C, while peaks at 260 and 530°C decrease slightly.

It is clear that the two endothermic peaks at 130-190°C and 235- 260°C and the exothermic peak at 840°C increase in relation to each other with increasing periods of hydration. The peak at 130-190°C is always of greater intensity than the peak occurring in the range 235-260°C. The endothermic effect at 290°C after 10 days of hydration is probably due to the conversion of the hexagonal hydroaluminates to the cubic C_3AH_6.

The exothermic peak between 800 and 900°C could be attributed to the formation of a compound from the dehydrated C_2AH_8 and/or C_4AH_{13}. The endothermic peak at 500-530°C registered along with the peaks at 150 and 240°C does not seem to be characteristic of hexagonal or cubic hydrates. This endotherm may be caused by a small amount of $Ca(OH)_2$ formed during the hydration of C_3A at low temperatures. Calcium hydroxide, however, exhibits an endothermal effect at 450°C in the original C_3A preparation.

Fig. V.1.3 shows the DTA thermograms for the samples hydrated at 12°C. The endotherms with peak temperatures in the range 130-190°C and 235-250°C are again evident. They represent the hexagonal compounds C_2AH_{7-9} and C_4AH_{12-14} respectively. The endotherm at 135-190°C has a characteristic temperature below 100°C; for a 6-hr hydration a shoulder is observed on the peak, and this seems to develop into an endothermic peak at about 105°C at 5 days. An endothermic effect shifting from 500 to 550°C and an exotherm at 840°C are also indicated.

All these effects grow steadily from 6 min to 10 days. At 6 hr an endothermic dent at 290°C gives evidence of the first appearance of C_3AH_6. It is interesting to note that this peak, which develops from 6 hr to increase in prominence at 24 hr, 5 days, and 10 days, is accompanied by a continued development of the 185 and 250°C peaks up to 5 days. Another endothermic peak at 490°C continues to increase in intensity up to 10 days.

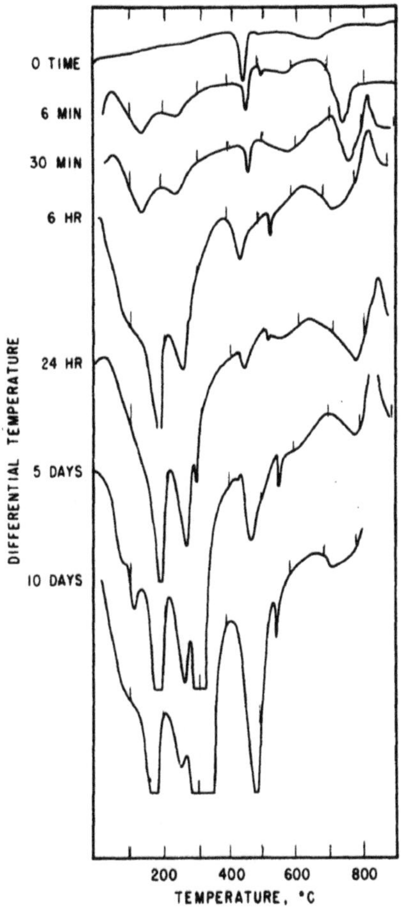

Fig. V.1.3: Hydration of C_3A at 12°C (*28*).

DTA thermograms of hydrated products were also obtained at 23°C for hydration times of 0.5, 6, and 30 min, 2 and 6 hr, and 1, 5, and 10 days. On the thermogram for 0.5 min (Fig. V.1.4), the endothermic peaks at 130 and 230°C were clearly observed. These peaks, together with the exotherm at 800-850°C can be seen clearly at 0.5 min, 6 min, 30 min, 2 hr, and 6 hr. At 30 min a very faint endothermic dent appears to be present at 285°C, suggesting that at 23°C C_3AH_6 may be present at 30 min.

At 2 hr the thermogram indicates a further increase of the hexagonal hydroaluminate as shown by the intensity of the peaks at 170 and 245°C. The endothermic peak at 285°C and another at 490°C suggest that C_3AH_6 has increased in amount at this period. At 6 hr there is an intensification of the peaks at 310 and 490°C for C_3AH_6, while there is a diminution of the peak

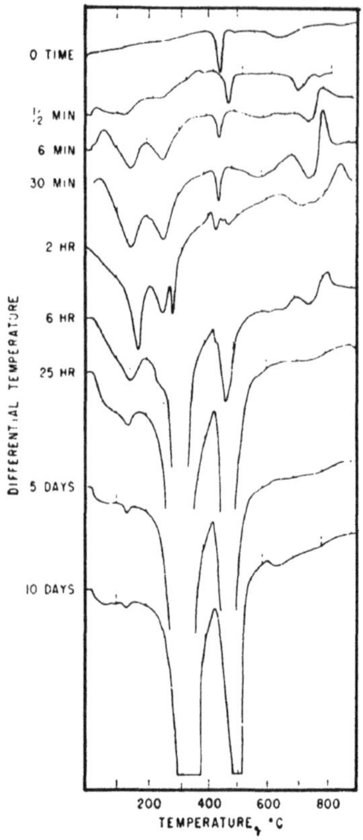

Fig. V.1.4: Hydration of C_3A at 23°C (*28*).

areas at 170, 240, and 810°C; the peak at 245°C is concealed by the peak at 310°C. This clearly shows the conversion of the hexagonal hydroaluminates to the cubic C_3AH_6.

After 24 hr of hydration the C_3AH_6 peaks at 320 and 495°C are much larger than those at 6 hr and the endothermic effect at 140°C is the only evidence that hexagonal hydroaluminates may still exist. At 10 days the peak at 140°C is still distinct although it is extremely small. Apparently the DTA characteristics of C_3AH_6 do not include an endothermic effect at 140°C.

It is also observed that at 23°C the thermograms do not exhibit a peak at 510°C even at early times when the hexagonal aluminate hydrates are present.

DTA thermograms obtained at 52° and at times 0.5, 6 and 30 min, 2 and 6

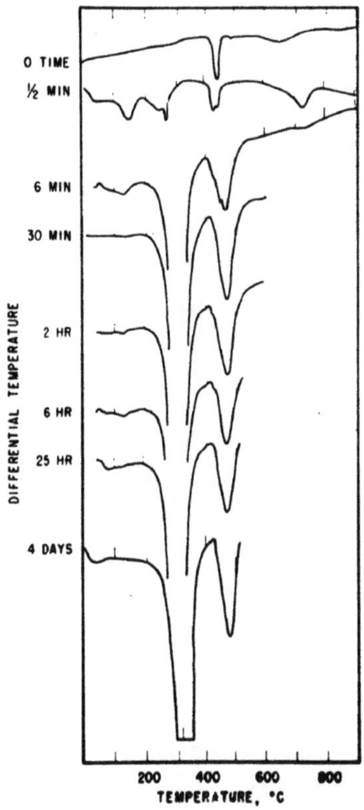

Fig. V.1.5: Hydration of C_3A at 52°C (*28*).

hr, 1 and 4 days are shown in Fig. V.1.5. Even at 0.5 min the endotherms at 140 and 240°C are quite prominent, the endothermic effect at 140°C being more intense than that at 240°C. At 270°C there is a large endotherm equal or greater in intensity than the peaks at 140° and 240°C. A doublet appears at 450-460°C. This should mean that a considerable amount of C_3AH_6 has already formed, the 270 and 460°C endotherms being indicative of this compound, and the other peak of the doublet indicating $Ca(OH)_2$.

At 6 min the hexagonal hydroaluminates decrease considerably, while the C_3AH_6 has formed in a large quantity. At 30 min the hexagonal hydroaluminates have virtually disappeared, while the C_3AH_6 peaks have shown little increase after 6 min. The thermograms at 2 and 6 hr and 1 and 14 days

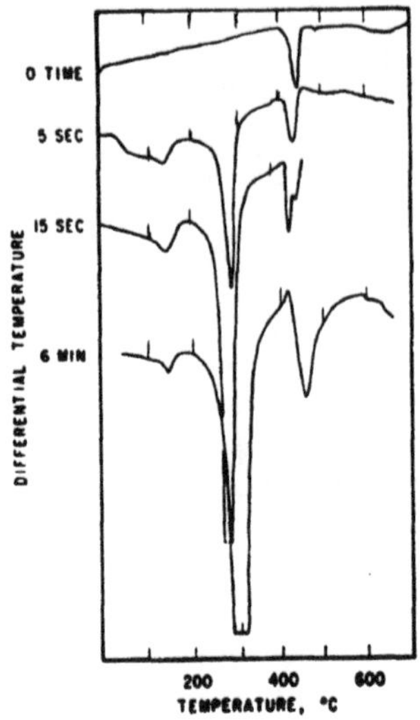

Fig. V.1.6: Hydration of C_3A at 80°C (*28*).

show almost no further increase in the intensity of the peaks and the hexagonal hydroaluminates are absent.

At 4 days the endothermic effect at 300 and 490°C is clearly demonstrative of the presence of the cubic C_3AH_6.

DTA thermograms were obtained at 80°C only at 5, 15, and 360 sec (Fig. V.1.6). At 5 sec a small endotherm at 130°C and a large effect at 270°C are present, apparently obscuring any effect at 240°C. The endotherm for $Ca(OH)_2$ at 440°C may be reinforced by the second endotherm for C_3AH_6 at 490°C. At 15 sec the 130°C peak has increased slightly, but the C_3AH_6 peak at 290°C has grown considerably. At 6 min the 130°C peak has decreased slightly, and the 300 and 470°C peaks for C_3AH_6 have increased further.

From a study of the thermograms, it is clear that the reactions involved are very temperature dependent; the rate of hexagonal hydrate formation increases as the temperature increases. At 6 min for 52 and 80°C, only a small amount of hexagonal hydroaluminate is present, C_3AH_6 being present at the earliest times examined (30 sec at 52° and 5 sec at 80°C).

There is evidence of a small amount of C_3AH_6 after 10 days at 2°C, after 6 hr at 12°C, and after 30 min at 23°C. At 12 and 23°C, at certain times of hydration, both the hexagonal hydrates and C_3AH_6 were increasing in quantity, but in general the disappearance of hexagonal hydrates indicated that they were converting to C_3AH_6, and at a faster rate as the temperature increased.

The probability of C_3AH_6 forming directly, especially at the higher temperatures, is very high. There is no evidence that the C_2AH_8 converts to C_4AH_{13} as an intermediate phase in the conversion to C_3AH_6 (*35*).

The DTA shows that at 52°C little hydration occurs after 30 min, whereas at 80°C, considering the rapid rate of hydration in the first 15 sec, the rate has been greatly reduced at 6 min; for 23°C hydration the product forms steadily up to 25 hr and slowly beyond that. The degree of hydration determined by TGA after 25 hr at 23°C is 52.5%, whereas after 4 days at 52°C it is 17.5%.

The endothermic peak at 500-530°C that developed with the other peaks at about 150 and 240°C when hydration occurred at 2 and 12°C was absent at the higher temperatures and thus does not seem to be characteristic of the hexagonal or cubic hydrates. This endotherm may be due to a small amount of $Ca(OH)_2$ formed during the hydration of C_3A at low temperatures even though the temperature of the endotherm is somewhat higher than the $Ca(OH)_2$ (*cf.* endotherm at 450°C in the original C_3A preparation). The latter endotherm was probably formed, however, by vapor phase hydration of CaO at more accessible regions, whereas the former endotherm may be the result of $Ca(OH)_2$ embedded in the hydroaluminate.

Alternatively, the hydration of CaO in the C_3A can explain the endotherm but it should produce a thermogram for the specimen hydrated at 23°C similar to that at 2 and 12°C.

The complete disappearance of the peak at 140°C for hydration at 52 and 80°C and almost complete at 23°C indicates that the thermogram for C_3AH_6 has only two endotherms—at about 320 and 500°C. It is evident that for systems that may produce several compounds, and where transformations of various types may occur, DTA thermograms must be interpreted from a series obtained at different stages as the reaction progresses. Also, in these reactions, where products may form on the surface of an unhydrated grain and progress inward and where the crystal size or habit may change, the peak temperatures will shift with an increase in degree of hydration.

Specimens used for X-ray diffraction patterns were those in which larger amounts of product were formed. In general the lines for C_4AH_{13} and C_2AH_8 were diffuse and not clear, in contrast to those for C_3A and C_3AH_6. Consequently the results are useful only in a general sense and show the advantage of using DTA for this system.

Length changes, measured at four temperatures, are shown in Fig. V.1.7. These results are reproducible to $\pm5\%$ except at 23°C, where the rate and

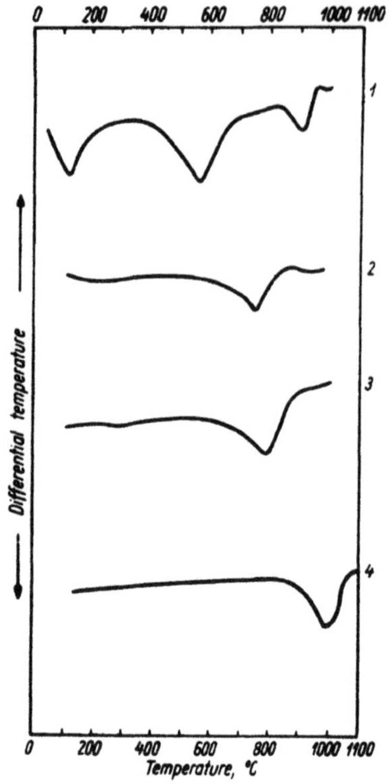

Fig. V.1.7: Length changes of C_3A compact at four temperatures (*28*).

degree of hydration and expansion were very great, with the result that in some cases the specimens broke. These results are used for qualitative and semi-quantitative analysis of the system. At each temperature, expansion began as soon as water contacted the specimen. The results corresponded to those obtained from the DTA.

The expansion rate at 52°C at short times is very high relative to the other temperatures, but after 5 min it decreases rapidly. After 1 day the expansion ceases completely after attaining a value of 0.66%, even though the specimen was kept for 3 days at 52°C and for 3 more days at 23°C. This specimen, autoclaved at 214°C and 300 lb/in.^{2} expanded by 21% and was completely hydrated. In results not shown here, C_3A after hydrating for 29 days at 12°C had expanded 14.4% and the rate of expansion was slow.

The pore size distribution and total pore volume results are shown in Fig. V.1.1. Specimen A is the unhydrated compact of C_3A. Specimen B is a C_3A compact hydrated at 52°C for 4 days (17.5% hydrated); the volume of pores above 0.5μ is still about half the original.

Thus, hydration of C_3A at 52 and 80°C involves formation of C_3AH_6, which acts as a very impermeable barrier localized around the C_3A grain, despite the fact that adequate space is available and the reaction has a high potential for expansion. The C_3AH_6 formed at a different region than that from the reaction at 23°C, the mechanisms of the respective reactions probably differing.

Specimen C in Fig. V.1.1, a compact of C_3A hydrated at 23°C for 25 hr (52.5% hydrated), shows a large reduction in volume of large pores and an increase in volume of smaller pores relative to specimens A and B. Specimen D hydrated at 2°C for one week (26.6% hydrated by estimate) gave a total pore volume of 10.9%.

By calculation, assuming the average composition of the hexagonal hydroaluminates as $C_3AH_{10.5}$ with a density of 2.04 (*41*), the pore volume is 9.13%. Errors in the degree of hydration determination and use of $C_3AH_{10.5}$ as the formula for the mixed product may account for this difference. These results show that the increase in solid volume of 1 g of the solid to be accommodated in the body is 0.1190 cm³ for specimen C and 0.1345 cm³ for specimen D.

The approximate expansions would be 6-8% and 4% respectively, suggesting that under the conditions of the reactions the hexagonal hydroaluminates can be accommodated in the voids with the development of less stress than the C_3AH_6. The rate of the respective reactions may also have an influence.

It is significant, however, that although the formation of the hexagonal hydroaluminates is the more accommodative, considerable hydration pressures are exerted as shown by the 4% expansion. Mikhailov (*42*) stated that formation of plate-structured compounds like the hexagonal hydroaluminates does not lead to expansion, but this may not be the case. It is indicated that the stresses created by the rapid formation of C_3AH_6 at 23°C and subsequent rupture of bonds weaken the structure under these conditions. This is in contrast to the views of Segalova *et al* (*13*) who consider that the conversion to the cubic form occurs by a solution mechanism resulting in dissolution of the bonds. It is not likely that all the bonds would break simultaneously; during the conversion more bonds can form because of the crystallization of C_3AH_6 and further hydration of the C_3A.

The exact mechanism of the C_3A and water reaction is difficult to prove conclusively, but the following explains most of the experimental evidence based on DTA and length change measurements. Water at 2, 12, and 23°C reacts immediately, forming poorly crystalline hexagonal hydroaluminates by direct reaction on the surface of the C_3A grain. These soluble products are a source of ions in the solution and also contribute to retardation of the reaction by limiting the supply of water to the C_3A. This stage of the reaction is consistent with the immediate expansion observed on first wetting.

As the hydration of C_3A continues, large crystals of the hexagonal aluminates may form. At a certain stage in the reaction the conversion to C_3AH_6 occurs, probably by a "through-solution mechanism." This conversion removes the hexagonal hydroaluminates from the surface of the grain and the reaction accelerates. It is significant that for a time at these three temperatures the quantity of both C_3AH_6 and hexagonal hydroaluminates increases. The conversion begins earlier at the higher temperature because the energy conditions at this temperature permit earlier nucleation. The rate of conversion is governed also by the rate of dissolution.

Energy conditions at 52 and 80°C permit C_3AH_6 to form not only by direct reaction of C_3A with water, but without going through the intermediate stage of the hexagonal hydroaluminate. This may account for the complete elimination of reaction between C_3A and H_2O; the C_3AH_6 grows epitaxially

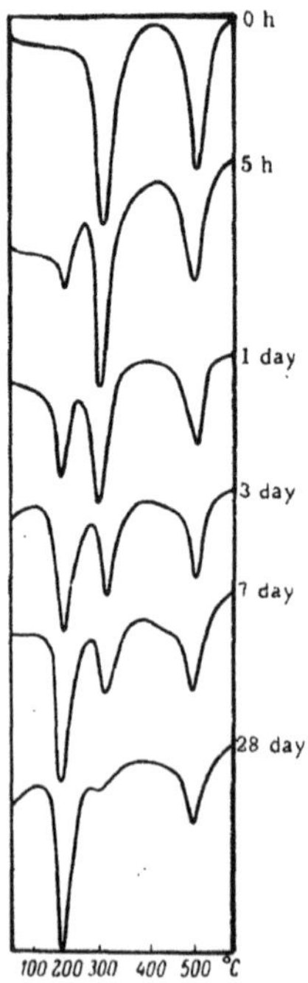

Fig. V.1.8: DTA of the system $Ca(OH)_2$:C_3AH_6 (*24*).

on the C_3A grain, completely sealing it off. A small amount of hexagonal hydroaluminate may be evidence, however, for a very rapid conversion of most of it to the C_3AH_6, but the possibility exists of C_3AH_6 forming by both mechanisms.

Gorshkov and Khmelevskaya (*31*) carried out DTA and dimensional studies at 22-25°C on the C_3A pastes pressed to 20.25 kg/cm^2. The sample exhibited an expansion of 0.38% at 210-230°C and a shrinkage of 0.68% at about 440°C, which was attributed to the dehydration of calcium aluminate hydrate. An expansion of 1.06% at 555°C corresponded to the decomposition of the hydrate to $C_{12}A_7$ and CaO.

Segelova, Solov'eva, and Rebinder (*13*) examined the strengths of a mix containing 20% C_3A and 80% sand cured at 20 or 65°C for different periods. The DTA showed that the strengths depended on the type of hydroaluminate present at a particular period. The specimens cured at earlier periods invariably showed the presence of hexagonal hydroaluminates. The decreased strengths at higher temperatures could be assigned to the presence of the cubic C_3AH_6 formed by the conversion of hexagonal hydroaluminates.

Segalova, Tulovskaya, and Brutskus (*24*) believe that in concentrated suspensions and in absence of $Ca(OH)_2$, C_3A gives the hexagonal $C_3A \cdot aq$, which ultimately converts to C_3AH_6. There is much evidence that $C_3A \cdot aq$ does not exist. They found that $C_4A \cdot aq$ is stable in $Ca(OH)_2$ solutions in concentrations higher than 1.08g CaO/liter.

A DTA study of the system $Ca(OH)_2$: C_3AH_6=5:1 molar ratio clearly demonstrates the stability of C_4AH_{13} up to 28 days by a peak at 180-200°C (Fig. V.1.8).

REFERENCES

1. Bogue, R.H., *The Chemistry of Portland Cement*, Reinhold Publishing Corporation, New York, p. 32, 1955.
2. Brandenberger, E., *Schweiz. Arch. angew. Wiss. u. Tech.*, *2*, 45, 1936.
3. Roberts, M.H., *J. Appl. Chem.* (*London*), *7*, 543, 1957.
4. Seligman, P., and Greening, N.R., *J. Portland Cem. Assoc. Res. Div. Labs.*, *4*, 2, 1962.
5. Johnson, H., and Thorvaldson, T., *Can. J. Res.*, *21B*, 236, 1943.
6. Schneider, W.G., and Thorvaldson, T., *Can. J. Res.*, *19B*, 123, 1941.
7. Buttler, F.G., Glasser, L.S., and Taylor, H.F.W., *J. Am. Ceram. Soc.*, *42*, 121, 1959.
8. Buttler, F.G., *Ph.D. Thesis*, Aberdeen, 1958.
9. Aruja, E., *Acta. Cryst.*, *13*, 1018, 1960.
10. Aruja, E., *Acta. Cryst.*, *14*, 1213, 1961.
11. Jones, F.E., and Roberts, M.H., *Res. Note, Building Res. Stn.*, London, 1959.
12. Kalousek, G.L., Davis, C.W., and Schmertz, W.E., *Proc. Am. Concrete Inst.*, *45*, 693, 1949.

13. Segalova, E.E., Solov'eva, E.S., and Rebinder, P.A., *Kollodn. Zh.*, *23*, 194, 1961.
14. Lefol, J., *Thesis*, *Paris*, 1937.
15. Young, J.F., *Mag. Concrete Res.*, *14*, 137, 1962.
16. Majumdar, A.J., and Roy, R., *J. Am. Ceram. Soc.*, *39*, 434, 1956.
17. Jones, F.E., *Proc. 4th Intern. Symp. Chem. Cements*, Washington, p. 204, 1960.
18. Feldman, R.F., and Ramachandran, V.S., *Mag. Concrete Res.*, *18*, 185, 1966.
19. Budnikov, P.P., Kolbassov, W.M., and Pantelejev, A.S., *Silikat. Tech.*, *11*, 271, 1960.
20. Sauman, Z., *Silikaty*, *3*, 46, 1959.
21. Govorov, A.A., *Akad. Fur Bau Architektur Ukr.*, *USSR*, *Kijev*, 1958.
22. Brutskus, T.K., and Segalova, E.E., *Kollodn. Zh.*, *26*, 11, 1964.
23. Tulovskaya, Z.D., and Segalova, E.E., *Zh. Prikl. Khim.*, *37*, 2, 1964.
24. Segalova, E.E., Tulovskaya, Z.D., Brutskus, T.K., and Rebinder, P.A., *Zh. Prikl. Khim.*, *37*, 1227, 1964.
25. Okorokov, S.D., and Golynkovolfson, S.L., *Trudy Leningradskogo tekhnologicheskogo Instituta imeni Lensoveta*, *Vypusk No. 56*, Leningrad, pp. 20-25, 1960.
26. Petzold, A. Von., and Göhlert, I., *Tonind.-Z. Keram. Rundschau*, *86*, 228, 1962.
27. Rey, M., *Silicates Ind.*, *22*, 533, 1957.
28. Feldman, R.F., and Ramachandran, V.S., *J. Am. Ceram. Soc.*, *49*, 268, 1966.
29. Kind, V.A., Okorokov, S.D., and Khodikel, E.P., *Tsement*, *6*, 32, 1939.
30. Govorov, A.A., *Trudy Khar'kov Inst. Inzhenerov. Zhelez nodorozh. Transporta im S.M. Kirova*, No. 39, 11, 1960.
31. Gorshkov, V.S., and Khmelevskaya, T.A., *Sb. Tr.*, *Vses. Nauchn-Issled. Inst. Novykh Stroit. Materialov*, No. 4, 77, 1961.
32. Turriziani, R., *Ric. Sci.*, *24*, 1709, 1954.
33. Turriziani, R., and Schippa, G., *Ric. Sci.*, *24*, 2645, 1954.
34. Midgley, H.G., and Rosaman, D., *Proc. 4th Intern. Symp. Chem. Cements*, Washington, p. 259, 1960.
35. Steinour, H.H., *Portland Cem. Assoc. Bull.* No. 34, 100 pp, 1951.
36. Assarsson, G., *Zement*, *26*, 293; 311; 327; 1937.
37. Bessey, G.E., *Proc. 2nd Intern. Symp. Chem. Cements*, Stockholm, pp. 178-215, 1938.
38. Feldman, R.F., Sereda, P.J., and Ramachandran, V.S., *Highway Res. Board, Highway Res. Rec.* No. 62, 106, 1964.
39. Feldman, R.F., Ramachandran, V.S., and Sereda, P.J., *J. Am. Ceram. Soc.*, *48*, 25, 1965.
40. Ramachandran, V.S., Feldman, R.F., and Sereda, P.J., *Mater. Res. Std.*, *5*, 510, 1965.
41. Lea, F.M., and Desch, C.H., *Chemistry of Cement and Concrete*, 2d ed., Edward Arnold, London, p. 227, 1956.
42. Mikhailov, V.V., *Proc. 4th Intern. Symp. Chem. Cements*, Washington, pp. 927-955, 1960.

V.2. HYDRATION OF C_3S AND β-C_2S

The two phases, tricalcium silicate (C_3S) and β-dicalcium silicate (β-C_2S), together forming 70-85% of the cement clinker, play a dominant role in the hardening of Portland cement. At room temperature the hydration products of the C_3S and β-C_2S phases are expected to be similar to the calcium silicate hydrates constituting the hydrated Portland cement. The hydration of C_3S and β-C_2S is one of the widely studied reactions in cement chemistry.

The unit cell of C_3S is triclinic and that of β-C_2S is monoclinic. The β-C_2S occurs in nature as larnite, while C_3S does not occur naturally. In the clinker, C_3S exists as alite with some Al and Mg in solid solution. The unit cell of alite is monoclinic and has the composition $C_{54}S_{16}AM$. The β-C_2S is usually associated with some K_2O in the clinker. Tricalcium silicate reacts faster than β-C_2S with water and this is attributed to the distorted calcium co-ordination and holes in the structure (*1*).

The hydration of C_3S and β-C_2S has been studied by three methods, *viz.* paste hydration, ball mill hydration, or bottle hydration. The paste hydration at a water/solid ratio of about 0.5-0.7 simulates the conditions of hydration of Portland cement in practice. A small steel ball mill can be used to hydrate the C_3S or C_2S at a water/solid ratio of 9.0 (*2*). In bottle hydration also a high water/solid ratio is maintained, but the reaction is carried out in a polyethylene bottle rotated on a wheel (*3*).

If the hydration of C_3S is taken to completion by any of the three methods, the reaction can be represented as follows:

$$2Ca_3SiO_5+6H_2O=Ca_3Si_2O_7\cdot 3H_2O+3Ca(OH)_2$$

The paste as well as the bottle-hydrated product is a calcium silicate hydrate very nearly related to the tobermorite mineral. Both C_3S and β-C_2S form the same type of calcium silicate hydrate but the yields are different.

Brunauer and Greenberg (*4*) showed that in paste hydration at room temperature C_3S gives a poorly crystallized calcium silicate hydrate of CaO/SiO_2 ratio 1.5, while β-C_2S gives a product of CaO/SiO_2 ratio 0.65. These hydrates were designated as tobermorite (G). Opinions differ on the exact CaO/SiO_2 ratio in the hydrated product of C_3S and C_2S. The tobermorite (G) should be designated as CSH(I) according to Taylor's nomenclature.

Taylor (*5*) and van Bemst (*6*) found that the hydrated C_3S had a CaO/SiO_2 ratio of nearly 2 and this corresponded to CSH(II). Graham, Spinks, and Thorvaldson (*7*) observed the products of hydration of both C_3S and β-C_2S to contain a CaO/SiO_2 ratio of 1.5, in agreement with the results of Brunauer, Kantro, and Copeland (*3*). The compound with a CaO/SiO_2 ratio of 1.5 is considered as a lime-poor CSH(II) by Kalousek and Prebus (*8*).

The ball mill hydration of C_3S yields afwillite ($C_3S_2H_3$), while the same treatment to C_2S results in the formation of a poorly crystallized mineral. Electron microscope studies demonstrate that the paste hydrated products are fibers made of rolled sheets and the bottle-hydrated products are cigar shaped.

A significant difference is obvious in the rates of hydration of C_3S and β-C_2S. Ershov (*9*) has discussed the rates of hydration of C_3S and β-C_2S in pastes. The C_3S hydrates to an extent of 15-20%, 35-50%, 45-70%, 68-75%, and 85-94% in 1 day, 3 days, 7 days, 1 month, and 6 months respectively.

Bottle hydration is more effective than paste hydration (*3*). For example, it takes only 48 days for the bottle-hydrated C_3S to get hydrated to an extent of 97%, whereas the paste form may take more than 180 days to get hydrated to the same extent. Similarly, the bottle-hydrated C_2S takes only 162 days to get hydrated to the same extent as the C_2S paste in 17 months.

An increase in the rate of hydration with increase in water/silicate ratio has been reported by Work and Lasseter (*10*). They showed that the rate of hydration at 3 months was 18.2, 19.0, and 20.4% at a water/silicate ratio of 0.45, 0.60, and 1.00 respectively. The rate of hydration is enhanced by increasing the temperature and is also reflected in the strengths of C_3S and C_2S pastes.

Bogue and Lerch (*11*) determined the compressive strengths of hydrated C_3S pastes, and taking the strength at 2 years as 100% found the strengths for C_3S pastes at 1 day, 3 days, 7 days, 28 days, 3 months, 6 months, and 1 year to be 12.8, 24.8, 52.7, 62.8, 62.8, 85.8 and 91.2% respectively. The corresponding strengths for the C_2S pastes were 0.0, 0.4, 1.0, 6.3, 36.0, 52.7 and 71.4%.

Differential Thermal Analysis

In 1939, the Russian workers Kind, Okorokov, and Khodikel (*30*) investigated by thermal analysis the dehydration behaviour of hydrated C_3S, C_3A, and C_2F. The hydration reactions of C_3S and β-C_2S have been followed using DTA by Petzold and Göhlert (*12*), Weigmann and Horte (*13*), Barta and Satava (*14*), Sauman (*15*), van Bemst (*6*), Berkovich *et al* (*16*), Mtschedlov-Petrosian *et al* (*17*), Gorhkov and Khmelevskaya (*18*), Tamas (*19*), Midgley (*20*), and Szekely and Vajda (*21*). There is some variation in the nature of the thermograms reported by different workers.

The thermograms of hydrated C_3S and C_2S show similarity. Tamas (*19*) compared the DTA results of hydrated C_3S and β-C_2S compounds with those of TGA and DTG (Figs. V.2.1 and V.2.2.). The weight loss at 170-210°C in DTG corresponds to an endothermic effect in DTA, attributed to

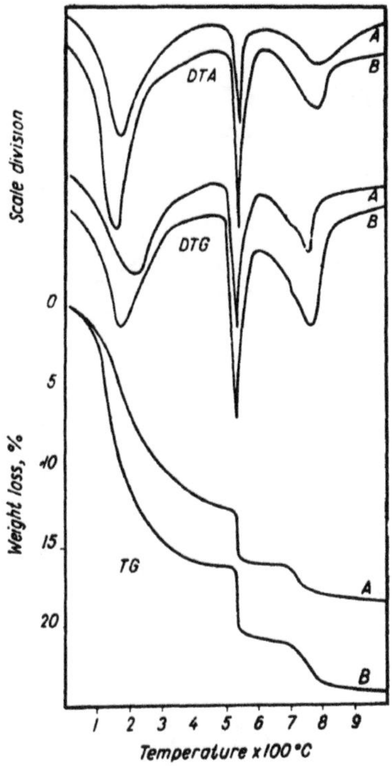

Fig. V.2.1: Derivatograms of C_3S (*19*).

the loss of interlayer water from the tobermorite-like phase. It is known that in montmorillonitic clay mineral also the interlayer water is lost at about 200°C. An endothermic effect at about 535°C involves a loss in weight in TGA, caused by the decomposition of $Ca(OH)_2$ formed in the hydration.

An endothermal doublet has also been reported in C_3S pastes indicating a stepwise dehydration of $Ca(OH)_2$. This may be due to the different modes of fixation of $Ca(OH)_2$ with the calcium silicate hydrate product. The higher temperature endothermic effect at 770°C is ascribed to the decomposition of calcium carbonate. However, the peak at about 800°C is attributed by Barta and Satava (*14*), Van Bemst (*22*), Bozhenov and Suvorova (*23*), and Petzold and Göhlert (*12*) to the dehydration of the calcium silicate hydrate.

In Fig. V.2.2 the DTA behaviour of hydrated β-C_2S is presented. The sensitivity of the DTA equipment had to be increased several times to realise discernible peaks in the samples hydrated for 21 or 200 days. The endothermic effect corresponding to the dehydration of $Ca(OH)_2$ has not been ob-

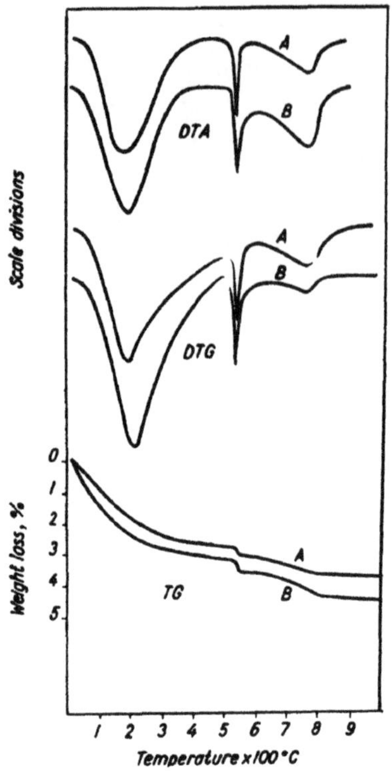

Fig. V.2.2: Derivatograms of β-C_2S (*19*).

tained in C_2S pastes (*12*), probably due to the low sensitivity of DTA adopted for the purpose. The rate of hydration of β-C_2S is much slower than that of C_3S. This can be seen from the TGA results.

At 200 days β-C_2S paste shows a weight loss less than 4%, whereas the C_3S hydrated at 21 days indicates a loss of about 13%. The difference in the rates of hydration is dictated by the extent to which C_3S and β-C_2S particles are enveloped by layers of the tobermorite gel and $Ca(OH)_2$ (*19*). By a hydration of 1 cc of C_3S, 1.3 cc of tobermoritic gel and 0.7 cc of $Ca(OH)_2$ are formed, whereas 1 cc of β-C_2S yields 1.86 cc of tobermoritic gel and 0.3 cc of $Ca(OH)_2$, the amount of the gel enveloping β-C_2S, being larger, impedes the diffusion of water into the mass.

The bottle-hydrated C_3S differs from the paste-hydrated C_3S in yielding an additional endothermic peak at about 335°C (*4, 6*). This peak probably represents the formation of a hydrate identical with Taylor's $C_2SH(II)$.

There is evidence of an exothermic peak at 800-900°C in pastes or suspensions (*12*) caused by the formation of wollastonite from CSH(I). Tamas (*19*) failed to get an exothermic peak at high temperatures.

The effect of mechanical vibrations on the rate of hydration of C_3S has been studied by Mchedlov-Potrosian *et al* (*17*) using DTA. This treatment did not affect the rate of hydration of C_3S but had a considerable influence on the reaction of C_3A with gypsum.

Gorshkov and Khmelevskaya (*18*) tried to correlate the dimensional changes of the hydrated C_3S or C_2S at 3 days with the thermal effects obtained at different temperatures. In C_3S an endothermic effect corresponding to a loss of 17.3% showed a shrinkage of 0.976%. The dehydration of $Ca(OH)_2$ involved a shrinkage of 0.674%. However, hydrated C_2S lost the zeolitic water with a slight expansion.

The CaO/SiO_2 ratio of the product of hydration of C_3S or β-C_2S is determined by estimating the uncombined lime by X-ray quantitative analysis (*24*), Franke's chemical method or its modifications (*25*, *26*), or the calorimetric method of Bessey (*27*). The chemical extraction methods are known to attack the calcium silicate hydrates, hence Kurczky and Schwiete (*28*) and Schwiete and Tan Tik-Ien (*29*) suggested the application of Dynamic Differential Calorimetry (DDC) as an alternative method for estimating free lime.

The DDC technique is similar to the DTA except that the temperature measurements are taken on platinum cups (Fig. V.2.3a) and not in the specimen (Fig. V.2.3b). The method also enables weighing of samples under dynamic conditions of heating.

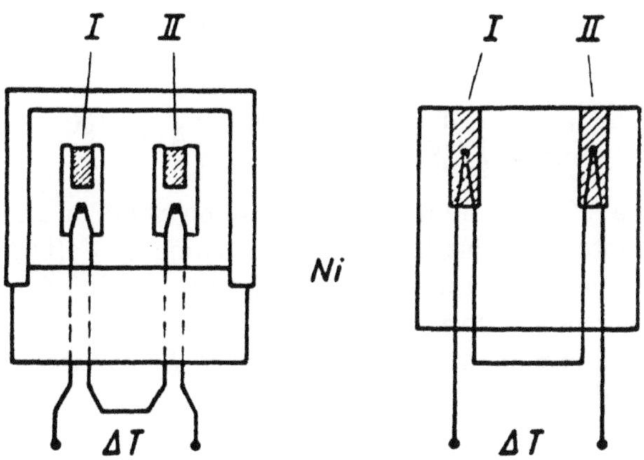

Fig. V.2.3: Arrangement of DDC and DTA (*28*).

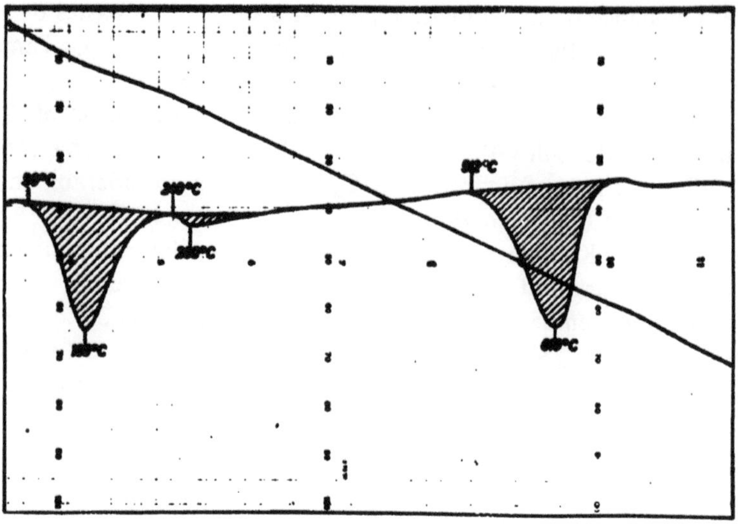

Fig. V.2.4: DDC plot of hydrated C_3S (hydration period, 1 month) (*28*).

The DDC plot of the C_3S paste hydrated for one month is shown in Fig. V.2.4. The area of the shaded portion of the endothermal peak with a characteristic temperature at 512°C indicates the amount of free $Ca(OH)_2$ present in the paste. The quantity of CaO liberated in the hydration of C_3S or β-C_2S can be calculated from the equations:

$$\frac{\text{moles CaO}}{\text{moles } C_3S} = \frac{0.757(F/K)}{E_2}\, 4.04\, \frac{100}{H}$$

and

$$\frac{\text{moles CaO}}{\text{moles } \beta C_2S} = \frac{0.757(F/K)}{E_2}\, 3.08\, \frac{100}{H}$$

where F=area of the $Ca(OH)_2$ decomposition peak in cm^2
K =calibration value for the determination of $Ca(OH)_2$ in cm^2/g
E_2=final weight of the sample after heating to 1000°C in the dynamic differential calorimeter
H=percentage of hydration of C_3S or β-C_2S obtained by X-ray diffraction

0.757, 3.08, and 4.04 are conversion factors.

In the paste hydration of C_3S the DDC method gave a CaO/SiO_2 value of 1.88, 1.87, and 1.92 after 1, 2, and 3 months respectively. The corresponding CaO SiO_2 values for β-C_2S were 1.86, 1.80, and 1.85.

It is an interesting fact that the C_3S or C_2S pastes with a high CaO/SiO_2

ratio of 1.8-1.9 are morphologically similar to the mineral tobermorite, which has a CaO/SiO_2 ratio of only 0.83. It is possible that the excess CaO is accommodated in the interlayer space of the tobermorite-like phase. Kurczyk and Schwiete have denoted the structural formula for this phase by

$$Ca^x\ [Ca_4Si_6O_{16}(OH)_2]\ [Ca^x(OH)_2]n$$

where Ca^x = atoms with unknown lattice constants
n = 5-7

REFERENCES

1. Bernal, J.D., Jeffery, J.W., and Taylor, H.F.W., *Mag. Concrete Res.*, *4*, 49, 1952.
2. Brunauer, S., Copeland, L.E., and Bragg, R.H., *J. Phys. Chem.*, *60*, 112, 1956.
3. Brunauer, S., Kantro, D.L., and Copeland, L.E., *J. Am. Chem. Soc.*, *80*, 761, 1958.
4. Brunauer, S., and Greenberg, S.A., *Proc. 4th Intern. Symp. Chem. Cements*, Washington, pp. 135-165,, 1960.
5. Taylor, H.F.W., *J. Chem. Soc.*, 3682, 1950.
6. Van Bemst, A., *Bull. Soc. Chim. Belges*, *64*, 333, 1955.
7. Graham, W.A.G., Spinks, J.W.T., and Thorvaldson, T., *Can. J. Chem.*, *32*, 129, 1954.
8. Kalousek, G.L., and Prebus, A.F., *J. Am. Ceram. Soc.*, *41*, 124, 1958.
9. Ershov, L.D., *Reports of Symp. on the Chem. Cem.* State Publication of the Literature on Structural Materials, Moscow, pp. 264-278, 1956.
10. Work, L.T., and Lasseter, F.P., *Chemical Reaction in the Setting of Cement, Concrete (Mill Section)*, *38*, 81-86; 89-92; 79-84, 1931.
11. Bogue, R.H., and Lerch, W., *Ind. & Eng. Chem.*, *26*, 837, 1934.
12. Petzold, A. and Göhlert, I., *Tonind.-Z. Keram. Rundschau*, *86*, 228, 1962.
13. Weigmann, J., and Horte, C.H., *Z. Anorg. Allgem. Chem.*, *286*, 268, 1956.
14. Barta, R., and Satava, V., *Stavivo*, *31*, 15, 1953.
15. Sauman, Z., *Silikaty*, *3*, 46, 1959.
16. Berkovich, T.M., Khekar, D.M., Gracheva, O.I., and Kupreeva, N.I., *Dokl. Akad. Nauk SSSR*, *120*, 372, 1958.
17. Mtschedlov-Petrosian, O.P., Budnikov, A.G., Levchuk, N.A., Latyshev, F.A., and Strelkova, I.S., *Silikaty Sbornik Statei po khim i Teknol. Silikatov, Vsesouz Khim. Obshchestyo im D.I. Mendeleeva*, No. 2, 67, 1959.
18. Gorshkov, V.S., and Khmelevskaya, T.A., *Sb. Tr., Vses Nauchn.-Issled. Inst. Novykh Stroit Materialov*, No. 4, 77, 1961.
19. Tamas, F., *Proc. 6th Conf. Silicate Ind.*, Budapest, pp. 425-436, 1961.
20. Midgley, H.G., *Proc. 4th Intern. Symp. Chem. Cements*, Washington, pp. 479-490, 1960.
21. Szekely, I., and Vajda, K., *Epitoanyag*, *7*, 405, 1955.
22. Van Bemst, A., *Silicate Ind.*, *22*, 213, 1957.
23. Bozhenov, P.I. and Suvorova, G.F., *Zement*, *23*, 8, 1957.

24. Copeland, L.E., and Bragg, R.H., *Anal. Chem.*, *30*, 196, 1958.
25. Franke, B., *Z. Anorg. Allgem. Chem.*, *247*, 180, 1941.
26. Pressler, E.E., Brunauer, S., Kantro, D.L., and Weise, C.H., *Anal. Chem.*, *33*, 877, 1961.
27. Bessey, G.E., *Building Res. Tech. Paper*, Building Research Station (U.K.), No. 9, 1930.
28. Kurczyk, H.G., and Schwiete, H.E., *Proc. 4th Intern. Symp. Chem. Cements*, Washington, pp. 349-358, 1960.
29. Schwiete, H.E., and Tan Tik Ien, A., *Proc. 4th Intern. Symp. Chem. Cements*, Washington, pp. 545-550, 1960.
30. Kind, V.A., Okorokov, S.D., and Khodikel, E.P., *Tsement*, *6*, 32, 1939.

V.3. HYDRATION OF THE FERRITE PHASE

The average ferrite content calculated as C_4AF in the normal Portland cement is about 8%. Actually, the ferrite phase in the Portland cement clinker lies in the solid solution range C_6AF_2–C_6A_2F, and for simplicity C_4AF can be considered as a representative of the ferrite phase.

The hydration of C_4AF proceeds at a slower rate than that of C_3A and is much more complicated. A paste of C_4AF forms hexagonal platy crystals initially, these converting to C_3AH_6 at longer periods of curing. The iron forms an amorphous hydrated ferric oxide and is also present in solid solution with C_3AH_6. In lime water, C_4AF reacts slowly to form a hexagonal solid solution of $C_4F{\cdot}aq$ and $C_4A{\cdot}aq$, which ultimately converts to a solid solution of C_3AH_6 and C_3FH_6. In the presence of gypsum, C_4AF reacts just as C_3A to give an analogous sulfoferrite, sulfoaluminate, and the solid solution of the two.

A systematic investigation of the paste hydration of C_4AF in water—in the presence of $CaSO_4{\cdot}2H_2O$, $Ca(OH)_2$, or $CaSO_4{\cdot}2H_2O + Ca(OH)_2$—has been carried out by Chatterji and Jeffery (*1*). Table V.3.1 summarizes the various compounds identified by X-rays. All references to the aluminate compounds include a possibility of the substitution of Fe^{+3} for Al^{+3}.

Differential Thermal Analysis

Not much work has been done on the hydration of C_4AF employing DTA. Kalousek and Adams (*2*), Watanabe and Iwai (*3*), and Budnikov and Gorshkov (*4*) have followed the hydration of C_4AF or C_2F with or without the presence of gypsum.

Kalousek and Adams prepared a series of hydrogarnets ranging in composition from $3Al_2O_3{:}OFe_2O_3$ to $3Fe_2O_3{:}OAl_2O_3$. The DTA of this series containing the Fe_2O_3-free product exhibited the largest endothermic effect at about 380°C and that with only Fe_2O_3 showed a small endothermic bulge at about 330-340°C. A small endothermic effect wes also observed in all the samples at about 550°C. These results were utilized to interpret the thermograms of C_4AF hydrated with or without gypsum.

Fig. V.3.1 shows the DTA behaviour of C_4AF hydrated for 7, 28, 60, 90, and 180 days. The sample hydrated for 7 days shows the presence of the hydrogarnets and C_4AH_{13} by the endothermic peaks at 360 and 220°C respectively. This was also confirmed by microscopic studies.

The peak area for the hydrogarnets is maximum at 60 days, after which it decreases and completely disappears at 180 days. There is a notable decrease of the endothermic effect occurring at 500° and 600°C. The endother-

TABLE V.3.1
Paste Components Identified by X-Ray Diffraction* [Chatterji and Jeffery (1)]

Compounds identified	Series A (C_4AF+H_2O) 1 day	Series A 14 days	Series A 3 months	Series B ($C_4AF+CaSO_4\cdot 2H_2O+H_2O$) 1 day	Series B 14 days	Series B 3 months	Series C ($C_4AF+Ca(OH)_2+H_2O$) 1 day	Series C 14 days	Series C 3 months	Series D ($C_4AF+Ca(OH)_2+CaSO_4\cdot 2H_2O+H_2O$) 1 day	Series D 14 days	Series D 3 months
C_4AF	6	4	2	7	4	2	6	4	2	7	6	5
$Ca(OH)_2$							3	2		2	2	
$CaSO_4\cdot 2H_2O$				3						2	1	
C_4AH_x (hex)	tr.	1	1		2	3	2	2	3			2
$C_3A\cdot CaCO_3\cdot xH_2O$ (hex)	3	2	2				3	3	2			
C_3AH_6 (cubic) (hydrogarnet)		5	6			2	2	5	6			2
C_2AH_x (hex)	2	1										
$C_3A\cdot 3CaSO_4\cdot 32H_2O$ (ettringite)				4						2	3	
$C_3A\cdot CaSO_4\cdot 13H_2O$ (monosulfate)					5	5						3

*Numerals 1 through 7 represent the estimated relative proportions of different components at various intervals of time. Hex = hexagonal, tr = trace.

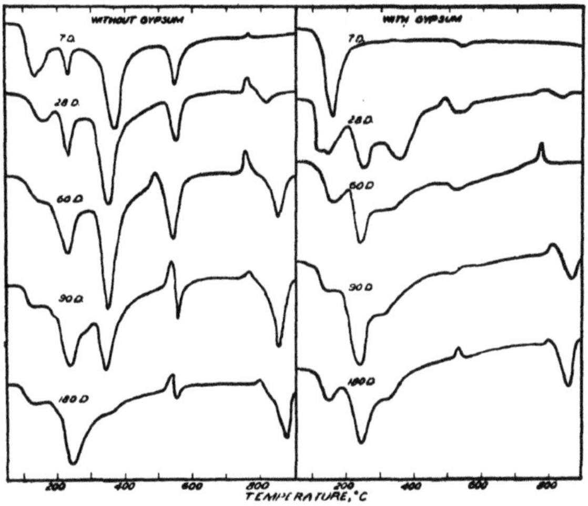

Fig. V.3.1: Thermal analysis curves of $4CaO \cdot Al_2O_3 \cdot Fe_2O_3$ with and without added gypsum ($0.5\ SO_3/R_2O_3$) processed at 25°C (*2*).

mic effect corresponding to C_4AH_{13} increases in amplitude, and at 180 days the sample contains only the hexagonal platy crystals of C_4AH_{13} with some Fe_2O_3.

The samples hydrated in gypsum exhibit thermograms different from those obtained without gypsum. At 7 days there is only one endothermic effect between 100 and 200°C, indicating the presence of high calcium sulfoaluminate and sulfoferrite. At 28 days two more endotherms appear and they are attributed to a solid solution of the Al_2O_3- and Fe_2O_3-bearing compound. The sample at 90 days contained a prominent endothermic effect at about 250°C, which may represent a mixture or a solid solution of $C_3A \cdot CaSO_4 \cdot 12H_2O$–$C_3A \cdot Ca(OH)_2 \cdot 12H_2O$ and $C_3F \cdot CaSO_4 \cdot 12H_2O$–$C_3F \cdot Ca(OH)_2 \cdot 12H_2O$. The 180-day sample represents a member of the solid solution.

No evidence was obtained for the presence of hydrogarnet in the hydrated C_4AF in the presence of gypsum. Kalousek and Adams (*2*) believe that the stable product in the C_4AF–H_2O system is C_4AH_{13} or a related phase containing Fe_2O_3, the hydrogarnets being metastable products.

REFERENCES

1. Chatterji, S., and Jeffery, J.W., *J. Am. Ceram. Soc.*, *45*, 536, 1962.
2. Kalousek, G.L., and Adams, M., *Proc. Am. Concrete Inst.*, *48*, 77, 1952.

3. Watanabe, K., and Iwai, T., *Cemento Gijutsu Nenpo*, *9*, 571, 1955.
4. Budnikov, P.P., and Gorshkov, V.S., *Zh. Prikl. Khim.*, *33*, 1246, 1960.

VI. Hydration of C_3A in Presence of Salts

VI.1. HYDRATION OF C_3A IN PRESENCE OF GYPSUM

Gypsum has long been used as a retarder in Portland cement. The retarding action is mainly based on the reaction of gypsum with the C_3A phase to form calcium sulfoaluminate hydrates. Jones (*1*, *2*) and D'Ans and Eick (*3*), in their studies on the equilibria in the quaternary system $CaO–Al_2O_3–CaSO_4–H_2O$, obtained a stable compound $C_3A \cdot 3CaSO_4 \cdot 31–32H_2O$ and evidence of a metastable compound of composition $C_3A \cdot CaSO_4 \cdot 12H_2O$. No C_3AH_6 was detected in the presence of sulfate ions.

A study of the two forms of calcium aluminate sulfate hydrates is of interest in cement chemistry as these are detected in the hydrated products of Portland cement.

The compound with 1 molecule of $CaSO_4$ is called the monosulfoaluminate or the low sulfate calcium sulfoaluminate, though the preferred name ought to be tetracalcium aluminate monosulfate hydrate. The sulfate-rich compound is related to the mineral ettringite and is referred to as the high sulfate calcium sulfoaluminate or trisulfoaluminate, but it would be more correct to designate it as hexacalcium aluminate trisulfate hydrate.

The monosulfoaluminate may be considered as a compound formed by replacement of (OH) groups by SO_4 groups in the compound $C_3A \cdot Ca(OH)_2 \cdot 12H_2O$. There is evidence of formation of the solid solution series, C_4AH_{13}-$C_3A \cdot CaSO_4 \cdot 12H_2O$. The low sulfate is a reaction product of calcium aluminate hydrate, gypsum and calcium hydroxide. The gypsum and Al_2O_3 in the reaction mixture should be in equimolar proportions. Heating to 110°C expels 5 molecules of water from the low sulfate form, and in DTA it is identified by an endothermic effect at about 200°C.

The high sulfate form is prepared by reacting together a gypsum-rich mixture of lime, gypsum, and calcium aluminate solution. Berman and Newman (*4*) carried out the thermal dehydration studies on the trisulfate form and registered losses corresponding to about 11-12, 23-24, 25-26, and

32 molecules of water at 60, 110, 145, and 200°C respectively. No solid solution exists between the compounds $C_3A \cdot 3CaSO_4 \cdot 31\text{–}32H_2O$ and $C_3A \cdot CaSO_4 \cdot xH_2O$ (*1*, *3*, *5*).

Midgley and Rosaman (*6*) discussed the possibility of formation of a solid solution, at least part of the way, between $C_3A \cdot 3CaSO_4 \cdot aq$ and $C_3A \cdot 3Ca(OH)_2 \cdot aq$. The compound $C_3A \cdot 3Ca(OH)_2 \cdot aq$ requires a very high concentration of lime for its formation and hence is not expected in cement–water mixes.

The formation of calcium sulfoaluminate hydrates by the reaction of C_3A with gypsum and $Ca(OH)_2$ has been a subject of study for many years. Recently, Stein (*7*) investigated these reactions by means of an isothermal calorimeter and Lawrence (*8*) followed the concentration of sulfate ions in a suspension of C_3A, while Schwiete *et al* (*9*) followed the reaction by chemical, X-ray, and electron optical methods.

Mechanism of Retardation. Many theories have been proposed to explain the retarding effect of gypsum in Portland cement, and these have been discussed in considerable detail by Steinour (*10*) and Bogue (*11*).

There are four main theories explaining the action of gypsum. The theory of reduced solubility of C_3A was put forward by Lerch (*12*) and Klein and Phillips (*13*), who considered that the sulfate solution reduces the solubility of C_3A and thus the reaction rate. Lerch maintained that the initial rapid reaction indicated in his calorimetric studies was due to the formation of a calcium aluminate hydrate. This reaction was retarded in sulfate solutions and the formation of sulfoaluminate was incidental.

Roller (*14*) proposed that the retardation of C_4AH_{13} is due to the action of gypsum with free alkalies in the clinker. This means that hexagonal hydroaluminate C_4AH_{13} is responsible for retarding the hydration of C_3A.

An impermeable layer of the low-form of sulfoaluminate, responsible for the retarding action, has been proposed by some workers (*10*, *15*). This has been criticized on the ground that it does not explain the rapid reaction at later periods. It has been suggested that the low form might revert to a more permeable high-sulfoaluminate form. But there is evidence that the high sulfoaluminate is the first product to be formed.

The formation of high sulfoaluminate on the surface of anhydrous C_3A crystals, diffusion of sulfate ions through this layer, formation of more sulfoaluminate, and breaking up of the layer are the processes that have received more support in recent years (*8*, *9*).

Differential Thermal Analysis

In the hydration of C_3A in the presence of gypsum, metastable compounds may form and exist for long periods. At any moment of hydration the nature and the amount of these compounds depend on the concentration of gypsum in the C_3A, the temperature, and the method of hydration. The technique of DTA is well suited to follow the formation of various products of hydration at different periods. A DTA study of the products of hydration in the system $C_3A–CaSO_4{\cdot}2H_2O–H_2O$ has been reported by Feldman and Ramachandran (*16*), Brutskus and Segalova (*17*, *18*), Murakami, Tanaka and Suzuki (*19*), Segalova and Brutskus (*20*), and Govorov (*21*).

The compounds that might form in the system $C_3A–CaSO_4{\cdot}2H_2O–H_2O$ are C_2AH_8, C_4AH_{13}, C_3AH_6, the low and high sulfoaluminate hydrates, and their solid solutions. The peak temperatures reported for these compounds vary, due to many factors. Hence it is always pertinent to determine the peak temperatures under a particular set of conditions and follow the changes as a function of any parameter.

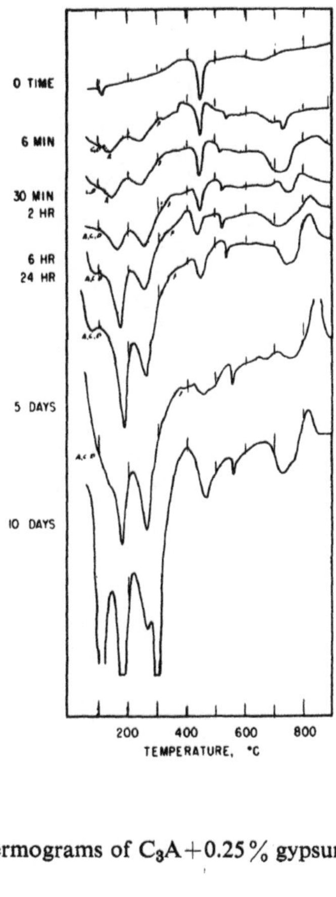

Fig. VI.1.1: Thermograms of C_3A+0.25% gypsum hydrated at 2°C (*16*).

In the hydration of C_3A two endothermal effects almost always appear together in the range 140-190°C and 230-260°C and hence they have been attributed to C_2AH_8 and C_4AH_{13} (*25*). Some investigators have assigned the endotherms at 90-100°C and 180-200°C to the dehydration of hexagonal C_3AH_x (*17*, *18*, *20*, *22*, *23*). Doubts have been expressed on the existence of the so-called hexagonal C_3AH_x.

Opinion is more in favour of treating the hexagonal C_3AH_x as a carboaluminate or as an equimolar mixture of C_2AH_8 and C_4AH_{13} formed by syncrystallization (*15*, *24*). The interpretation of the peak temperature for high and low sulfoaluminates has to be made with caution. The high sulfoaluminate may exhibit an endothermal peak anywhere in the range 120-180°C, while the low sulfoaluminate may indicate the peak in the range 200°-250°C.

The endothermal effect of the low sulfoaluminate occurs at about the

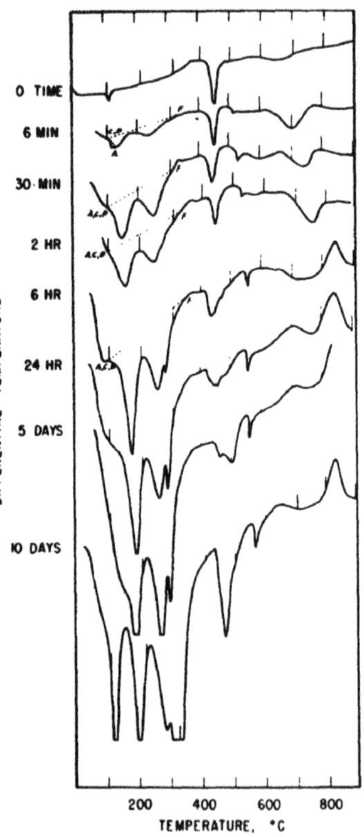

Fig. VI.1.2: Thermograms of C_3A+0.25% gypsum hydrated at 12°C (*16*).

same temperature at which C_4AH_{13} and the solid solution $C_3A \cdot Ca(OH)_2 \cdot 12H_2O$–$C_3A \cdot CaSO_4 \cdot 12H_2O$ exhibit endothermic effects, and hence it is difficult to differentiate between them.

The sequence of hydration reactions on the C_3A–gypsum mixtures can be illustrated with the thermograms obtained by Feldman and Ramachandran (*16*). In their studies these workers used compacts of C_3A containing 0.25-20% gypsum and cured them at temperatures of 2, 12, 23, and 50°C for periods ranging from ½ min to 34 days. The results are shown in Figs. VI.1.1-1.12.

Based on the thermal data of C_3A hydration, the endothermal effects were assigned to different compounds. The endothermic peak in the range 170-200°C was attributed to high sulfoaluminate, the endothermic effect in the range 140-180°C and 240-285°C to the hexagonal C_2AH_8 and C_4AH_{13} com-

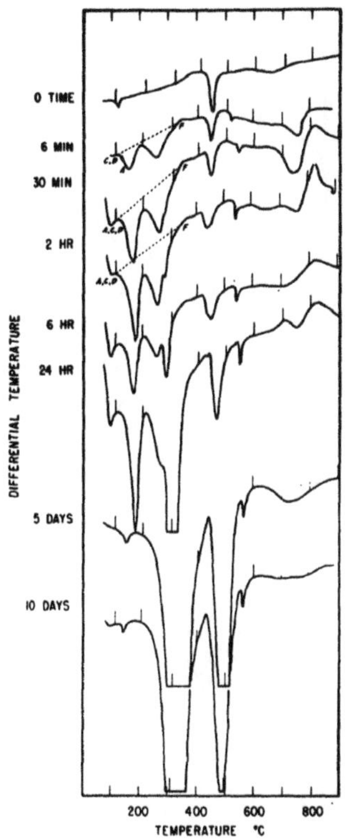

Fig. VI.1.3: Thermograms of C_3A+0.25% gypsum hydrated at 23°C (*16*).

pounds respectively, and the two endothermic effects at 290-300°C and 460-500°C to the cubic C_3AH_6. The compound C_4AH_{13} may be associated with the low sulfoaluminate in solid solution.

Thermograms of the hydration products of C_3A containing 0.25% gypsum and cured at 2, 12, and 23°C are shown in Figs. VI.1.1, 1.2 and 1.3 respectively. At 2°C the hexagonal hydroaluminates C_2AH_8 and C_4AH_{13} appear at 6 min and increase in quantity up to 5 days, after which a significant proportion of C_3AH_6 is indicated. The reaction proceeds at a faster rate at 12 and 23°C.

Thermal curves clearly show that the rate of formation of hexagonal hydroaluminates and their conversion to cubic C_3AH_6 are accelerated with increasing temperatures. A comparison of these with the corresponding thermograms of hydration of pure C_3A described in Chapter V reveals that

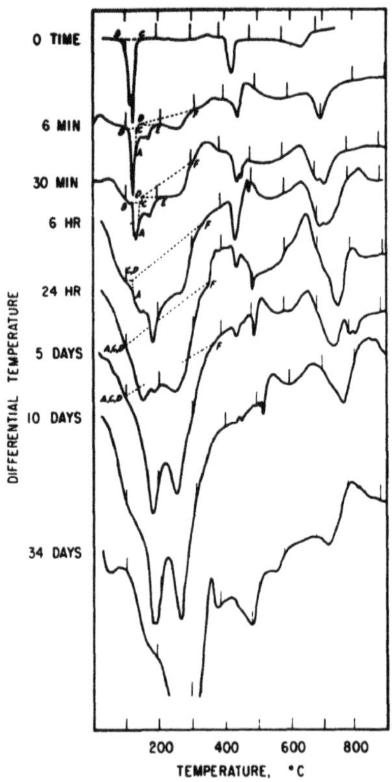

Fig. VI.1.4: Thermograms of C_3A+2.5% gypsum hydrated at 2°C (*16*).

addition of gypsum retards the rate of formation of hexagonal hydroaluminates and their conversion to C_3AH_6. The addition of 0.25% gypsum has negligible effect on the thermal curves at higher temperatures. With 0.25% gypsum there is no evidence of a peak corresponding to the sulfoaluminate.

In Figs. VI.1.4, 1.5, 1.6 and 1.7 are presented the thermal curves of C_3A + 2.5% gypsum hydrated at 2, 12, 23, and 52°C respectively. The rate of consumption of gypsum can be studied by its characteristic endothermal effects at about 120 and 135°C. At 2 and 12°C these peaks persist up to 6 hr, whereas at 23°C they disappear within about 30 min. The hexagonal hydroaluminates and the high sulfoaluminate coexist.

There is little evidence of formation of C_3AH_6 at 2 and 12°C up to 10 days, but at 23°C this compound is formed after 24 hr. There is a definite indication that hexagonal hydroaluminates continue to form even when gypsum is

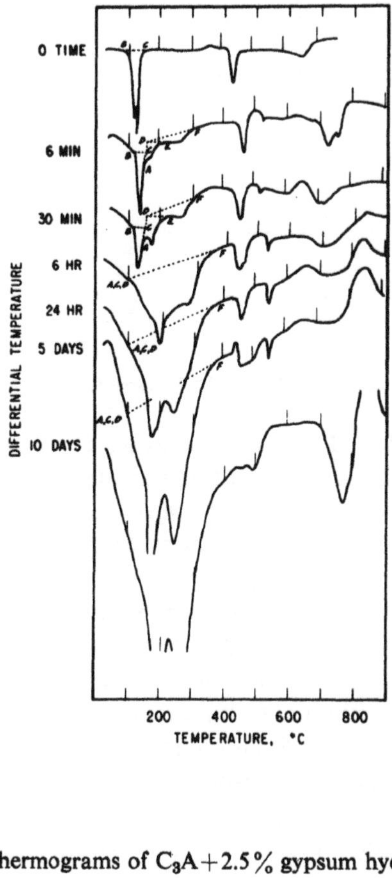

Fig. VI.1.5: Thermograms of C_3A+2.5% gypsum hydrated at 12°C (*16*).

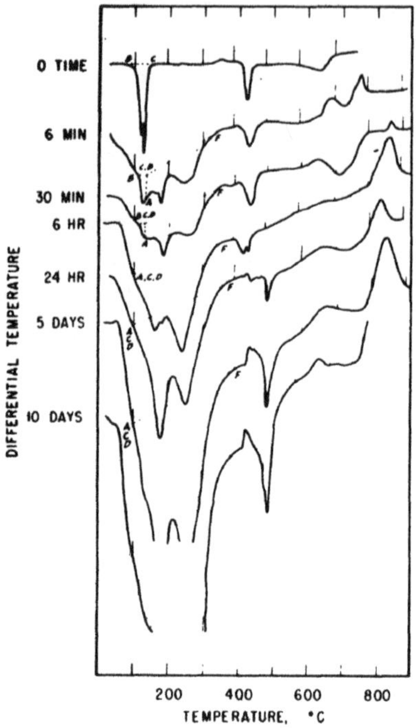

Fig. VI.1.6: Thermograms of C_3A+2.5% gypsum hydrated at 23°C (*16*).

not consumed completely. The hydration at 52°C proceeds at a faster rate, with the formation of C_3AH_6, high sulfoaluminate, and hexagonal hydroaluminates. The rates of conversion are however, lower compared with those without gypsum addition. The hexagonal hydroaluminates continue to form after all the gypsum has apparently reacted.

DTA results for the hydration of C_3A containing 10% gypsum at 2, 12, 23, and 52°C are shown in Figs. VI.1.8, 1.9, 1.10, and 1.11 respectively. The hexagonal hydroaluminates and the high sulfoaluminate appear at 6 min. The quantity of the sulfoaluminate increases with increasing temperature, but the quantity of hexagonal hydroaluminates remains relatively unchanged.

The intensities of the peaks corresponding to hexagonal hydroaluminate and the high sulfoaluminate seem to increase after 24 hr, 6 hr, and 2 hr at 2, 12, and 23°C respectively. Gypsum disappears after 24 hr at 23°C, where-

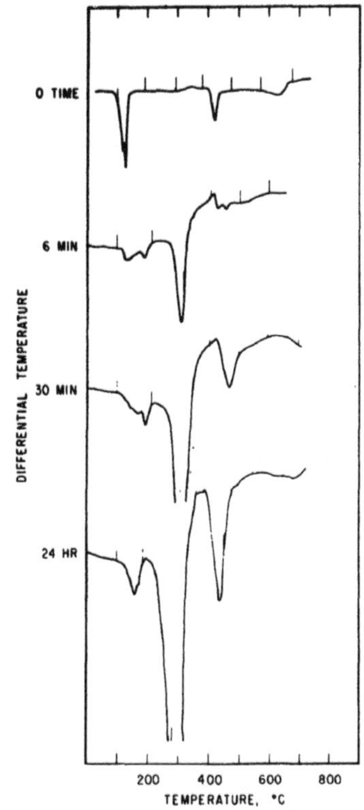

Fig. VI.1.7: Thermograms of $C_3A+2.5\%$ gypsum hydrated at 52°C (*16*).

as it takes 5 days to disappear at 2 or 12°C.

Addition of 10% gypsum retards the onset of C_3AH_6 considerably. At 23°C a small amount of C_3AH_6 is detected after 10 days and at 52°C it is detected after 24 hr. The hydration rate is enhanced at 52°C, considerable amounts of both hexagonal hydroaluminates and sulfoaluminate being formed after 30 sec.

Thermograms of C_3A containing 20% gypsum are shown in Fig. VI.1.12. The thermal behavior is similar to that of C_3A with 10% gypsum at 23°C, except that more sulfoaluminate is formed at the expense of hexagonal hydroaluminates.

A comparison of the hydration reaction at any temperature for different amounts of gypsum reveals interesting facts (Figs. VI.1.1, 1.4 and 1.8). With 0.25% gypsum, hydration at 2°C is retarded, though there is no indication

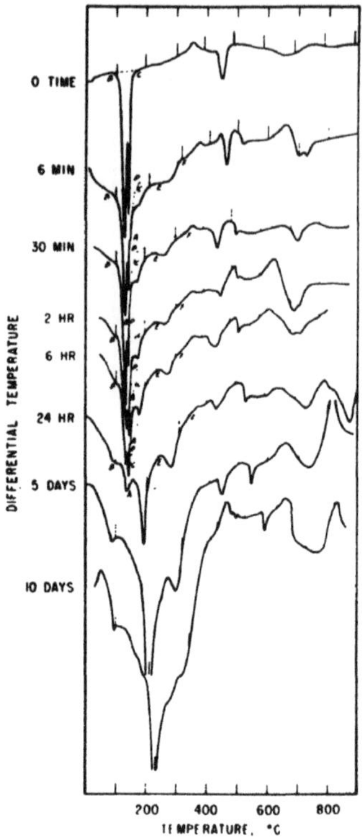

Fig. VI.1.8: Thermograms of C_3A+10% gypsum hydrated at 2°C (*16*).

for the presence of the sulfoaluminate. At 2.5% gypsum the high sulfoaluminate is detected. The formation of C_3AH_6 is prevented more efficiently at higher gypsum contents. The C_3A with 10% gypsum suppresses more effectively the formation of hexagonal hydroaluminates than that with lower amounts. At 10% gypsum no C_3AH_6 forms even at 10 days.

At 12°C (Figs. VI.1.2, 1.5, and 1.9) the mixture with 2.5% gypsum has a much greater retarding effect than that with 0.25% gypsum. Larger amounts of products seem to form with 10% gypsum.

At 23°C (Figs. VI.1.3, 1.6, and 1.12) there is a marked difference between the thermograms of mixtures containing 0.25% and 2.5% gypsum. At 2.5% the initial appearance and the rate of formation of hexagonal hydroaluminates and C_3AH_6 are delayed considerably. The samples containing 10 or 20% gypsum have a greater retarding effect than those containing less gypsum at

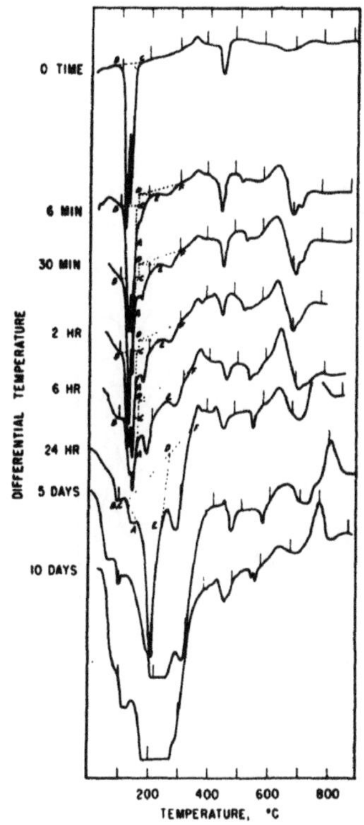

Fig. VI.1.9: Thermograms of C_3A+10% gypsum hydrated at 12°C (*16*).

52°C.

At 52°C (Figs. VI.1.7 and 1.11) with only 2.5% gypsum, C_3AH_6 forms relatively quickly at 6 min, even though some gypsum remains unreacted at this period. The mixture with10% gypsum slows down the overall reaction to a greater extent than that with 2.5% gypsum.

The DTA examination reveals that the hexagonal hydroaluminates are formed almost immediately on exposure of the C_3A–gypsum mixture to water. The formation of hexagonal hydroaluminates is retarded by gypsum; the conversion of the hexagonal hydroaluminates to C_3AH_6 is also delayed. In general the rate of hydration of C_3A and the rate of reaction of gypsum increase with temperature. As the temperature is increased higher gypsum contents are required for equal suppression of the hexagonal hydroaluminates or for the prevention of conversion to C_3AH_6.

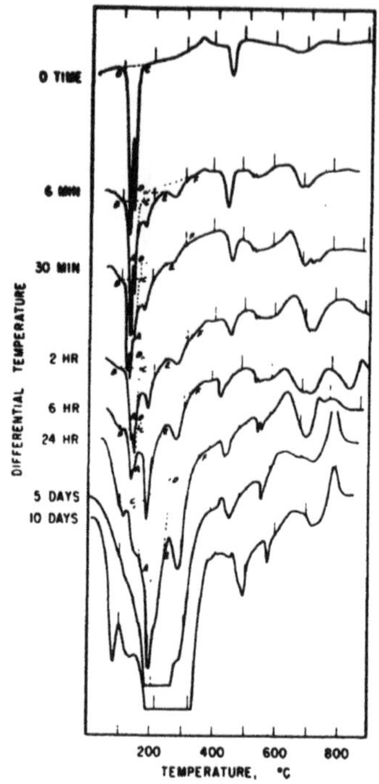

Fig. VI.1.10: Thermograms of $C_3A + 10\%$ gypsum hydrated at 23°C (*16*).

Segalova and Brutskus (*20*) demonstrated by DTA the formation of the solid solution of calcium monosulfate form with $C_4A \cdot aq$ in the hydrating mixtures of trisulfate form and C_3A. They investigated two samples of high sulfoaluminate containing 5 or 8% C_3A by hydrating them for periods ranging from 30 min to 1 month. The results are presented in Fig. VI.1.13. With 5% C_3A the solid solution appears after 30 min and is associated with an endothermic effect at 180°C. At 8% C_3A the solid solution is indicated after 8 hr. Segalova and Brutskus also found that with the initial ratio of $CaSO_4$:$C_3A \geq 3$, only the trisulfate formed, while with $CaSO_4$:$C_3A < 3$, the solid solution of monosulfate form with $C_4A \cdot aq$ was the main reaction product.

Butt, Maier, and Varshal (*26*) subjected the pastes containing C_3A, $CaSO_4 \cdot 2H_2O$ and sand to a temperature of 50, 90, 125, 175, or 200°C for 24 hr and

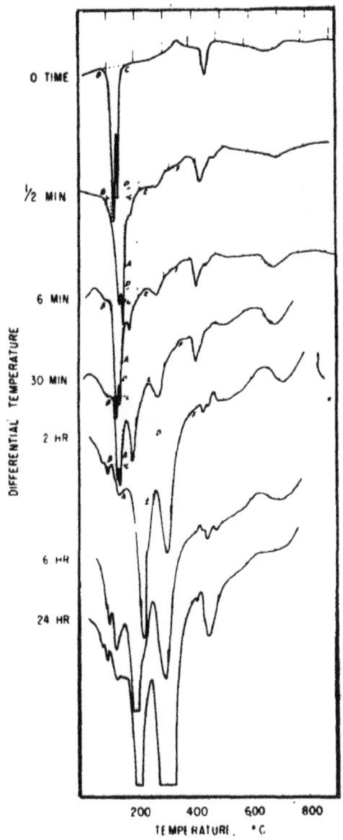

Fig. VI.1.11: Thermograms of $C_3A + 10\%$ gypsum hydrated at 52°C (*16*).

studied the products by DTA. The monosulfate form existed as a stable phase at 50-200°C. The presence of silica was found to influence the thermal transformations of trisulfate to monosulfate, by forming a hydrogarnet with a peak in the range 300-400°C.

Murakami, Tanaka, and Suzuki (*19*) studied the hydration product of C_3A containing $CaSO_3$ and detected a compound of composition $C_3A \cdot CaSO_3 \cdot 7H_2O$. DTA showed that at 190°C 4 molecules were lost, while the remaining water was lost at 260°C.

Mechanism of Retardation. Feldman and Ramachandran (*16*) presented a quantitative interpretation of DTA results by hydration at 2, 12, 23, and 52°C of C_3A–gypsum mixtures containing 0, 0.25, 2.5, 10, and 20% gypsum, and compared these with length changes in compacts hydrated under similar conditions. The estimation of the relative reaction rates from DTA peak

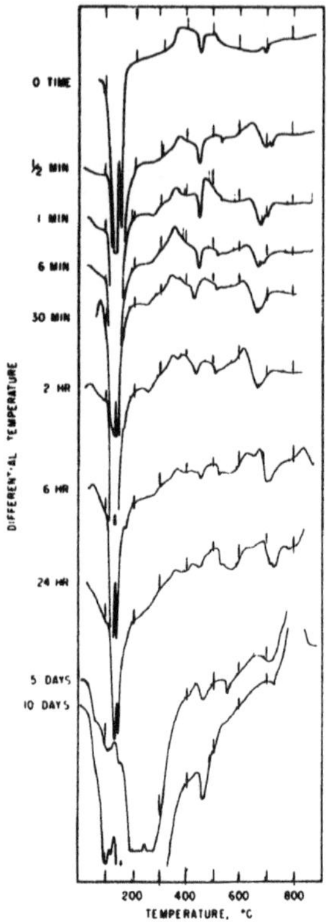

Fig. VI.1.12: Thermograms of $C_3A+20\%$ gypsum hydrated at 23°C (*16*).

areas and measurement of length changes during these reactions showed considerable agreement. These studies also enabled them to formulate a mechanism of the retarding action of gypsum.

Gypsum decreases the reactivity of C_3A by sorption of $(SO_4)^{--}$ ions, thus hindering the operation of certain dislocation sources. Gypsum also reduces the rate of conversion of hexagonal hydroaluminates to the cubic hexahydrate by sorption of $(SO_4)^{--}$ ions. This is a second mechanism by which the $(SO_4)^{--}$ ions can retard the normal hydration of C_3A. The large volume of sulfoaluminate product may contribute to a general decrease in the porosity of C_3A but there is no evidence to suggest that this product forms a particularly impermeable layer.

A sufficiently large addition of gypsum affects hydration at higher temperatures by preventing the impermeable growth of C_3AH_6 over the C_3A layer.

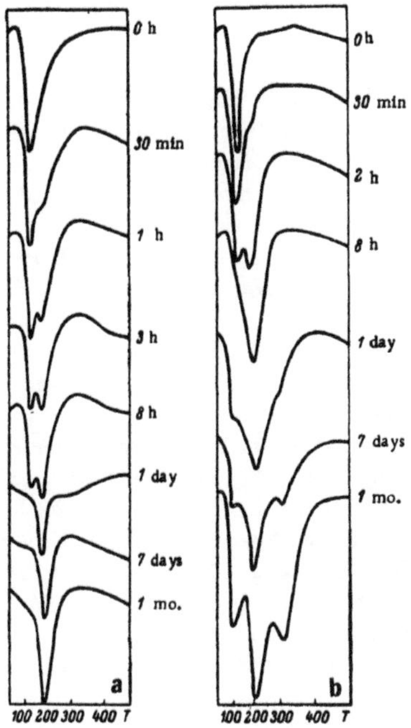

Fig. VI.1.13: Thermograms of products with 4.5% $C_3A \cdot CaSO_4 \cdot 31H_2O$ and 95.5% mix of hydrated Al_2O_3 and C_3A (w/s=0.5). (a) 5% C_3A, (b) 8% C_3A (*20*).

Under these conditions gypsum may be termed an accelerator. The degree of retardation of C_3A with a given amount of gypsum at any particular temperature is dependent on (i) concentration of $(SO_4)^{--}$ ions on and around the surface of C_3A, (ii) rate of reaction of $(SO_4)^{--}$ with the hexagonal hydro-aluminates, and (iii) thickness of the hexagonal hydroaluminate layer around the C_3A grains.

REFERENCES

1. Jones, F.E., *J. Phys. Chem.*, *48*, 311, 1944.
2. Jones, F.E., *Trans. Faraday Soc.*, *35*, 1484, 1939.
3. D'Ans, J., and Eick, H., *Zement-Kalk-Gips*, *6*, 302, 1953.
4. Berman, H.A., and Newman, E.S., *Proc. 4th Intern. Symp. Chem. Cements*, Washington, p. 247, 1960.

5. Turriziani, R., and Schippa, G., *Ric. Sci., 24*, 2356, 1954.
6. Midgley, H.G., and Rosaman, D., *Proc. 4th Intern. Symp. Chem. Cements*, Washington, p. 259, 1960.
7. Stein, H.N., *Rec. Trav. Chim., 81*, 881, 1962.
8. Lawrence, C.D., *Special Report 90*, Highway Res. Board, Washington, 1966.
9. Schwiete, H.E., Ludwig, U., and Jager, P., *Zement-Kalk-Gips, 17*, 229, 1964.
10. Steinour, H.H., *Portland Cement Assoc. Bull. 34*, Feb., 1951.
11. Bogue, R.H., *The Chemistry of Portland Cement*, 2nd Ed., Reinhold Publishing Corp., New York, 1955.
12. Lerch, W., *Proc. Am. Soc. Testing Mater., 46*, 1252, 1946.
13. Klein, A.A., and Phillips, A.J., *Nat. Bur. Std., Tech. Paper No. 43*, p. 71, 1914.
14. Roller, P.S., *Ind. & Eng. Chem., 26*, 669, 1934.
15. Lea, F.M., and Desch, C.M., *The Chemistry of Cement and Concrete*, Edward Arnold Publishers, London, 1956.
16. Feldman, R.F., and Ramachandran, V.S., *Mag. Concrete Res. 18*, 185, 1966.
17. Brutskus, T.K., and Segalova, E.E., *Zh. Prikl. Khim.*, **37**, 1126, 1964.
18. Brutskus, T.K., and Segalova, E.E., *Kollodn. Zh., 26*, 8, 1964.
19. Murakami, S., Tanaka, H., and Suzuki, P., *Yogyo Kyokai Shi., 67*, 8, 1959.
20. Segalova, E.E., and Brutskus, T.K., *Zh. Prikl. Khim., 38*, 1954, 1965.
21. Govorov, A.A., *Trudy Kharikov. Inst. Inzhenerov Zheleznodorozh. Transporta im. S.M. Kirova, 39*, 11, 1960.
22. Segalova, E.E., Solov'eva, E.S., and Rebinder, P.A., *Kollodn. Zh., 23*, 94, 1961.
23. Govoroff, A., *Rev. Mater. Construct. Ed. C.*, No. 490/491, 181, 1956.
24. Wells, L.S., Clarke, W.F., and McMurdie, H.F., *J. Res. Nat. Bur. Std., 30*, 367, 1943.
25. Feldman, R.F., and Ramachandran, V.S., *J. Am. Ceram. Soc., 49*, 268, 1966.
26. Butt, Y.M., Maier, A.A., and Varshal, B.G., *Dokl. Akad. Nauk SSSR, 136*, 398, 1961.

VI.2. HYDRATION OF C_3A IN PRESENCE OF CALCIUM LIGNOSULFONATE

Addition to concrete of small quantities of surface-active agents such as lignosulfonic acids and their salts or hydroxylated carboxylic acids and their salts lowers the water requirement and retards the setting. A reduced water/cement ratio causes strength development in concretes. This can be clearly seen in Fig. VI.2.1 in which the strength development in a concrete containing lignosulfonate additive is compared with that without any additive. The control sample always shows lower strengths than that containing the active agent.

The most widely used surface active agents are based on lignosulfonic acid or its Ca and NH_4 salts. These agents have a molecular weight between 2000 and 100,000 and are absorbed on the cement particles causing deflocculation and initial retardation of the hydration.

Adsorption data have demonstrated that of the various phases in Portland cement, C_3A absorbs calcium lignosulfonate to the maximum extent. The silicate phases absorb the lignosulfonate to a negligible extent (*2*) (Fig. VI.2.2). From these results it is evident that the adsorption of calcium lignosulfonate

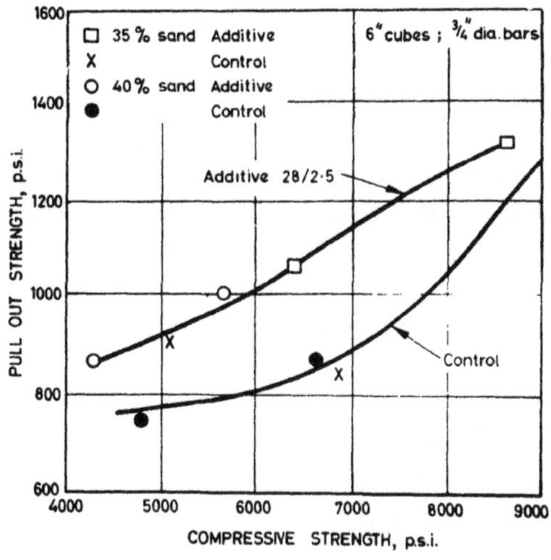

Fig. VI.2.1: Effect of lignosulfonate on bond and compressive strength (*1*).

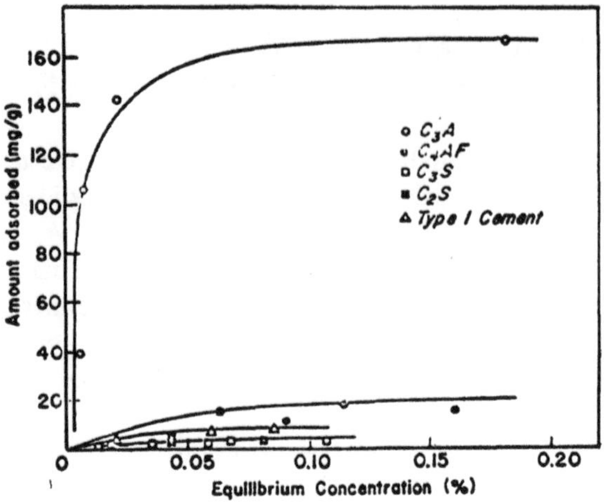

Fig. VI.2.2: Adsorption from aqueous solution of calcium lignosulfonate on cement compounds (*2*).

on the aluminate phase is mainly responsible for the set-retarding and water-reducing action in cements. For details of the mechanism of the action of calcium lignosulfonate the work of Hansen (*3*), Steinour (*4*), Suzuki and Nishi (*5*), Ernsberger and France (*6*, *7*), Young (*8*) and Lukyanova, Segalova and Rebinder (*9*) should be consulted.

Differential Thermal Analysis

The strength development in concretes in the presence of surface active agents may not be solely due to the lowering of the water/cement ratio but may also be influenced by the changes in the hydration products. By applying the methods of DTA, X-ray diffraction, and microscopy Young (*8*) demonstrated the changes in the hydration products of C_3A in the presence of lignosulfonates.

In Fig. V1.2.3 are shown the thermograms of pure C_3A hydrated for 6 min, 15 min, 1 hr, 1 day, or 7 days. The C_3A paste exhibits an intense endothermic effect between 300 and 400°C in all the mixes. Another endothermic peak of smaller magnitude also appears in the temperature interval 500-600°C. The two effects are indicative of the presence of C_3AH_6 and increased amounts of this compound are formed at longer periods of hydration.

Thermograms of C_3A samples hydrated in the presence of lignosulfonate (lignosulfonate/ C_3A ratio, 0.05) show different characteristics at lower tem-

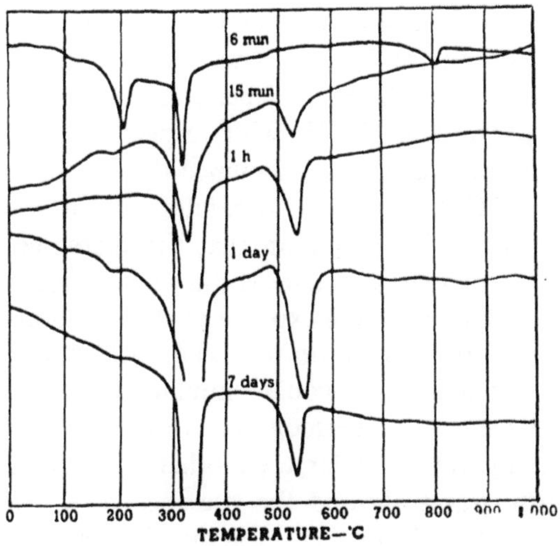

Fig. VI.2.3: Thermograms of hydrated C_3A (8).

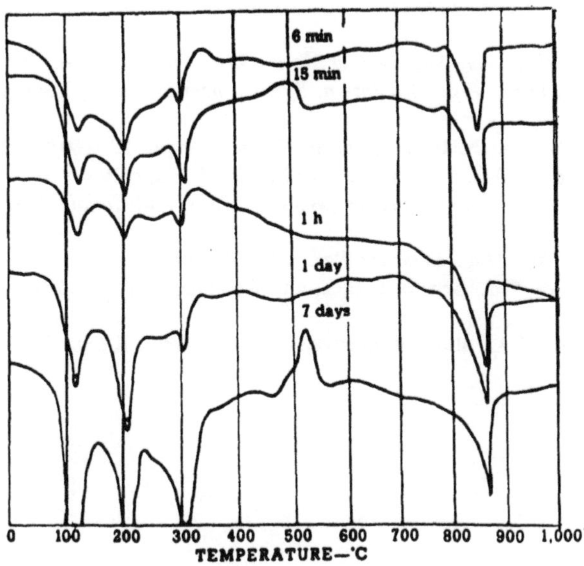

Fig. VI.2.4: Thermograms of C_3A pastes hydrated in presence of lignosulfonate (8).

peratures (Fig. VI.2.4). Three endothermic peaks occur in the vicinity of 100, 200, and 300°C, the intensities of which increase progressively for curing periods ranging from 6 min to 7 days. These endotherms could be assigned to a mixture of C_4AH_{13} and C_2AH_8. Hence it can be inferred that lignosulfonate stabilizes C_4AH_{13} and C_2AH_8 with respect to the cubic C_3AH_6.

Tikhonov and Tihomirova (*10*) have assumed that lignosulfonate alters the hydration products of C_3A but the DTA results are not in consonance with this observation. Young is of opinion that the increased strength of cements in the presence of lignosulfonate is due to the formation of acicular crystals of C_4AH_{13} or C_2AH_8 at the expense of the hexagonal plates.

Young carried out further work on the DTA of C_3A in presence of gypsum and $Ca(OH)_2$ because this would simulate the environment in the hydrating Portland cement. Thermograms revealed the presence of the high sulfoaluminate hydrate by a large endotherm at 150°C. After 3 hr of hydration the thermograms showed the presence of C_2AH_8 and C_4AH_{13}. There is a clear indication from the DTA that lignosulfonate retards conversion to C_2AH_8 and C_4AH_{13}.

REFERENCES

1. Hobbs, C., *Chem. and Ind. (London)*, 526, 1964.
2. Blank, B., Rossington, D.R., and Weinland, L.A., *J. Am. Ceram. Soc., 46*, 395, 1963.
3. Hansen, W.C., *Am. Soc. Testing Mater. Spec. Tech. Publ. No. 266*, 265, 1960.
4. Steinour, H.H., *Proc. 3rd Intern. Symp. Chem. Cements*, London, 261-333, 1952.
5. Suzuki, S., and Nishi, S., *Semento Gijutsu Nenpo, 13*, 160, 1959.
6. Ernsberger, F.M., and France, W.G., *Ind. & Eng. Chem., 37*, 598, 1945.
7. Ernsberger, F.M., and France, W.G., *J. Phys. and Colloid Chem., 52*, 267, 1948.
8. Young, J.F., *Mag. Concrete Res., 14*, 137, 1962.
9. Lukyanova, O.I., Segalova, E.E., and Rebinder, P.A., *Kollodn. Zh., 19*, 82, 1957.
10. Tikhonov, V.A., and Tihomirova, L.A., *Zh. Prikl. Rhim., 27*, 1067, 1954.

VI.3. HYDRATION OF C_3A IN PRESENCE OF CALCIUM CHLORIDE

Salts such as sodium chloride and calcium chloride (about 2%) are added to Portland cement to accelerate the development of strength during the winter months in some countries, At lower concentrations the salts act as retarders of the set. Forsen (*1*), from his extensive studies on the accelerators and retarders, concluded that the set is regulated by the formation of complex calcium salts, analogous to the calcium aluminate sulfate hydrates. Two calcium aluminate chloride hydrates exist—$C_3A \cdot CaCl_2 \cdot xH_2O$ and $C_3A \cdot 3CaCl_2 \cdot 30H_2O$—the latter being formed by the hydration of C_3A in 23% $CaCl_2$ at −10°C (*2*).

Differential Thermal Analysis

Not much work has been reported on the use of DTA in a study of the hydration of C_3A in presence of accelerators, though this appears to be a promising field of application.

Heller and Ben-Yair (*3*) studied the expansion of cements in $CaCl_2$ solutions. They prepared a synthetic calcium chloroaluminate hydrate that exhibited a large endothermic effect at 215°C and four endotherms of smaller magnitude at 110, 265, 330 and 550°C.

Andreeva, Segalova, and Volynets (*4*) obtained evidence of the complex $C_3A \cdot CaCl_2 \cdot xH_2O$ in hydrated mixes containing C_3AH_6, $CaCl_2$, and 90-98% quartz or $CaCO_3$. The reaction rate was much higher if C_3A was used in place of C_3AH_6.

Serb-Serbina, Savvina, and Zhurina (*2*) detected by DTA and other analytical methods both forms of chloroaluminates, *viz.* $C_3A \cdot CaCl_2 \cdot 10H_2O$ and $C_3A \cdot 3CaCl_2 \cdot 30H_2O$, while hydrating C_3A in presence of $CaCl_2$. The frost resistance of cements containing $CaCl_2$ was explained by the formation of an impervious layer of these complex chloroaluminates.

REFERENCES

1. Forsen, L., *Zement*, *19*, 1130, 1930.
2. Serb-Serbina, N.N., Savvina, Y.A., and Zhurina, V.S., *Dokl. Akad. Nauk* SSSR, *111*, 659, 1956.
3. Heller, L., and Ben-Yair, M., *J. Appl. Chem.* (*London*), *16*, 223, 1966.
4. Andreeva, E.P., Segalova, E.E., and Volynets, E.E., *Dokl. Akad. Nauk* SSSR, *123*, 1052, 1958.

VI.4. HYDRATION OF C_3A IN PRESENCE OF CALCIUM CARBONATE

The reaction of calcium carbonate with tricalcium aluminate in the presence of water results in the formation of calcium aluminate carbonate hydrate. Two complex carboaluminates are known, analogous to the high and low calcium sulfoaluminate hydrate (*1*-*5*). The study of the formation of the carboaluminate complexes is of interest because they may be formed in hydrating cements in the presence of carbon dioxide or calcium carbonate aggregate.

The two carboaluminates are usually prepared by the introduction of carbonate ions into aluminate solutions, either by means of an alkali carbonate or by the bubbling of CO_2 into the solution. The compound that was assigned a composition C_3AH_{12} earlier is in fact a low carbonate complex form $C_3A \cdot CaCO_3 \cdot H_n$, where n has values between 10.5 and 11.0 (*6*-*10*). The low carbonate complex has also been formed from the reaction of $CaCO_3$ and C_3A by stirring techniques (*11*-*13*).

In the paste hydration of C_3A with 0.5% $CaCO_3$ Budnikov and co-workers observed the formation of the low form of carboaluminate (*14*-*16*). They also reported that the formation of the complex slowed down the hydration. It has been reported by Seligman and Greening (*13*) that a 5% addition of $CaCO_3$ to Portland cement forms a considerable quantity of carboaluminate.

Farran (9), working with calcite as an aggregate in a high alumina cement, observed that the interfacial film consisted of a crystalline phase of a solid solution of low carboaluminate in C_4AH_{13}. The presence of the low form of carboaluminate has also been observed by Hadley (*17*) in cement pastes containing carbonate as the aggregate.

The phenomenon of false set has been reported in cements exposed to an atmosphere containing CO_2. Manabe (*18*) found that the carbonation of C_3A decreased its reactivity to gypsum and promoted the false set. Seligman and Greening (*13*) also obtained evidence of formation of the low form of carboaluminate in cements and explained the decreased reactivity of C_3A to sulfate ions due to the coating of the complex.

Differential Thermal Analysis

The influence of $CaCO_3$ additions on the hydration behaviour of C_3A has been studied by DTA by Feldman, Ramachandran, and Sereda (*19*), Butt, Varshal, and Maier (*20*), and Budnikov, Kolbasov, and Penteleev (*16*).

Feldman, Ramachandran, and Sereda prepared compacts of C_3A containing 2.5, 10, 20, and 50% $CaCO_3$ and exposed them to water in liquid or vapor phase. The hydration reaction was examined by DTA, infrared absorption, and X-ray diffraction. These results were correlated with those of dimensional changes in the compacts.

In Fig. VI.4.1 are presented the thermograms of precipitated $CaCO_3$, the C_3A preparation, hydrated C_3A, and the hydrated C_3A–$CaCO_3$ mixtures. Precipitated $CaCO_3$ exhibits an intense endothermic peak at 845°C owing to decarbonation (curve a). The C_3A thermogram shows three endothermic peaks at 275, 450, and 760°C caused by the partially hydrated and carbonated portion of the material. Hydrated C_3A indicates four endothermic effects (curve c).

The peak at 330°C is of greater intensity than the others. The two endo-

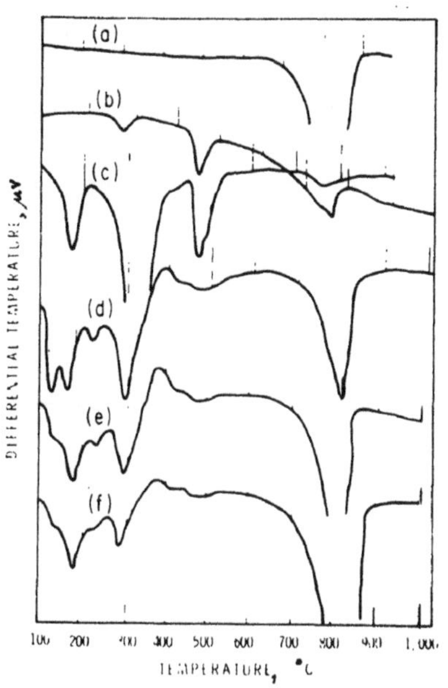

Fig. VI.4.1: Differential thermal curves. (a) $CaCO_3$, (b) C_3A preparation unhydrated, (c) C_3A preparation hydrated, (d) C_3A+10% $CaCO_3$ hydrated, (e) C_3A+20% $CaCO_3$ hydrated, (f) C_3A+50% $CaCO_3$ hydrated (*19*).

therms at 330 and 450°C represent the stepwise dehydration of C_3AH_6. The endothermic peak at 175°C may be due to the preeence of C_4AH_{13} or C_2AH_8 or both. The endothermic dip at 760°C is due to the decomposition of a small amount of $CaCO_3$ present in the sample.

The DTA behaviour of $CaCO_3$–C_3A mixtures containing 10, 20, and 50% $CaCO_3$ is shown by curves (d), (e), and (f). The decarbonation peaks of these samples occur at 800, 815, and 840°C respectively. Additions of $CaCO_3$ result in a corresponding increase in the intensity of the $CaCO_3$ decomposition peak. It is interesting to note that additions of $CaCO_3$ to C_3A enormously decrease the peak at about 300°C for C_3AH_6 dehydration. With higher $CaCO_3$ contents the peak area decreases further. Increased amounts of $CaCO_3$ progressively inhibit C_3AH_6 formation; in addition, there will be a decrease in the peak intensity with decrease in C_3A content. These results show that hydration of C_3A is suppressed by $CaCO_3$ additions.

The hydrated C_3A–$CaCO_3$ mixtures show endothermic peaks at about 180°C, and this effect is also present in the hydrated C_3A sample. Whereas the peak at 330°C decreases in intensity, the peak at 180°C persists at all $CaCO_3$ concentrations and is perhaps due to the carboaluminate complex. Additions of $CaCO_3$ result in the development of two endothermic peaks at about 140 and 230°C. Both seem to decrease in intensity with higher $CaCO_3$ content. The endothermic peaks may represent the formation of calcium

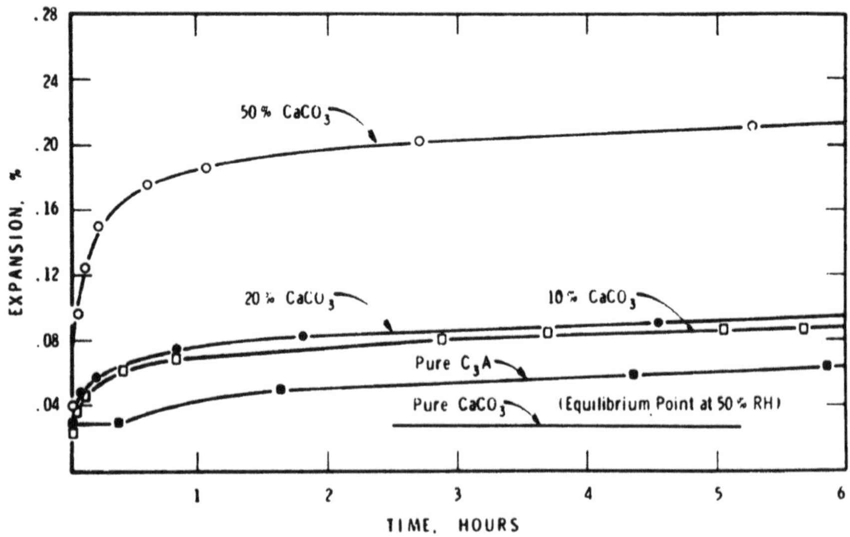

Fig. VI.4.2: Expansion of C_3A and $C_3A/CaCO_3$ mixtures on exposure to 50% relative humidity (*19*).

carboaluminates reported by earlier workers. All the $CaCO_3$-treated samples exhibit an exothermic bulge below 400°C, and this has also been observed by Carlson and Berman (*4*).

Thermograms of carboaluminates reported by different workers are not identical and hence it is difficult to interpret these results by comparison. By employing DTA it is difficult to conclude whether the low or the high or the mixture of carboaluminates is present.

The length change results of Feldman, Ramachandran, and Sereda (*19*) are presented in Fig. VI.4.2. Pure C_3A expands to about 4% within a few hours of hydration. The addition of $CaCO_3$ reduces the expansion in all cases, the expansion for the sample containing 50% $CaCO_3$ being less than 1%. In the DTA of the C_3A–$CaCO_3$ mixtures, addition of 50% $CaCO_3$ inhibits the formation of C_3AH_6 to the greatest extent.

Butt, Varshal, and Maier (*20*) studied the stability of hydrogarnets in CO_2 atmosphere by DTA. The treatment of C_3AH_6 by CO_2 for various periods shows that $CaCO_3$ is formed with an endotherm at 650-800°C by a corresponding decrease in the endotherm at 300-400°C.

REFERENCES

1. Greene, K.T., *Proc. 4th Intern. Symp. Chem. Cements*, Washington, pp. 359-374, 1960.
2. Bessey, G.E., *Proc. 2nd Intern. Symp. Chem. Cements*, Stockholm, pp. 178-215, 1938.
3. Jones, F.E., *Proc. 2nd Intern. Symp. Chem. Cements*, Stockholm, pp. 231-245, 1938.
4. Carlson, E.T., and Berman, H.A., *J. Res. Nat. Bur. Std.*, *64A*, 33, 1960.
5. Stein, H.N., *J. Appl. Chem.* (*London*), *13*, 228, 1963.
6. Turriziani, R., and Schippa, G., *Ric. Sci.*, *26*, 2792, 1956.
7. Roberts, M.H., *J. Appl. Chem.* (*London*), *7*, 543, 1957.
8. Buttler, F.G., and Taylor, H.F.W., *J. Appl. Chem.* (*London*), *9*, 616, 1959.
9. Farran, J., *Silicates Ind.*, *23*, 595, 1958.
10. Lhopitallier, P., *Proc. 4th Intern. Symp. Chem. Cements*, Washington, pp. 1007-1033, 1960.
11. Manabe, T., Kawada, N., and Nishiyama, M. *Semento Gijutsu Nenpo*, *15*, 54, 1961.
12. Roberts, M.H., *Proc. 4th Intern. Symp. Chem. Cements*, Washington, pp. 1033-1034, 1960.
13. Seligmann, P., and Greening, N.R., *Proc. 4th Intern. Symp. Chem. Cements*, Washington, pp. 408-410, 1960.
14. Budnikov, P.P., and Azelitskaya, R.D., *Zh. Prikl. Khim.*, *32*, 1181, 1959.
15. Budnikov, P.P., Kolbasov, V.M., and Penteleev, A.S., *Silikat. Tech.*, *11*, 271, 1960.
16. Budnikov, P.P., Kolbasov, V.M., and Penteleev, A.S., *Dokl. Akad. Nauk SSSR*, *129*, 1104, 1959.

17. Hadley, D.W., *Highway Res. Board, Proc.*, *40*, 462, 1961.
18. Manabe, T., *Proc. 4th Intern. Symp. Chem. Cements*, Washington, pp. 404-08, 1960.
19. Feldman, R.F., Ramachandran, V.S., and Sereda, P.J., *J. Am. Ceram. Soc.*, *48*, 25, 1965.
20. Butt, Y.M., Varshal, B.G., and Maier, A.A., *Zhur. Vsesoyuz. Khim. Obshchestva im D.I., Mendeleeva*, *6*, 355, 1961.

VII. Autoclaved Cementitious Products

The hydrothermal treatment of cement or lime or both with silica-bearing materials results in the formation of cementitious solids. In the U.S.A. and Canada autoclaved blocks are in great use and are made by autoclaving cement and either dense or lightweight aggregate.

Autoclaving is employed to produce articles of high early strength and reduced shrinkage. Many types of materials can be mixed with lime or cement and processed in high-pressure steam to produce concrete products of good strength. The materials that have been used are silica flour, pumice, scoria, tuff, cinder, calcined clay, fly ash, and blastfurnace slag.

Autoclaved products include sand-lime and flint-lime bricks, lightweight cellular concrete blocks, panels and blocks such as 'Silicate Concrete' and 'Silicalcite', and asbestos cement sheets and pipes. In all these products the cementing agent is a hydrous calcium silicate. In practice the autoclaving cycle may vary, but a typical treatment may involve an autoclaving pressure of 140-150 $lb/in.^2$ for 8 hr with a rapid release of pressure within about 30 min.

High strength calcium silicate materials in the form of blocks and panels, plain or reinforced, are gaining increasing importance in many countries, including U.S.S.R., Germany, U.K., Holland, Sweden, Switzerland, Denmark, Finland and Norway. In the U.S.S.R. the calcium silicate industry is expanding at such a rate that the production of calcium silicate products will be about 20-25% of all the building materials made in 1970 (*1*).

A wider use of the autoclaved products may be predicted because of many advantages these possess. One of the most desirable aspects is that products of required strength can be made under controlled hydrothermal conditions. The strengths attained in 24 hr are equivalent to those obtained after 28 days of normal moisture curing.

By control of the starting materials, the forming, and autoclaving pressure, calcium silicate products with a strength of the order of 30,000 $lb/in.^2$ can be obtained (*2*). The building units can be produced in white or black or any other color and size. The autoclaved products have good resistance to

TABLE VII.1

The Normal Range of Properties of Building Bricks and Blocks [Taylor (3)]

Property 1	*Calcium silicate bricks* 2	*Clay bricks* 3	*Concrete bricks (dense)* 4	*Aerated concrete block (autoclaved)* 5	*Lightweight aggregate concrete block* 6
Strength					
Compressive strength ($lb/in.^2$) tested in a wet condition	1500-8000	1500-12000	1000-5000	300-800	300-2500
Volume stability					
Drying shrinkage (% of length)	0.01-0.035	0.00-0.015*	0.02-0.05	0.05-0.10	0.04-0.08
Thermal conductivity (k)					
($Btu/ft^{-2} \cdot hr^{-1} \cdot in^{-1} \cdot deg \cdot F^{-1}$)	8-11	5-9	7-12	0.7-1.5	1.0-5.0
Durability under severe natural exposure	Good if strength above 2000 $lb/in.^2$	Excellent to very poor†	Good to poor†	Good	Good to poor†
Fire resistance	Good	Very good to moderate†	Good to poor†	Very good	Very good to poor†
Properties affecting laying and plastering	Good	Varied; good to poor†	Good	Good	Good
Liability to efflorescence (soluble salts)	None	Vary from none to severe†; fairly common	None or slight†	None or slight†	Generally none
Density (lb/ft^3)	105-130	90-150	110-140	25-55††	40-100

* Clay bricks have sometimes a small but significant progressive expansion, more particularly during their early life.

† According to quality and type.

†† Aerated concretes are normally made to a closely controlled density at a specified value within this range. The most widely made grades have densities of 45-50 $lb/in.^2$

sulfate action also, do not normally exhibit efflorescence, and have low moisture contents.

Table VII.1 compares the properties of calcium silicate bricks with those of other well known building bricks and blocks.

Chemistry

Cement Compounds. In the manufacture of autoclaved products, cement or lime or both and SiO_2-bearing materials form the raw mix. The usual approach to the study of the autoclaved products consists in gaining an understanding of the possible products of hydration of the individual cement minerals, *viz.* C_3S, C_2S, C_3A, the ferrite phase, and neat cement paste, as well as of the mixtures of cement or lime with various types of aggregates.

Buckle and Taylor (*4*) studied the phases developed in C_3S pastes at 18-250°C by X-ray and optical methods. At temperatures below 180°C the products comprised an ill-crystallized tobermoritic mineral, CSH gel, and calcium hydroxide. Higher temperatures promoted the formation of α-dicalcium silicate. Tricalcium silicate hydrate (TSH) was also detected in pastes at lower temperatures, while in suspensions this hydrate could be detected only above 200°C. Higher mechanical strengths of the pastes autoclaved at temperatures below 200°C were attributed to the presence of a gelatinous matrix.

The hydration products of β-C_2S very, depending on the type of stabilizer used. The products may be CSH gel or α or γ-dicalcium silicate hydrate. Funk (5, 6) found that CSH gel was formed along with $Ca(OH)_2$ with stabilizers such as 0.5% B_2O_3 or 1% Cr_2O_3. A modified γ-C_2S, more reactive than the normal γ-C_2S was indicated in products containing 0.2% Na_2O or Na_2SO_4 as a stabilizer. The CSH gel formed at 100°-120°C had a Ca:Si molar ratio of 1.88-1.89.

In excess of water and at temperatures above 100°C, α-dicalcium silicate hydrate was observed. α-C_2S is normally regarded as an inert material, but under hydrothermal conditions it reacts with water. The hydration product of γ-C_2S at 170°-250°C is variously described as γ-C_2S hydrate, $C_2SH(C)$, $C_2SH(III)$, or $C_2SH(CI)$.

Speakman and Taylor (*7*) have described the product of hydration of γ-C_2S as a submicroscopic intergrowth of kilchoanite with ill-crystallized calciochondrodite. Kilchoanite or phase Z is a polymorph of rankinite C_3S_2 and calciochondrodite has the composition C_5S_2H.

In the high-pressure steam curing, C_3A hydrates to form C_3AH_6, and above about 215°C a compound of composition $C_4A_3 \cdot 3H_2O$ forms (*8*).

Cement Paste. The cement paste is expected to yield the same compounds as those obtained from its mineral constituents under hydrothermal conditions. The compounds expected to form are CSH gel, α-C_2S hydrate, TSH, $Ca(OH)_2$, C_3AH_6, and a modified form of γ-C_2S.

Kalousek and Adams (*9*) made a systematic investigation of the hydration products of cement pastes cured at temperatures between 25 and 175°C. The products of hydration at about 100°C resembled those at 25°C in indicating a successive formation of ettringite, a solid solution of the composition C_4AH_{13}–$C_3A \cdot CaSO_4 \cdot 12H_2O$ and the phase X.

The phase X refers to a single phase containing all the components of the cement. This phase forms within 4 hr of treatment at 100°C. The treatment at 175°C, however, gave α-C_2S hydrate with the composition 1.8-2.4 CaO: 1.00 SiO_2:1.00-1.25 H_2O and $Ca(OH)_2$. No evidence was found for the presence of C_3AH_6 or other hydrogarnets.

Kalousek (*10*) and Kalousek and Prebus (*11*) investigated the neat cements according to the autoclaving schedules in practice and identified hillebrandite, lime-rich fibrous and platy tobermoritic materials, α-C_2S hydrate, and $Ca(OH)_2$.

Cement- or Lime-Aggregate Reactions. The chemistry of the reactions in the system CaO–SiO_2–H_2O is of direct interest in the manufacture of sand–lime bricks, lightweight insulation materials, and other related products. Extensive work has been done on the hydrothermal reaction of $Ca(OH)_2$ with SiO_2 and the relevant results have been reviewed by Taylor (3, 12, 13).

At different conditions of autoclave treatments the products are semicrystalline phases CSH(I) and CSH(II), 11 Å tobermorite, α-C_2SH, xonotlite, gyrolite, afwillite, hillebrandite, and 14 Å tobermorite. The reactions in the lime-quartz system may be summarized as follows (*12*):

1. The initial reaction gives CSH(II) or CSH gel, with a Ca:Si ratio of at least 1.5. In pastes of low overall Ca:Si ratio (*e.g.* 0.8-1.0), the lime therefore disappears before the quartz.
2. At low overall Ca:Si ratios, the CSH(II) reacts with more silica to give CSH(I). When first formed, this probably has a Ca:Si ratio of 1.25. The value drops gradually to 0.8 as the reaction proceeds.
3. The CSH(I) recrystallizes to tobermorite. The rate of this process depends greatly on the temperature. At 175°C tobermorite can be detected within 12 hr, but at 125°C (the lowest temperature at which it has been found in a paste) several days' reaction is needed. This process occurs more quickly if reactive aluminous materials are present, but more slowly if silica gel is substituted for the quartz.
4. If the temperature is high enough, the tobermorite is changed into xonotlite. This phase has been detected in pastes cured at 165°C, but its formation under these conditions is slow. It is not formed to a detectable extent in less than 24 hr below 180-190°C.

5. At overall Ca:Si ratios below 0.83, unreacted quartz remains after all the CSH(II) and CSH(I) have been converted into tobermorite. Reaction of tobermorite or CSH(I) with residual quartz can occur, giving gyrolite. This reaction, however, is very slow, probably because it can only take place at the interface of the quartz and the tobermorite. Gyrolite can nevertheless sometimes be detected in pastes with Ca:Si $\leq$0.67 after one or more days' autoclaving at 165-200°C. Its formation under these conditions is not very reproducible. Xonotlite is sometimes formed when gyrolite might be expected, presumably because the tobermorite alters to xonotlite more quickly than it can react with the quartz.
6. If the overall Ca:Si ratio exceeds 1.2-1.3, there is not enough quartz to convert all the CSH(II) into CSH(I). A different set of reactions thus occurs, which begins with recrystallization of the CSH(II). The usual product of this recrystallization is α-C_2S hydrate, which is formed slowly at 125°C and more quickly at higher temperatures.
7. Phase F, of possible composition $C_5S_3H_2$, seems to be an alternative recrystallization product of CSH(II), but its formation is irreproducible; it is formed most often at around 165°C.
8. At sufficiently high temperatures, C_2S α-hydrate or phase F can alter to hillebrandite. This phase is formed to a detectable extent in pastes cured for 1-2 days at 165-200°C.

In Table VII.2 the compounds detected in autoclaved lime-silica pastes at different temperatures are given.

The behavior of the cement–silica system is analogous to that of the lime–silica system, the cement contributing CaO and SiO_2 in the reaction. It is an interesting fact that under autoclaving conditions cement–silica pastes do not yield hydrated alumina or ferric oxide phase but may enter the crystal

TABLE VII.2
Compounds Detected in Autoclaved Lime-Silica Pastes [Taylor (14, 15)]

Compound	*Overall Ca:Si ratios of pastes in which detected*	*Chemical composition*	*Range of temp. in which formed, °C*
Gyrolite	0.33-0.67	$C_3S_4H_3$ (?)	165-200
CSH	0.33-3.00	$C_{0.33\text{-}3.0}SH_x$	90-175
Tobermorite	0.33-1.08	$C_5S_6H_5$	125-200
Xonotlite	0.67-3.00	C_6S_6H	165-200
Phase F	1.00-3.00	$C_5S_3H_2$ (?)	155-200
α-C_2S hydrate	1.50-3.00	C_2SH	125-175
Hillebrandite	1.67-2.00	C_2SH	165-200

lattice of the hydrous calcium silicate (*16*).

Extensive data have been published by Kalousek and coworkers (*9-11*, *16-18*) on the autoclaved products in the cement-silica system. The amount of α-C_2S hydrate formed depends on the quartz content in the cement, increasing with quartz content up to 9% and further additions promoting the formation of tobermorite. Highest strengths correspond to the formation of large amounts of tobermorite. These results have been confirmed by Sanders and Smothers (*19*). Neese, Spangenberg and Weiskirchner (*20*) found CSH(I), tobermorite, and nekoite in pastes containing cement:quartz =3:1.5.

Some work has also been reported on the products of reaction between cement and pumice or other alumina-bearing compounds and these show somewhat different behaviors.

It is generally observed that the development of strengths in autoclaved products is due to the crystalline tobermorite, while low strengths are ascribed to the presence of α-C_2S hydrate. Taylor *et al* (*2*) however, found that these conclusions are not valid for the hydrothermal products of lime–silica flour mixes. They observed that with molding pressures of 0 to 15,000 $lb/in.^2$ the compressive strength of the autoclaved products increased from 3500 to 30,000 $lb/in.^2$ The products of high strengths contained a nearly amorphous compound with a CaO/SiO_2 ratio of 1.05. The degree of compaction was shown to be an important factor in the development of strengths.

Differential Thermal Analysis

Lime–Silica System. The hydrothermal products generally appear amorphous to microscopic observations, and a detailed study by X-rays also is not easy. The technique of DTA has been found to be particularly useful in the study of the hydrothermal products of lime or cement with various types of aggregates. In addition to clearly providing indications on the rate of consumption of $Ca(OH)_2$, the technique can be used to ascertain the progressive formation of different products, and in some instances the results can be correlated with strength development.

Kalousek (*16*) was probably the first cement chemist to draw attention to the potentialities of the DTA method for investigating the phases developed in autoclaved products. Figs. VII.1, 2, and 3 obtained by him for the system CaO–SiO_2–H_2O illustrate the manner in which thermograms differ, depending on the CaO/SiO_2 molar ratio, the retention time, and the temperature and pressure of autoclaving.

The raw materials consisted of $Ca(OH)_2$ and SiO_2 with CaO/SiO_2 ratios of 1.6, 1.7, and 1.75, autoclaving temperatures 130, 140, 150, or 175°C, and

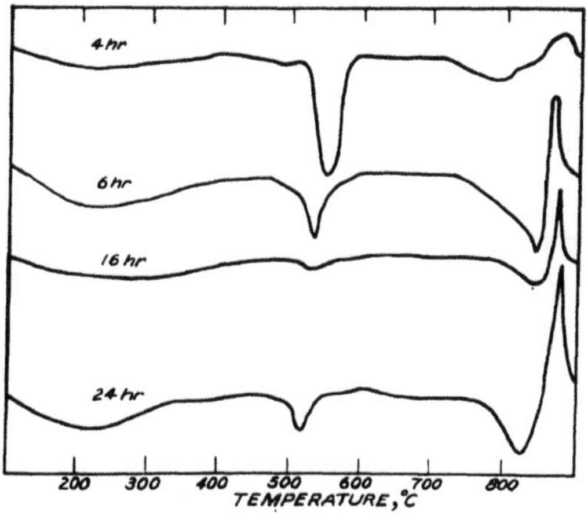

Fig. VII.1: DTA curves of 1.6 C:S products autoclaved at 130°C (*16*).

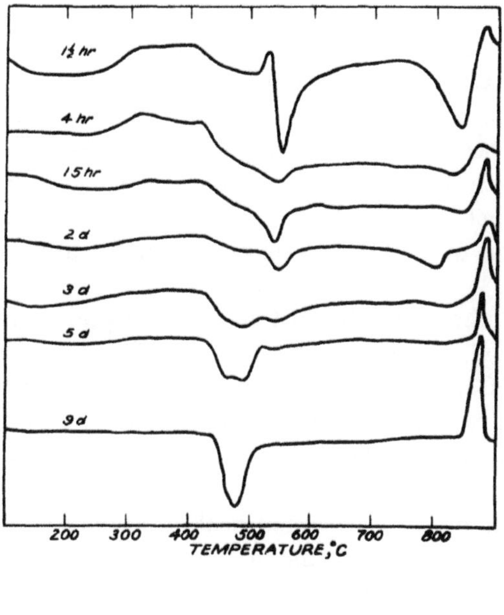

Fig. VII.2: DTA curves of 1.6 C:S products autoclaved at 150°C (*16*).

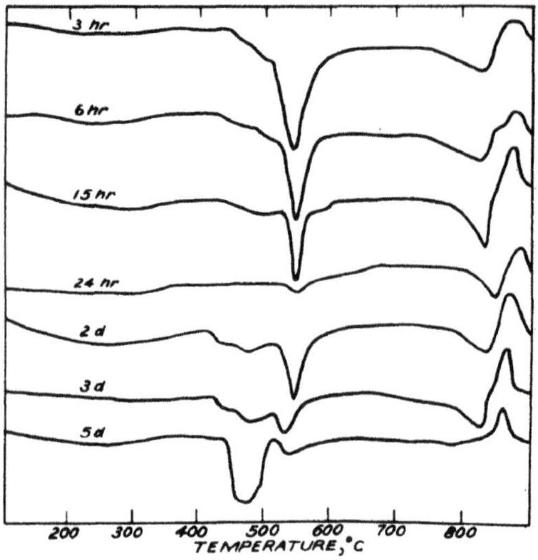

Fig. VII.3: DTA curves of 1.75 C:S products autoclaved at 175°C (*16*).

retention periods ranging from 1½ hr to 9 days.

In all the mixes the endothermal effect between 500 and 600°C indicates the presence of $Ca(OH)_2$, which gradually decreases in intensity and finally disappears. At shorter periods of autoclaving all the curves exhibit the endothermic peak between 500 and 600°C for the decomposing $Ca(OH)_2$, the intensity of which decreases to a considerable extent even at 16 hr, 4 hr, and 24 hr for products autoclaved at 130, 150, and 175°C respectively. The rate of consumption of $Ca(OH)_2$ is maximum in the mix containing CaO/SiO_2 ratio of 1.6.

Further, the endothermal bulge of $Ca(OH)_2$ reappears sooner in the samples treated at 150°C. Thermograms also reveal that at higher CaO/SiO_2 ratios depletion of $Ca(OH)_2$ takes longer periods. The curves show unmistakable evidence of exothermal dents between 350 and 550°C and an endothermal valley at about 800-890°C, the exact causes of which could not be explained.

All the curves exhibit an exothermal peak at about 860-870°C of varied intensity, and this is attributed to the presence of a microcrystalline gel-like compound of the formula $C_5S_4H_n$. A comparison of the exothermic peak intensities at 800-900°C for 1.6 CaO/SiO_2 and 1.75 CaO/SiO_2 products shows that the intensities decrease as the CaO/SiO_2 ratio increases. In other words, this should imply the presence of increased quantities of $C_5S_4H_n$ in

1.6 CaO/SiO_2 products.

X-ray studies also confirmed these observations by showing lines of decreased intensities at 3.03 and 1.82 Å for products with higher CaO/SiO_2 ratios.

The compositions having a CaO/SiO_2 ratio higher than 1.25 and treated at temperatures of 130°C or beyond always contained the α-hydrate, which could be identified in the thermograms by an endothermic effect at about 470-480°C. In 1.75 CaO/SiO_2 products the α-hydrate formed after 2 days.

The intensity of the α-hydrate increased further at 5 days at the expense of three effects, *viz.* the endothermic peak due to $Ca(OH)_2$, an endothermic valley at 800-840°C, and the exothermic peak denoting the presence of $C_5S_4H_n$. It is evident from Fig. VII.3 that in the 1.75 CaO/SiO_2 mix the α-hydrate should be the predominant product at 5 days.

The X-ray results confirmed these observations by showing the absence of lines corresponding to $Ca(OH)_2$ and the gel-like product but exhibiting strong lines at 4.20, 3.90, 3.26, and 2.42 Å for the presence of α-hydrate.

Further data in the system $Ca(OH)_2$–silicic acid and $Ca(OH)_2$–quartz at 125-175°C were reported by Kalousek (*10*). In this work attention was mainly directed to the study of the products of molar composition 0.8–1.25 CaO/SiO_2. The hydrothermal products of the reaction ranged in composition from $C_4S_5H_n$ to $C_5S_4H_n$ and DTA was able to differentiate them. The compositions $C_4S_5H_n$, CSH_n, and $C_5S_4H_n$ showed similar X-ray behaviour, but DTA showed characteristic peaks for their identity.

The 0.8-1.0 CaO/SiO_2 hydrates exhibited a strong exothermic peak at 830±5°C, the intensity for 1.0 CaO/SiO_2 hydrate being twice that of 0.8 CaO/SiO_2 hydrate. The 1.25 CaO/SiO_2 product showed the exothermal reaction at 865±5°C, the 1.33 CaO/SiO_2 hydrate, at 900±5°C.

The surface area results showed some interesting features. The 0.8 CaO/SiO_2 mix autoclaved at 175°C for about 4 hr contained a lime-rich phase and $C_5S_4H_n$ with a surface area of 120 m^2/g. At about 8 hr the formation of $C_4S_5H_n$ involves an increase in surface area of about 170 m^2/g. Thereafter up to 19 hr there is only $C_4S_5H_n$ without any contribution to surface area. The formation of $C_4S_5H_5$ at 24 hr decreases the surface area to about 70 m^2/g.

The sequence of reactions can be followed conveniently by DTA. In Fig. VII.4 the thermal curves of the mix containing an CaO/SiO_2 ratio of 0.8 and autoclaved at 175°C for different periods is shown. At 1 hr, only $Ca(OH)_2$ is present, which is obvious from the large endothermic peak between 500 and 600°C. A poorly-crystallized α-hydrate is formed at 90 min and a decrease in the intensity of the $Ca(OH)_2$ peak is noted. After 3½ hr $Ca(OH)_2$ is almost completely absent and the composition of the product corresponds to 1.75 CaO/SiO_2 α-hydrate.

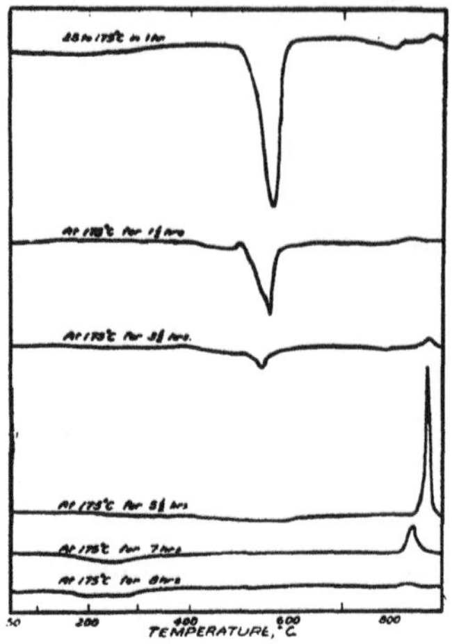

Fig. VII.4: DTA of products obtained by autoclaving 0.8 C/S mixture at 175°C (*10*).

At periods beyond 3½ hr, reaction occurs between the lime-rich phase and the unreacted quartz. The sequence of these reactions was formulated by Kalousek as follows:

$$7CaO + 4SiO_2 + aq. \rightarrow 7CaO \cdot 4SiO_2 \cdot nH_2O \quad (1)$$

$$5(7CaO \cdot 4SiO_2 \cdot nH_2O) + 8SiO_2 \rightarrow 7(5CaO \cdot 4SiO_2 \cdot nH_2O) \quad (2)$$

$$5CaO \cdot 4SiO_2 \cdot nH_2O + SiO_2 \rightarrow 5(CaO \cdot SiO_2 \cdot nH_2O) \quad (3)$$

$$4(CaO \cdot SiO_2 \cdot nH_2O) + SiO_2 \rightarrow 4CaO \cdot 5SiO_2 \cdot nH_2O \quad (4)$$

$$4CaO \cdot 5SiO_2 \cdot nH_2O \text{ (fibrous)} \rightarrow 4CaO \cdot 5SiO_2 \cdot 5H_2O \text{ (tobermorite)} \quad (5)$$

At 8 hr, well-crystallized tobermorite is present in the product and this is characterized by the absence of the high temperature peak.

The reaction of a lime–silicic acid mix having a 0.8 CaO/SiO_2 ratio differs from that of a lime–quartz mix. A comparison of the thermograms of lime–silicic acid mixtures (Fig. VII.5) can be made to study the different behaviors of these systems. There is almost complete absence of $Ca(OH)_2$ in the $Ca(OH)_2$–silicic acid system even at 40 min after hydration. The intermediate product $C_5S_4H_n$ forms rapidly and at 2 hr exhibits a strong exothermic peak beyond 800°C, denoting conversion to CSH_n. The trans-

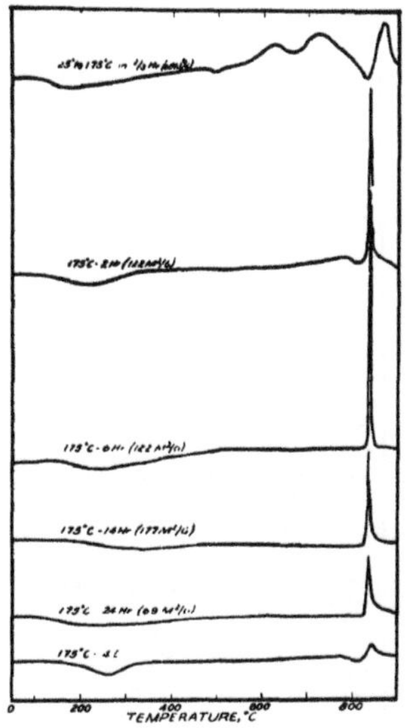

Fig. VII.5: DTA of products prepared at different times of autoclaving at 175°C from 0.8 C/S mixture made with silicic acid (*10*).

formation to tobermorite, however, takes longer than 18-19 hr, and complete conversion is not attained up to seven days.

Kalousek made the observation that in mixes containing silicic acid the reaction is rapid in the initial stages and CSH_n and $C_4S_5H_n$ persist for longer periods prior to the final transformation to tobermorite.

The phase equilibria in the system $CaO–SiO_2–H_2O$ were studied by Peppler (*21*) at 180°C with initial CaO/SiO_2 ratios up to 3. The DTA of the products and the X-ray diffraction data of the heated specimens in the DTA apparatus were utilized to identify the reaction products.

In Fig. VII.6 the DTA of solid reaction products obtained for the mixes containing CaO/SiO_2 ratios from 0.25 to 3.0 shows that the preparations designated No. 2 to No. 6 belong to one family and the rest to another. The samples numbered 2 to 6 contain unaltered silica gel, gyrolite ($C_2S_3H_2$), and xonotlite (C_5S_5H).

The endothermic peak in the range 100-300°C represents the dehydration of silica gel and the exothermic effect at 800-850°C represents the crystallization of wollastonite. Gyrolite is a metastable product and converts to more stable xonotlite. The proportion of xonotlite increases with CaO/SiO_2 ratio,

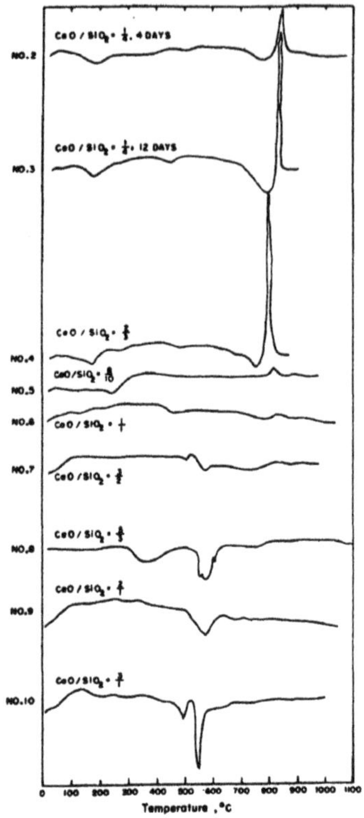

Fig. VII.6: DTA curves of solid reaction products obtained by an approach to equilibrium from undersaturation (*21*).

the sample No. 6 with CaO/SiO_2 ratio of 1 being mainly xonotlite.

Samples 7 to 10 are similar in having negligible silica gel and an endothermic effect at 500-600°C associated with the dehydration of hillebrandite (C_2SH) to β-C_2S. The intensity of this peak increases from sample No. 7 to No. 10, thereby indicating that the amount of hillebrandite increases with the CaO/SiO_2 ratio. Some $Ca(OH)_2$ with a break at about 500°C is present in sample 10. The optical and X-ray results are in agreement with those of DTA.

The three forms of dicalcium silicate hydrate synthesized under hydrothermal conditions have been designated α, β and γ-types. Kalousek, Logiudice, and Dodson (*22*) made a detailed study of the synthesized forms from lime and quartz or silicic acid by the 'dry mixing' or 'lime slurry' method.

The 'dry mix' method consisted of mixing the ingredients in a mortar in a dry state and then adding 4 parts of water to 1 part of solid. The α-form containing 1.8 to 2.4 CaO/SiO_2 could be synthesized from CaO and quartz by the dry mix method autoclaved at 175°C for 3 days. The α-type hydrate is characterized by an endothermic peak at 470-480°C.

The β-hydrate of CaO/SiO_2 ratio 1.8 to 2.4 was prepared from CaO and silicic acid dispersed in water or from quartz by treatment at 175-250°C for 5-12 days. The β-hydrate prepared from silicic acid exhibited an endothermic effect at 540-560°C, whereas that prepared from quartz exhibited a peak at 600°C. The β-hydrate is similar to the mineral hillebrandite.

The thermal curves of γ-hydrate are shown in Fig. VII.7. The γ-form has a composition of about 1.5 to 2.25 CaO/SiO_2 and is prepared from quartz and CaO dispersed in water at temperatures 200-300°C for 24 hr to 2 days. All the samples except that processed at 200°C show an endothermal valley at 730-800°C.

From the ignition loss, the compositions of the 1.5, 1.75, and 2.0 C/S hydrates were computed to be $C_{1.5}SH_{0.3}$, $C_{1.75}SH_{0.3}$, and $C_2SH_{0.4}$.

It was concluded that the β-hydrate is the conversion production of the α-type. There is a possibility that the α-type hydrate is the intermediate phase in the formation of the γ-type.

The mineralogical composition of the autoclaved products depends on the type of aggregate used with lime. Kalousek (*18*) used DTA to identify the changes in the thermal behaviors of hydrothermally treated lime in admixture with quartz, pumice, shale, or slag. Hydrated compounds with CaO/

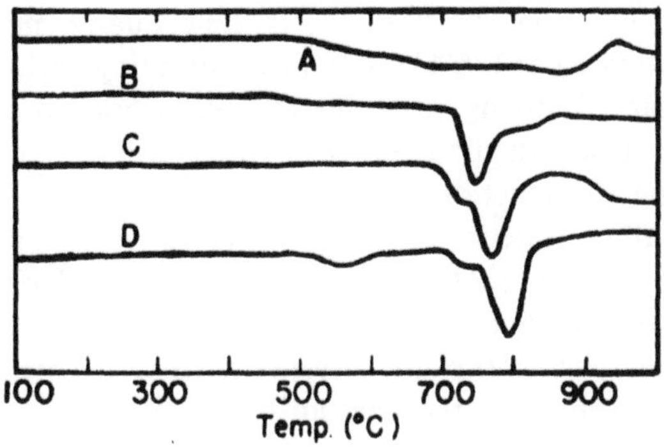

Fig. VII.7: Thermal analysis curves of γ-type hydrates made from silicic acid or quartz. (A) 1.5 C/S 200°C, 24 hr; (B) 1.75 C/S, 225°C, 2 days; (C) 2.0 C/S, 225°C, 2 days; (D) 2.25 C/S, 300°C, 2 days (*22*).

SiO_2 ratios 0.9 to 1.3 consisted of a poorly crystallized lime-rich phase in mixes containing quartz. Pumice and shale gave products with similar structure, but containing CaO/SiO_2 ratio in excess of 1.3. DTA showed differences in curves for different phases formed from a given aggregate though X-ray diffraction patterns were similar.

An attempt was also made by Midgley and Chopra (*23*) to find a relation between the mineralogical composition and compressive strengths in the autoclaved products containing various proportions of lime and aggregates such as pulverized fuel ash, expanded colliery shale, ground quartz, granulated slag, or foamed blastfurnace slag. The DTA technique in conjunction with X-ray diffraction analysis confirmed the presence of various cementitious minerals at different periods of autoclaving from 2 to 48 hr at a pressure of 160 lb/in.^{2}

The lime–quartz mixtures contained tobermorite and xonotlite, the lime-pulverized fuel ash or shale contained tobermorite and a hydrogarnet of approximate composition $3CaO \cdot Al_2O_3 \cdot SiO_2 \cdot 4H_2O$, and lime-slag gave a poorly crystalline tobermorite, α-dicalcium silicate hydrate, and a hydrogarnet.

It has already been stated (p.168) that the nature of the products in the autoclaved products in the system $CaO\text{-}SiO_2\text{–}H_2O$ varies with the temperature, retention period, the type of aggregate, and the initial CaO/SiO_2 ratio.

Tun Soo-Lee and Loo Gon-Chen (*24*) re-examined these factors by applying DTA, TGA, X-ray, and microscopic techniques. In addition to confirming the observations of other workers, they found that the method of mixing and the initial water/solid ratio are important factors determining the nature and amount of the hydration products.

Fig. VII.8 represents the thermograms of lime–quartz mixes with CaO/SiO_2 ratio 0.8 and treated at 175°C for 16 hr by adopting various mixing procedures such as manual mixing, grinding, and wet vibration milling for 1 or 1½ hr. It can be seen that the manual mixing is the least effective of methods in promoting the reaction.

The cryptocrystalline tobermorite ($C_5S_6H_5$) is shown by an endotherm at 170°C, the CSH(I), by an exothermic peak at 825°C. Grinding and wet vibration milling for 1 hr accelerates the reaction, and the products consist of cryptocrystalline tobermorite and crystalline tobermorite ($C_5S_6H_5$) characterized by a small endothermal dent at 280°C in curves 2 and 3.

The wet vibration milling for longer periods promotes the formation of well crystallized tobermorite. The increased activity of mixes on wet vibration milling is ascribed to the deformation of crystal lattice under the forces of impact.

The rate of reaction also changes with the initial water/solid ratio and

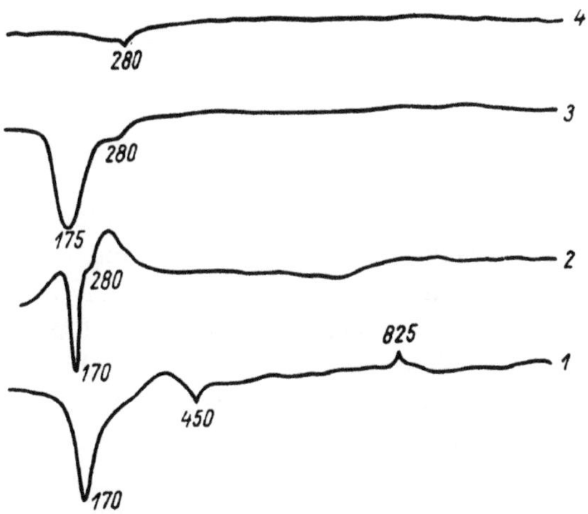

Fig. VII.8: DTA curves showing the effect of mixing methods. (1) Manual mixing, (2) grinding, (3) wet vibration milling for 1 hr, (4) wet vibration milling for 1.5 hr (*24*).

this is demonstrated in Fig. VII.9. At a water/solid ratio of unity, an endothermic peak at 235°C due to CSH(I) and an exothermic peak at 830°C due to C_2SH_2 are registered. At a water/solid ratio of 5 or 20, the presence of well crystallized $C_5S_6H_5$ is indicated by the endothermic peaks at 270°-280°C.

Improved crystallization of products may be achieved by subjecting the raw materials to ultrasonic waves prior to the hydrothermal treatment (*63*). A

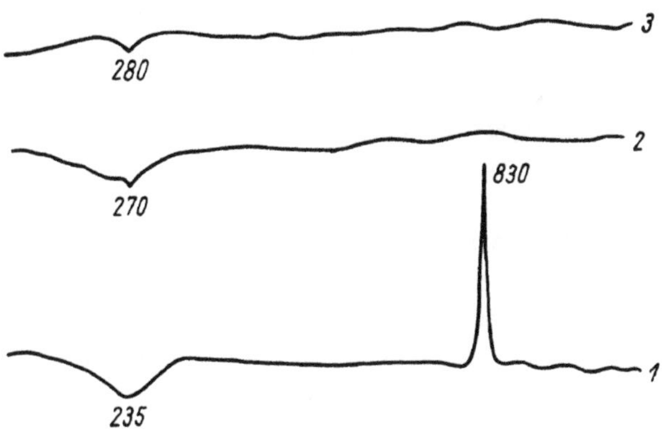

Fig. VII.9: Influence of w/s on phase changes. (1) w/s=1, (2) w/s=5, (3) w/s=20 (*24*).

mix containing $CaO–SiO_2$ gel in ratio 2:1 was treated with ultrasonic waves of frequency 2 Mc for 30-40 min, and placed in an autoclave at 180°C for 250 hr, and the resulting product compared with an untreated mix. DTA of the treated sample showed intense peaks corresponding to the formation of well crystallized hillebrandite.

It is generally believed that the binding matrix in autoclaved products is a platy form of CSH. However, according to Cole and Moorehead (*25*) the high strengths in pressed and autoclaved specimens may be caused more by physical than chemical bonds. They compared cast lime–silica specimens with those pressure-molded at 40,000 lb/in.^{2} and autoclaved at different temperatures from 175 to 220°C for periods ranging from 1 to 12 hr.

The autoclaved specimens were subjected to X-ray, chemical, and electron microscopic examinations. The DTA curves of cast and pressed specimens at 185°C for ¾, 5¾, and 12 hr showed an exothermic reaction at about 850°C for the presence of fibrous calcium silicate hydrate except in the pressed specimen cured for ¾ hr. The low temperature endothermic effect at about 200°C, especially in the paste specimens, was caused by the tobermoritic mineral.

The unreacted quartz and calcium hydroxide showed endothermal effects at about 575 and 480°C respectively. The endotherm at 780°C was attributed to the well crystallized $CaCO_3$, while that in the range 600-750°C was attributed to the poorly crystallized calcite. An exothermic peak at about 325°C may possibly have been due to a delayed transition. The high temperature exothermic peak provided an idea of the CaO/SiO_2 ratio of the specimens.

The exothermic peak at 820°C for the sample cured at ¾ hr corresponded to a compound with a CaO/SiO_2 ratio of about 1.2. The specimens cured for 12 hr exhibited the exothermic peak at a lower temperature of 790°C that could be ascribed to a compound of CaO/SiO_2 ratio 0.9. Curing at longer periods increased the proportion of platy to fibrous crystals, but a decrease in strength was also observed beyond 8 hr.

These observations led Cole and Moorehead (*25*) to conclude that the formation of platy tobermorite may only partly explain the strength development in autoclaved products. Since the conclusions of these workers are essentially based on the specimens formed at high molding pressures not normally used by others, it may be questioned whether they are valid for specimens molded at lower pressure ranges.

In the $CaO–SiO_2–H_2O$ system containing small amounts of impurities it has been observed that the X-ray diffraction patterns do not alter, though the DTA results indicate variation in thermal behaviors. The DTA is sensitive to additions of even 1% Al_2O_3 to low lime hydrous calcium silicates, though the X-ray diffraction may not indicate any change (*26*). This shows that additions of small amounts of Al_2O_3 and Fe_2O_3 modify the nature of the hydrous calcium silicates.

In Fig. VII.10 thermograms of hydrothermal products containing CaO/SiO_2 ratio 0.8 and Al_2O_3 from 0.2 to 20.7% are shown. Curve 1 refers to the thermogram of 0.8-1.0 CaO/SiO_2 tobermorite. In the presence of 5.7% Al_2O_3 the sample exhibits a strong exothermic peak at 860°C. Increasing amounts of Al_2O_3 shift the exothermic peak to a higher temperature. This effect is attributed to the progressive substitution of Al_2O_3 in the lattice of tobermorite. At higher Al_2O_3 contents the formation of C_3ASH_4 is established by an endothermal effect at 320 to 360°C.

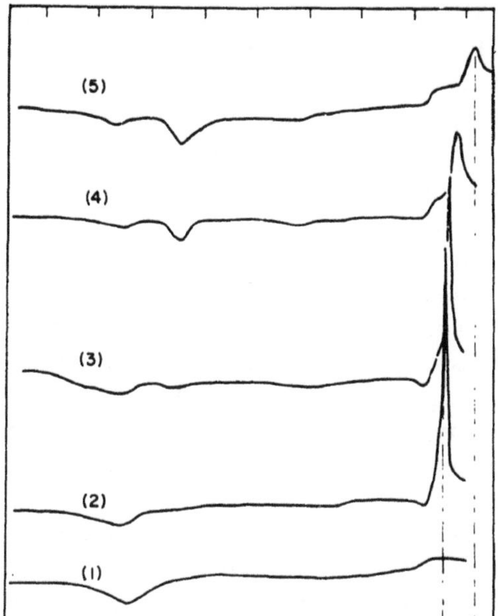

Fig. VII.10: DTA of 0.8 C/(S+A) reaction products. (1) pure tobermorite; (2), (3) tobermorite and C_3ASH_4 with 5.7 and 9.3% combined Al_2O_3 respectively; (4), (5) products made from mixtures with 14.4% and 20.7% Al_2O_3 (*26*).

Incorporation of certain accelerators may increase the strength of calcium silicate products. Addition of about 1% sodium chloride enhances the strength of autoclaved products containing a CaO/SiO_2 ratio of 0.5 (*27*). The DTA of untreated sample shows the dehydration peak at 500°C for the presence of $Ca(OH)_2$, while the sample treated with NaCl exhibits an endothermic effect at about 200°C for the presence of calcium silicate hydrate. The NaCl-treated samples do not indicate the presence of an endothermic effect for $Ca(OH)_2$.

The addition of clay confers a beneficial effect on calcium silicate bricks. It is deliberately added to make the pressing easier. Under hydrothermal conditions clays may react with lime to form cementitious compounds (*58*). A hydrothermal treatment of lime with kaolin or bentonite at 8 atm for 4 to 10 hr results in the formation of calcium hydrosilicate and several calcium hydroaluminates with endothermal effects at 183, 784, and 910°C. Evidence has also been obtained through DTA for C_4AH_x in the hydrothermally treated calcined kaolin or bentonite with lime (*28*).

The system $CaO–Al_2O_3–H_2O$ has been studied at 100-1000°C at water pressures up to 3000 atm and the stable hydrates reported are $Ca(OH)_2$, C_3AH_6, and $C_4A_3H_3$ (*29*). The DTA of C_3AH_6 shows two peaks for the stepwise dehydration of 4.5 and 1.5 molecules of water. The $C_4A_3H_3$ exhibits an endothermic peak at about 700°C.

Many other workers have applied DTA for the study of hydrothermal products involving lime and silica under different conditions (*30-41*, *64*, *65*).

Individual Cement Compounds. The chemistry of the reactions involving cement and aggregates is complex. A study of the reaction of individual cement compounds such as C_3S, β-C_2S, and C_3A, or the ferrite phase alone or in presence of various types of aggregates under hydrothermal conditions, should provide a better understanding of the chemical processes occurring in the hydration of cement.

Tricalcium silicate converts to hydrated silicates under autoclave treatment at a faster rate than β-C_2S. The strength of C_3S paste autoclaved at 8 atm is almost double that of C_2S paste treated under similar conditions. The technique of DTA could be used to distinguish the hydrated calcium silicate products obtained with C_3S or C_2S. Higher strengths generally correspond to increased intensities of the DTA peaks and increased amounts of the hydrated calcium silicate with CaO/SiO_2 ratio 0.9 (*42*).

The products obtained in the C_3S–quartz system need not be exactly similar to those formed without the aggregate. A C_3S specimen treated at 16 atm for 8 hr may contain α-C_2S hydrate, whereas in combination with quartz, calcium silicate hydrate with CaO/SiO_2=0.9 may form in addition (*43*).

The α-C_2S hydrate is characterized by an endothermic peak at about 450°C

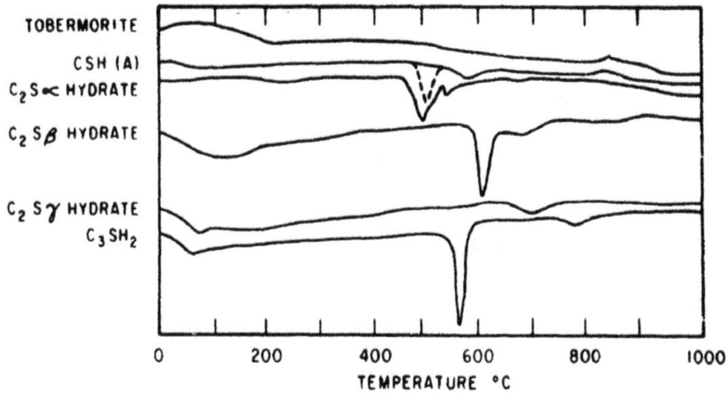

Fig. VII.11: Differential thermograms of various calcium silicate hydrates (*40*) [Crown Copyright, reproduced by permission of Her Britannic Majesty's Stationery Office].

and the calcium silicate hydrate of CaO/SiO_2 ratio 0.9 by an exothermic peak at about 800°C. It has also been observed that with an initial mix containing a CaO/SiO_2 molar ratio of 3 prepared with different starting materials the products are not the same (*40*). At an autoclaving temperature of 180°C, if lime and silica are the starting materials, α-C_2S hydrate forms initially and then converts to β-C_2S hydrate; with β-C_2S and lime as the starting materials α-C_2S hydrate is formed.

A mixture of calcium silicate hydrate (CSH) and lime gives Flint's CSH(A) and α-C_2S hydrate. C_3S is converted to $C_3S \cdot 1\frac{1}{2}H_2O$. Typical thermograms of some of the hydration products are given in Fig. VII.11.

A systematic investigation of the hydration of β-C_2S alone or in presence of quartz of different specific surface areas, silica gel, $Ca(OH)_2$, or C_3S has been reported (*44*). The DTA of β-C_2S cured at 175°C for periods from 2 hr to 7 days showed a considerable quantity of α-C_2S hydrate at 8 hr, as shown by two endothermic peaks, at about 350° and 400-450°C.

In mixtures containing 35% β-C_2S and 65% quartz of specific surface 3100 cm^2/g an endothermic effect at 575°C representing $\alpha \rightleftharpoons \beta$ quartz transition could be seen for periods up to 72 hr. An exothermal peak appeared just above 800°C at 16 hr, denoting CSH(I), and the intensity of this peak increased at longer curing periods. The DTA also indicated a slow down in the reaction when quartz with a specific surface of 4750 cm^2/g was used.

Further experiments with different quartz fractions showed that with particles in the size range 0.3-0.5 mm large amounts of α-C_2S hydrate along with tobermorite formed. Increased amounts of tobermorite were detected by using the quartz fraction of size $<40\ \mu$. In DTA, mixes with coarser particles showed an exothermal peak of large intensity while those with finer fractions failed to exhibit a peak.

If $Ca(OH)_2$ or C_3S were added to β-C_2S–quartz mixtures the hydration products were CSH(I) and tobermorite. In β-C_2S pastes cured at 180°C for periods up to 90 days Midgley and Chopra (*40*) found γ-C_2S hydrate but with added lime α-C_2S hydrate formed preferentially. The effect of SiO_2 or CaO on the hydrothermal reaction of β-C_2S at pressures of 100 atm has been reported by Bozhenov *et al* (*45*). β-C_2S reacts with SiO_2 and CaO additives and exhibits an endothermic peak at 100-200°C due to the expulsion of adsorbed water and the decomposition of phase X.

An exothermic peak at 780-840°C with SiO_2 additions increases in intensity with the amount of SiO_2. Additions of CaO to β-C_2S shows the 520-580°C peak for $Ca(OH)_2$ decomposition and another endothermic peak at 720-730°C (*46*).

The effect of prolonged treatment of C_3S or C_2S and their mixtures with C_3A, C_4AF, and C_2F at 16 atm for 100 hr has been studied by DTA (*47*). C_3A yields C_3AH_6 and C_3S converts mainly to C_3SH_2 and C_2SH accompanied by an endothermic peak at 730°C. Addition of small amounts of silicates to C_3A results in the formation of hydrogarnet. Failure to observe the hydrogarnet peak at 520-540°C in steam-cured Portland cements is ascribed to the superposition of $Ca(OH)_2$ peak. Addition of C_2F to the silicates prevents the appearance of the endothermic peak for the formation of hydrogarnet.

Hydrothermal reactions in calcium aluminate and tetracalcium aluminoferrite have been reported by Krylov and Kantsepol'skii using DTA (*48*). DTA and microscopic methods have been applied by Tikhonov and Shpynova (*49*) to differentiate the normally cured products of hydration of cement minerals from those cured in steam.

Synthesis of various hydrated compounds such as tobermorite, C_6S_6H, C_2SH_2, C_2SH, and C_3SH_2 was accomplished at temperatures up to 300°C and 100 atm by Butt and Rashkovich (*50*). These compounds were also examined by DTA.

Cement and Cement Aggregates. The DTA studies of the hydrothermal products of neat cement pastes have been reported by Kalousek (*9*) and Bozhenov *et al* (*51*, *52*). Certain trends in the variation of the intensity of endothermic peaks have been observed by Bozhenov in Portland cement subjected to increasing steam pressures.

Basically, Portland cement exhibits five endothermic effects—the first four at 120-135°C, 200-210°C, 420-430°C, and 565-580°C. With increasing pressure the low temperature endothermic effect at 120-130°C decreases and is

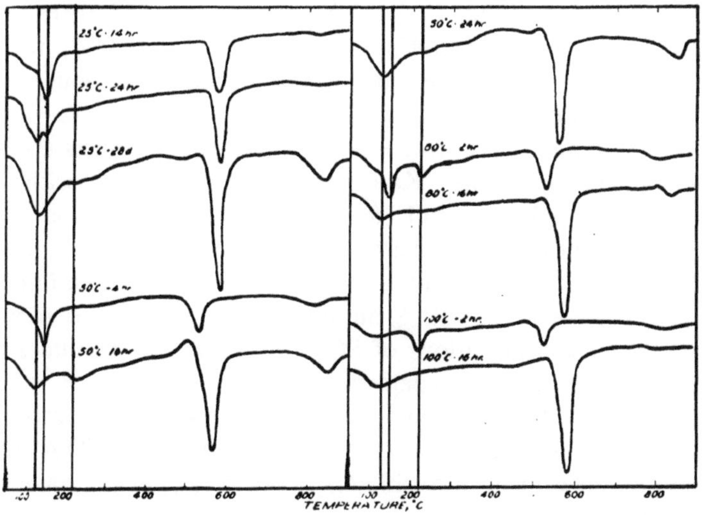

Fig. VII.12: Thermal analysis curves showing formation and transformations of SO_3-bearing products in hydrating cements at different temperatures (*9*).

completely absent at 200 atm, and the effect at 200-210°C representing ettringite also disappears. The endothermic peak for $Ca(OH)_2$ decomposition also decreases with pressure but that at 600-800°C increases.

The steam-cured cement exhibits additional endothermic peaks at 700-730° and 890-910°C, corresponding to the development of new hydrated compounds.

Kalousek (*9*) found that the thermograms of cements cured up to 100°C were similar to those formed at 25°C. In Fig. VII.12 the thermal curves of the cement pastes cured at 25°C are compared with those obtained at 50, 80, and 100°C for different periods. The paste cured at 25°C for 14 hr shows an endothermic dip at 150°C corresponding to the formation of the high calcium sulfoaluminate, the intensity of which decreases at higher autoclaving periods.

An endothermic peak observed at 130-140°C in the 24-hr sample is attributed to phase X (a transformation product of sulfoaluminate and sulfoferrites) and that at 220°C to the solid solution $3CaO \cdot Al_2O_3 \cdot CaSO_4 \cdot 12H_2O$–$3CaO \cdot Al_2O_3 \cdot Ca(OH)_2 \cdot 12H_2O$ or to the corresponding Fe_2O_3-containing products. At 50°C the trisulfate forms in larger amounts even at 4 hr, but at 16 hr phase X and ettringite are identified by the two endothermic peaks at about 130-140° and 220°C respectively. Phase X is the major product at 24 hr. An increase in the rate of reaction is noticed at 80°C.

The sample cured at 100°C for 2 hr exhibits an endothermic effect at about 220°C for the solid solution and another above 100°C for phase X. The DTA results show that the transformations are in the sequence trisulfate form→solid solution→phase X.

The formation and transformation of SO_3-bearing products are affected if gypsum is added to cement. Kalousek (*9*) mixed gypsum with cement in quantities of 2.75, 5.5, 11.0, and 16.5% (calculated as SO_3) and after curing them at 70 and 100°C studied the products by DTA. Phase X was detected within 24 hr in pastes containing 2.75% gypsum and the trisulfate form was stabilized to some extent at 5.5% addition. The presence of trisulfate was evident at higher concentrations of SO_3. However, the trisulfate form was not stabilized at a curing temperature of 100°C.

The products of hydrothermal curing of cement-aggregate mixes are not the same as those formed from neat cement pastes. The DTA of the products of Portland cement and different types of aggregates have also been reported (*16*, *18*, *53*, *54*).

In the system cement–lime–silica, for the product of a particular composition the thermograms show a similarity irrespective of whether the starting material is lime-bearing, lime, cement, or their mixtures. The compositions corresponding to 0.9-1.3 CaO/SiO_2 show an exothermic effect at 840-860°C and those with higher CaO/SiO_2 ratios show a valley in the range 400-500°C and depressed exothermic peaks. Prolonged hydrothermal treatment promotes the formation of α-type hydrate.

In pumice–cement mixes the sharpness of the exothermic peak decreases at higher CaO/SiO_2 contents. The presence of pumice seems to stabilize the lime-rich solids while preventing the formation of α-type hydrate. For any comparable composition the expanded shale-based products show the second exothermic peak at higher temperatures. Mixes with expanded shale fines exhibit an exothermal hump at 300-550°C in 2.1 CaO/SiO_2 products. The specimens with expanded slag fines have a flattened exothermal effect.

DTA yields typical thermograms to differentiate products of varying compositions prepared from a particular aggregate, though X-ray results are very similar.

It is possible by a careful examination of the thermal curves to explain the differences in compressive strengths of autoclaved cement–silica–water products containing quartz in amounts of 2-60% (*9*). Menzel (*55*) steam-cured cement–quartz mixtures at 210 and 182°C and plotted a curve showing the effect of the composition of raw mix on the compressive strength (Fig. VII.13).

The mix with 8-10% SiO_2 shows a minimum strength. With higher silica additions the compressive strength increases progressively and reaches a

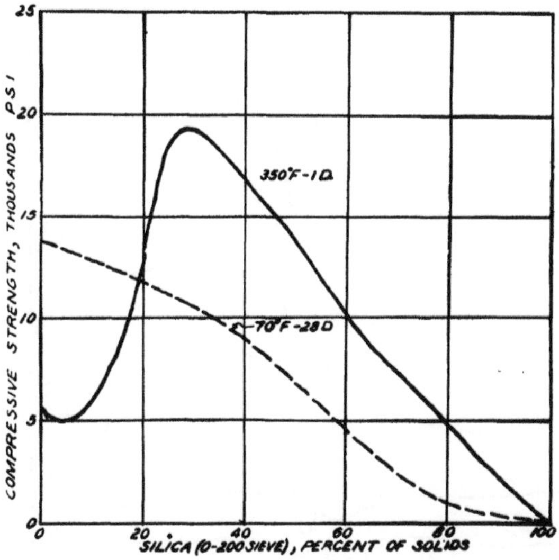

Fig. VII.13: Relation between compressive strength and amount of ground quartz by weight of cement and quartz (*9*).

maximum at a concentration of about 30% SiO_2. Further additions of silica result in a decrease of the strength. The thermograms are shown of cement–silica mixes containing different amounts of silica (Fig. VII.14).

The endothermic effect at about 560°C due to the dehydration of $Ca(OH)_2$ is absent in the specimen containing 9% SiO_2. Another endothermic effect at 470-480°C denoting α-C_2S hydrate is present in increasing proportions up to 9% SiO_2, decreasing thereafter and disappearing altogether at 30% SiO_2. An exothermic effect is evident at 800-810°C in the mixtures with 12% SiO_2 and is most intense at 44% SiO_2.

The DTA demonstrates that at 9% SiO_2 and beyond no $Ca(OH)_2$ is obtained as a reaction product. The maximum amount of α-C_2S hydrate at 9% SiO_2 corresponds to lower compressive strengths. Increased strengths above 9% SiO_2 additions are ascribed to the formation of lower amounts of α-C_2S hydrate and increased amounts of calcium silicate hydrate.

The calcium silicate hydrate is characterized by an exothermic peak at about 840°C. The maximum intensity of the exothermic peak at 840°C for the 40-45% SiO_2 mix coincides approximately with higher strengths.

Chebukov and Egorova (*54*) studied the DTA of products obtained by steam curing or exposing to natural atmospheric conditions a mixture of slag Portland cement, lime-ash cement, and unground ash in the ratio 25:30:45. Under normal curing at 2 days the DTA showed a small endothermic effect at 190-200°C due to the presence of calcium hydroaluminates. An endother-

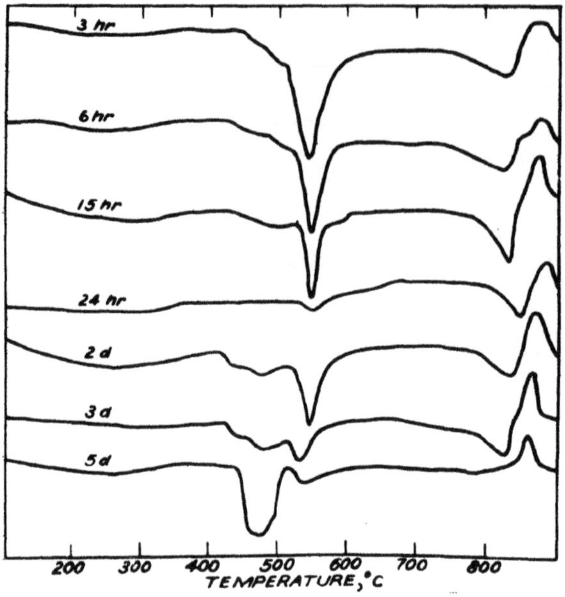

Fig. VII.14: Thermal analysis of cement–silica mixtures processed at 175°C for 24 hr (*9*).

mic effect at 470-500°C indicated the decomposition of $Ca(OH)_2$ and another endothermic effect at 780-800°C, the presence of calcium hydrosilicates.

The high temperature endothermic effect increased in intensity at 28 days of curing. Specimens steam cured at 98-100°C for 8 hr and subsequently stored in air for 2, 7, 28, or 540 days showed somewhat different thermal behavior. The endothermic effects in the range 450-500°C were attributed to hydrogarnets and calcium hydrosilicates of lower CaO/SiO_2 ratios. The endothermal effects at 680-710°C represented the decomposition of calcium hydrosilicates.

The specimen aged for 2 or 7 days showed the endothermic peak at 150-170°C corresponding to ettringite, while that aged for 540 days showed an endothermic peak at 190-210°C for the presence of low calcium sulfoaluminate. Some specimens of steam cured foam concrete were also subjected to DTA examination.

The development of products in cement–asbestos mixes is different from that of pure cement. Kalousek and co-workers (*53*) studied the binders in 24 samples comprising five laboratory-prepared, ten autoclaved, and nine normal-cured commercial asbestos–cement specimens. The five laboratory prepared samples had cement, silica, asbestos in the ratio 75:25:0, 62.5:37.5:0, 50:50:0, 50:30:20, and 50:30:20. They were either prepared as cast pastes or as pressed cakes after suspension (water/solid ratio of 6.0) and filtration.

In the DTA, endotherms were observed at 430, 475, and 550°C corresponding to $Mg(OH)_2$, α-C_2SH, and $Ca(OH)_2$ respectively. An exothermic peak

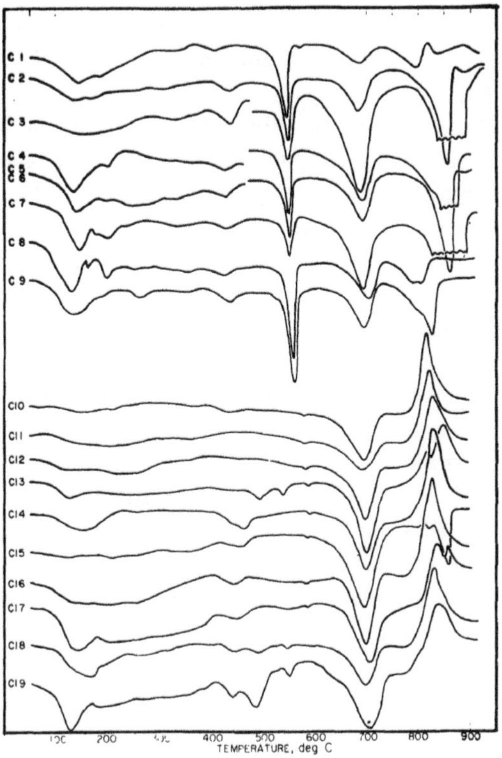

Fig. VII.15: DTA tracings at 50% sensitivity from 50-500°C and 600-950°C, and 25% sensitivity from 500-600°C on normal-cured commercial products and at 50% sensitivity on autoclaved products (*53*).

was also exhibited by the specimens in the range 830-880°C indicating tobermorite. The intensity of the peak increased with the period of autoclaving from 4 to 46 hr. The endothermic areas for $Ca(OH)_2$ decomposition decreased with curing time.

The filtered and pressed cakes contained better crystallized tobermorite and exhibited sharper exotherms than the pastes. At higher W/C ratios or in presence of asbestos the exotherms were more intense, but the endotherms corresponding to α-C_2SH decreased in amplitude.

The DTA of normal-cured commercial samples (C_1 to C_9) and the autoclaved samples (C_{10}-C_{19}) is shown in Fig. VII.15. The normal-cured specimens do not contain ettringite as they do not exhibit the characteristic endothermic peak at 170°C. Small amounts of either or both monosulfate ($C_3A \cdot CaSO_4 \cdot 12H_2O$) and C_4AH_{19} are indicated in samples C_4, C_5, C_6, C_7, and C_8. An endothermic effect at 270°C is noticed only in sample C_9. The peak at 700°C is ascribed to chrysotile asbestos.

Calcium carbonate is present in substantial amounts in samples C_3, C_4, C_5, and C_7. In the autoclaved samples, C_{13} and C_{18} have some $Ca(OH)_2$ and α-C_2SH, and C_{19} contains moderate quantities of α-C_2SH.

Some samples show quartz by an endothermic effect at 575°C. All the autoclaved specimens except C_{15} show an exotherm at 875°C for the presence of tobermorite. In C_{15} the exotherm is perhaps masked by the endothermic effect due to $CaCO_3$ decomposition.

DTA was used to estimate uncombined $Ca(OH)_2$ in the asbestos–cement samples and the results were compared with the method involving extraction with glycerol-ethanol solvent. In Table VII.3 the estimated values of uncombined $Ca(OH)_2$ obtained by these methods are compared.

At increased periods of autoclaving the amount of uncombined $Ca(OH)_2$ decreases. In almost all samples DTA showed very little or no $Ca(OH)_2$. In commercial autoclaved specimens, all except three samples showed little $Ca(OH)_2$. Normal-cured specimens had more $Ca(OH)_2$, and the solvent method showed higher values.

These data have shown that the glycerol–ethanol may extract adsorbed $Ca(OH)_2$ as well as partially decompose the hydrated calcium silicate.

Magnesium Hydrosilicates. Magnesium hydrosilicates such as talc, antigorite, and chrysotile have been synthesized by curing at high temperatures and pressures. By a hydrothermal treatment of the dehydrated natural magnesium hydrosilicate mineral, the original mineral can be reformed, giving it binding properties (*45*). For example, a dehydrated chrysotile asbestos mineral at 800°C shows a compressive strength of 6 kg/cm² cured under normal conditions, while an autoclaved sample of the dehydrated mineral shows a compressive strength of 108 kg/cm². It has also been observed (*56*) that

TABLE VII.3
Uncombined $Ca(OH)_2$ Determined by DTA and ASTM Method C 500 [Kalousek et al (53)]

Laboratory Samples				Commercial Products			
		$Ca(OH)_2$, %				$Ca(OH)_2$, %	
Mix No.	*Cure time, hr*	*Method C500*	*DTA*	*Product No.*	*Cure*	*Method C500*	*DTA*
1	*2*	*3*	*4*	*5*	*6*	*7*	*8*
L1	46	0.75	0.00	C1	normal	8.3	7.0
L1	16	1.23	0.43	C2	normal	8.1	6.1
L1	12	4.50	2.00	C3	normal	7.2	7.1
L2	46	0.34	0.00	C4	normal	8.1	6.6
L2	16	0.69	0.09	C5	normal	7.0	5.5
L2	12	0.82	0.00	C6	normal	7.3	4.5
L2	8	2.20	1.60	C7	normal	5.5	4.4
L3	46	0.28	0.09	C8	normal	10.1	7.6
L3	16	0.44	0.07	C9	normal	11.2	8.2
L3	12	0.64	0.00	C10	autoclave	0.25	0.00
L3	8	1.00	0.32	C11	autoclave	0.18	0.00
L4	46	0.41	0.00	C12	autoclave	0.34	0.00
L4	16	0.45	0.00	C13	autoclave	0.77	0.48
L4	12	0.51	0.09	C14	autoclave	0.31	0.00
L4	8	0.63	0.12	C15	autoclave	0.07	0.00
L5	46	0.24	0.09	C16	autoclave	0.34	0.00
L5	16	0.58	0.07	C17	autoclave	0.54	0.00
L5	12	0.82	0.00	C18	autoclave	0.75	0.31
L5	8	0.88	0.32	C19	autoclave	1.50	0.74

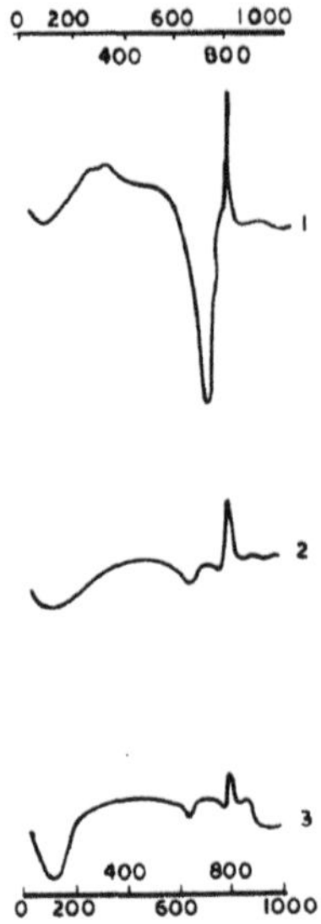

Fig. VII.16: DTA of chrysotile asbestos (*45*).

asbestos wastes of the asbestos–cement industries could be cured at 8 atm to achieve in the asbestos–magnesium slabs a strength of 310 kg/cm.²

In Fig. VII.16, curve 1 refers to the DTA behavior of untreated chrysotile asbestos. The sample dehydrated at 800°C and steam cured at 100 atm has the thermogram with characteristics similar to that of the untreated sample. The dehydrated mineral at 900°C subjected to autoclave treatment (curve 3) also exhibits similar characteristics to the original chrysotile but with diminished peak intensities.

Butt and Rashkovich (*57*) studied the composition of the products formed in the system MgO–quartz–H_2O under hydrothermal conditions. The starting material containing 3-70% MgO was subjected to temperatures of 150, 175, and 204°C from a few hours to 11 days. The thermograms of all the products were very similar but the intensities of the peaks differed. The composition of the products varied from 1.1 $MgO \cdot SiO_2 \cdot 0.85H_2O$ to $1.5MgO \cdot SiO_2 \cdot 1.2H_2O$. The DTA work of Kalousek and Mui (*59*), Yang (*60*) and Sabatier (*61*, *62*) also deals with $MgO \cdot SiO_2$–H_2O system.

REFERENCES

1. Bessey, G.E., '*Calcium Silicate Products in the U.S.S.R.*,' *Progress in Applied Chemistry*, Vol. XLVIII, Society of Chemical Industry, London, p. 627, 1963.
2. Taylor, W.H., Cole, W.F., and Moorehead, D.R., *Nature*, *201*, 918, 1964.
3. Taylor, H.F.W., *The Chemistry of Cements*, Academic Press, London and New York, Vol. 2, p. 105, 1964.
4. Buckle, E.R., and Taylor, H.F.W., *J. Appl. Chem.* (*London*), *9*, 163, 1959.
5. Funk, H., *Z. Anorg. Allgem. Chem.*, *291*, 276, 1957.
6. Funk, H., *Z. Anorg. Allgem. Chem.*, *297*, 103, 1958.
7. Speakman, K., and Taylor, H.F.W., *Intern. Symp. Autoclaved Calcium Silicate Building Products*, London, 1965.
8. Johnson, H., and Thorvaldson, T., *Can. J. Res.* (*B*)., *21*, 236, 1943.
9. Kalousek, G.L., and Adams, M., *J. Am. Concrete Inst.*, *48*, 77, 1951.
10. Kalousek, G.L., *J. Am. Concrete Inst.*, *51*, 989, 1955.
11. Kalousek, G.L., and Prebus, A.F., *J. Am. Ceram. Soc.*, *41*, 124, 1958.
12. Taylor, H.F.W., *Proc. 4th Intern. Symp. Chem. Cements*, Washington, pp. 167-190, 1962.
13. Taylor, H.F.W., *Intern. Symp. Autoclaved Calcium Silicate Building Products*, London, 1965.
14. Aitken, A., and Taylor, H.F.W., *J. Appl. Chem.* (*London*), *10*, 7, 1960.
15. Taylor, H.F.W., *Progress in Ceramic Science*, Pergamon Press, London, p. 138, 1961.
16. Kalousek, G.L., *Proc. 3rd Intern. Symp. Chem. Cements*, London, p. 334, 1952.
17. Kalousek, G.L., Davis, C.W., and Schmertz, W.E., *J. Am. Concrete Inst.*, *45*, 693, 1949.
18. Kalousek, G.L., *J. Am. Concrete Inst.*, *Proc.*, *50*, 365, 1954.
19. Sanders, L.D., and Smothers, W.J., *J. Am. Concrete Inst. Proc.*, *54*, 127, 1957.
20. Neese, H., Spangenberg, K., and Weiskirchner, W., *Tonind.-Z. Keram. Rundschau*, *81*, 325, 1957.
21. Peppler, R.B., *J. Res. Nat. Bur. Stad.*, *54*, 205, 1955.
22. Kalousek, G.L., Logiudice, J.S., and Dodson, V.H., *J. Am. Ceram. Soc.*, *37*, 7, 1954.
23. Midgley, H.G., and Chopra, S.K., *Mag. Concrete Res.*, *12*, 73, 1960.
24. Tun Soo-Lee and Loo Gon-Chen, *Proc. 7th Conf. Silicate Ind.*, *Publishing House of Hungarian Acad. Sci.*, Budapest, pp. 293-310, 1965.
25. Cole, W.F., and Moorehead, D.R., *Intern. Symp. Autoclaved Calcium Silicate Building Products*, London, 1965.
26. Kalousek, G.L., *J. Am. Ceram. Soc.*, *40*, 74, 1957.
27. Butt, Y.M., and Krzheminskii, S.A., *Dokl. Akad. Nauk SSSR*, *89*, 709, 1953.
28. Budnikov, P.P., *Proc. Symp. Chem. Cem. Sci. Tech. Soc. Industry of Bldg. Mat.*, Moscow, p. 294, 1956.
29. Majumdar, A.J., and Rustum Roy, *J. Am. Ceram. Soc.*, *39*, 434, 1956.
30. Butt, Y.M., and Parimbetov, B., *Sbornik Trudov Resp. Nauch-Issledovatel Inst. Mestnykh Stroitel. Materialov*, No. 8, 41, 1954.
31. Butt, Y.M., and Maier, A.A., *Ibid*, No. 13, 35, 1957.

32. Toropov, N.A., Nikogosyan, K.S., and Boikova, A.I., *Dokl. Mezvuz Konf. po. Izuchen Autoklavn. Materialov. i ikh Primenen v. Stroitel'stve*, Leningrad, p. 92, 1959.
33. Sauman, Z., *Silikaty*, *6*, 149, 1962.
34. Shigeo Akaiwa and Giichi Sudo, *Semento Gijutsu Nenpo*, *10*, 14, 1957.
35. Boikova, A.I., Nikogosyan, K.S., and Toropov, N.A., *Tr. Shestogo Sovesch po Eksperin i Tekhn. Mineralog i Petrogr., Akad. Nauk SSSR, Inst. Geol. Rudn. Mestorozhd., Petrogr., Mineralog i Geokhim., Inst. Khim. Silikatov*, Leningrad, p. 181, 1961.
36. Barrer, R.M., and Denny, P.J., *J. Chem. Soc.*, 983, 1961.
37. Gaze, R., and Robertson, R.H.S., *Mag. Concrete Res.*, *8*, 7, 1956.
38. Toropov, N.A., and Boikova, A.I., *Cement-Wapno-Gips*, *7*, 203, 1961.
39. Greenberg, S.A., *J. Phys. Chem.*, *58*, 362, 1954.
40. Midgley, H.G., and Chopra, S.K., *Mag. Concrete Res.*, *12*, 19, 1960.
41. Neese, H., *Tonind.-Z Keram. Rundschau*, *83*, 124, 1959.
42. Butt, Y.M., and Rashkovich, L.N., *Tsement*, *22*, 21, 1956.
43. Butt, Y.M., Rashkovich, L.N., and Danilova, S.G., *Dokl. Akad. Nauk SSSR*, *107*, 571, 1956.
44. Sauman, Z., *Intern. Symp. Autoclaved Calcium Silicate Building Products*, London, 1965.
45. Bozhenov, P.I., Kavalerova, V.I., Salnikova, V.S., and Sovorova, G.F., *Proc. 4th Intern. Symp. Chem. Cements*, Washington, p. 327, 1962.
46. Bozhenov, P.I., and Kavalerova, V.I., *Dokl. Mezhvuz. Konf. po Izuch. Autoklavn. Materialov i ikh Primeneniya v Stoit*, Leningrad, p. 187, 1959.
47. Butt, Y.M., and Rashkovich, L.N., and Volkov, V.V., *Izvest. Vysshykh Uchele Zavedenii, Khim. i Khim. Teknol.*, No. 3, p. 130-5, 1958.
48. Krylov, G.M., and Kantseepol'skii, I.S., *Uzbek. Khim. Zh. Akad. Nauk Uzbek, SSR*, No. 3, p. 37, 1958.
49. Tikhonov, V.A., and Shpynova, L.G., *Dokl. Mezhvuz. Konf. po Izuch. Autoklavn. Materialov i ikh Primeneniya v Stroit*, Leningrad, p. 102, 1959.
50. Butt, Y.M., and Rashkovich, L.N., *Sbornik Trudov Resp. Nauk-Issledovatel. Inst. Mestnykh Stroitel. Materialov*, No. 15, 13, 1959.
51. Bozhenov, P.I., and Suvorova, G.F., *Proc. Symp. Chem. Cements. Sci. Tech. Soc. Industry of Bldg. Materials*, Moscow, pp. 341-50, 1956.
52. Bozhenov, P.I., and Sovorova, G.F., *Tsement*, *23*, 8, 1957.
53. Kalousek, G.L., Camarda, F.V., Kopanda, J.E., and Jugovic, Z.T., *Mater. Res. Std.*, *6*, 169, 1966.
54. Chebukov, M.F., and Egorova, A.M., *Zh. Prikl. Khim.*, *37*, 225, 1964.
55. Menzel, C.A., *J. Am. Concrete Inst.*, *31*, 125, 1934.
56. Bozhenov, P.I., and Salnikova, V.S., *Coll. of Sci. Works on the Chem. and Tech. of Silicates, Promstroii Zdat.* 1956.
57. Butt, Y.M., and Rashkovich, L.N., *Akad. Nauk SSSR, Inst. Khim. Silikatov*, Leningrad, p. 322-9, 1956.
58. Butt, Y.M., and Parimbetov, B., *Sbornik Trudov Resp. Nauch-Issledovatel Inst. Mestnykh Stroitel Materialov*, No. 9, 95, 1955.
59. Kalousek, G.L., and Mui, D., *J. Am. Ceram. Soc.*, *37*, 38, 1954.
60. Yang, J., *J. Am. Ceram. Soc.*, *43*, 542, 1960.
61. Sabatier, G., *Compt. Rend.*, *230*, 1962, 1950.
62. Sabatier, G., *Bull. Soc. Franc. Mineral.*, *77*, 953, 1077, 1954.

63. Nikogosyan, K.S., *Khim. i Prakt Primenenie Silikatov, Inst. Khim. Silikatov. Akad. Nauk SSSR*, p. 107, 1960.
64. Vydrevich, E.Z., and Gal'perin, E.L., *Zh. Prkl. Khim.*, *34*, 1971, 1961.
65. Berkovich, T.M., Kheiker, D.M., Gracheva, O.I., Zerin, L.S., and Kupreeva, N.I., *Dokl. Akad. Nauk SSSR*, *120*, 853, 1958.

VIII. Non-Portland Cements Based on Silicates and Aluminates

VIII.1. SLAGS AND SLAG CEMENTS

In the manufacture of pig iron, limestone is used as a flux and is fed into the blast furnace along with coke and iron ore. Blast-furnace slag is formed by the combination of limestone flux with coke ash and aluminous and siliceous constituents of iron ore.

The slag is treated variously to obtain 'granulated slag,' 'air-cooled slag,' 'foamed glass,' or 'slag wool.' Quick cooling of the slag as it comes out of the blast furnace produces a light, porous product known as granulated slag. A hard and dense material formed on a slow cooling of the slag is known as air-cooled or 'lump' slag. Foamed slag is a very light material formed by a quick chilling of the slag with an amount of water just enough to entrap the steam in the mass. A very rapid cooling of the slag by a jet of air or a mixture of air and steam results in threads, which felt together to make slag wool.

Uses

Granulated slag finds application in making slag cement, Portland blast-furnace cement, and supersulfate cement. Slag cement is a mixture of granulated slag and hydrated lime, with added salts to accelerate the set. Portland blast-furnace cement is composed of an interground mixture of Portland cement clinker and granulated slag. The International Standards Organization describes supersulfate cement as a mixture consisting of a minimum of 75% blast-furnace slag, calcium sulfate, and a maximum of 5% lime, Portland cement clinker, or Portland cement.

The slag alone or in admixtures finds wide applications. Air-cooled slag is used as a stone in road construction, as an aggregate in concretes, and as a raw material in the manufacture of Portland cement. Foamed slag makes a good lightweight aggregate. Insulating materials are made from slag wool. Granulated slag is one of the constituents in expanding cements. The Portland blast-furnace cement conforms to the specifications of Portland cement. Slag cements have a good plasticity and are resistant to sulfate attack. Supersulfate cements can withstand severe conditions of exposure to sulfates and are used in underground works.

Chemistry

A chemical analysis of blast-furnace slag shows that it contains mainly CaO, SiO_2, Al_2O_3, and MgO, with small proportions of sulfide and ferrous and manganous oxides.

Granulated slags are glassy as they are formed by quick chilling, and hence are considered as supercooled liquids with considerable internal viscosity. There is controversy on the nature of the glassy phase. Some investigators have considered the slag as forming a random network, while others have used the crystallite theory to define the glass. The air-cooled slag contains crystalline materials, several of which have been identified.

The major mineral present in air-cooled slags is melilite, an isomorphous solid solution occurring between the end members gehlenite (C_2AS) and akermanite (C_2MS_2). Akermanite predominates at higher MgO contents. Other minerals present in smaller quantities include C_2S (α', β, and γ-forms), CS, C_3S_2, C_3MS_2, MA, CMS_2, CMS, CAS_2, and M_2S. Only the α, and β-forms of C_2S are known to be reactive towards water.

Granulated slag alone does not show any setting property with water unless an activator such as lime, Portland cement, alkali, or alkali sulfate is present. The activator is believed to open up the silicate structure in the glassy slag.

Kramer (*1*) considers that lime, in addition to releasing the latent hydraulicity of the slag, acts as a pozzolana. According to D'Ans and Eick (*2*) lime prevents the formation of a water-impervious layer of acidic gels over the grains, thus facilitating the opening up of SiO_2 framework of the glass.

The usefulness of slag depends on its hydraulicity. A number of methods have been suggested for evaluating the potential activity of slags. Keil (*3*) has reviewed the methods based on chemical indexes. Other methods include methylene blue staining, reaction with aluminum sulfate, and determination of the refractive index or fluorescence in ultraviolet light.

The hydration products of granulated slag activated with lime or Portland cement may consist of a tobermorite-like phase, ettringite, a hexagonal calcium aluminate hydrate, or a solid solution or an intergrowth with low calcium sulfoaluminate hydrate. The formation of $C_2AS \cdot aq$ cannot be ruled out. Free $Ca(OH)_2$ is formed in small quantities.

Supersulfate cement, also known as sursulfate cement, is extensively used in structures exposed to sulfate waters. This cement is made by grinding together a mixture of 80% slag, 5% Portland cement clinker, and about 15% gypsum or anhydrite.

In supersulfate cements ettringite is the initial product of reaction of $CaSO_4$ with lime and aluminum in the slag. At later stages tobermorite gel is detected. The hexagonal C_4AH_x is expected at high Al_2O_3 contents. An excess of Portland cement in supersulfate cements causes reduction in strength. This is attributed to the formation of the low form of calcium sulfoaluminate hydrate. The resistance of supersulfate cements to sulfate attack is ascribed to almost complete absence of $Ca(OH)_2$ and the presence of less basic calcium silicate hydrate.

Differential Thermal Analysis

Devitrification. Much success has been achieved in the estimation of glass content of slags by the application of DTA (*4-16*). The DTA method has yielded very valuable results in the identification of hydrated products of blast-furnace slag cements and supersulfate cements (*17-24*).

A granulated slag heated to temperatures of 800-900°C shows devitrification accompanied by an evolution of heat, and this is registered as one or more exothermic peaks in a thermogram. The temperature, the number and the intensity of the exothermic peaks differ from slag to slag. Hence, attempts have been made to correlate the DTA results with the slag content or hydraulicity of slags and also to study the mechanism and products of devitrification (*4-16*).

For control purposes there is a need to estimate the glassy material in granulated slags and also in Portland blast-furnace cements. The area of exothermic peaks in the DTA may be used conveniently to estimate the glassy phase. The ground Portland cement clinker does not exhibit any peak at 800-900°C and hence does not interfere with exothermic devitrification effects.

Krüger (*9*) took three granulated blast-furnace slags and with each of them he mixed unhydrated Portland cement from 10 to 90% by weight. In thermograms of the three undiluted slags the exothermic peaks beyond 800°C signified devitrification. The area under the exothermic peaks should be proportional to the extent of devitrification and hence to the glass content. An endothermal valley was also observed in each of the samples at a temperature before the onset of the exothermal effects.

The delineation of the peak area as carried out by Krüger is explained for sample A in Fig. VIII.1.1. The base line LM is extended till it intersects the curve at P_1. From P_1 a line parallel to the temperature axis is drawn to cut the line CD at P_2. The line CD is drawn from a point P_3 at an angle of 45° to the line running parallel to the temperature axis. The shaded portion

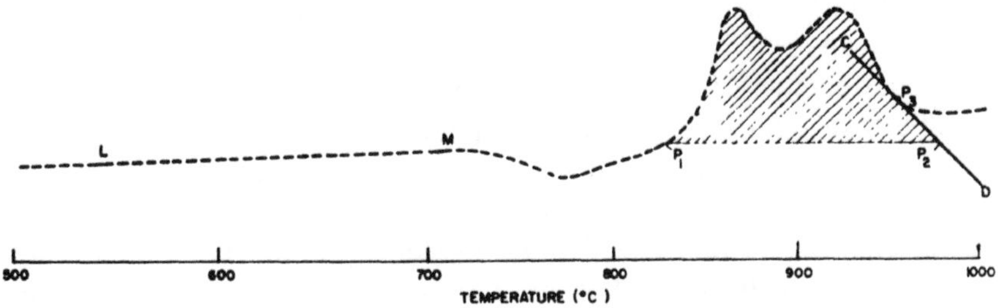

Fig. VIII.1.1: Method of delineation of the area under the devitrification peak of blast-furnace slag (*9*).

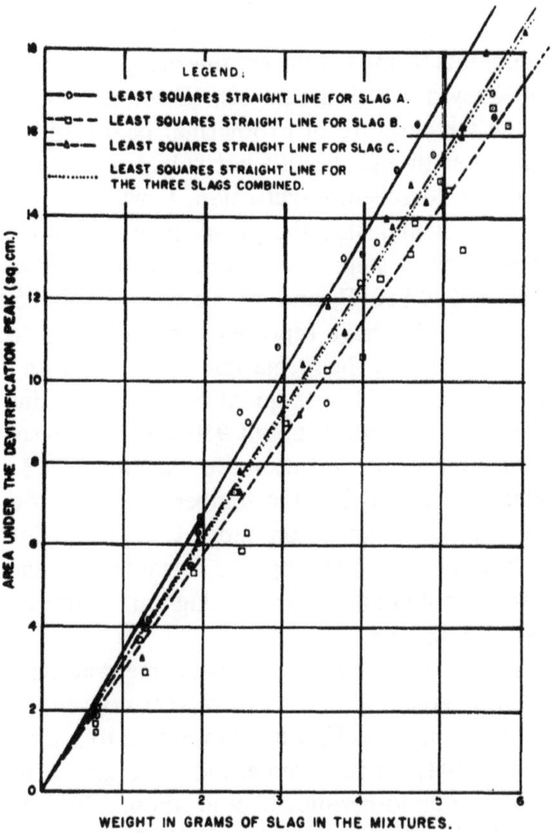

Fig. VIII.1.2: Relation between peak area and slag content of unhydrated Portland cement–blast-furnace slag mixtures (*9*).

denotes the exothermic peak area.

The relationship between the peak area and the slag content for the three slags is shown in Fig. VIII.1.2. A straight line relationship is obtained with correlation coefficients of 0.9959, 0.9914, and 0.9954. Further accuracy is possible with refinements in the technique.

Other investigators have also examined the exothermal effects in blast-furnace slags of different compositions (*6-8*, *11*).

The mechanism responsible for the existence of an endothermal valley preceding the devitrification peaks has been a subject of speculation. Schrämali (*13*) assigns the endothermal effect to an absorption of heat for creating the required mobility conditions for subsequent crystallization. Krüger *et al* (*15*) believe that sintering and partial melting occur at the temperature of the endothermal effect. Dehydration and decarbonation are also considered as causes that lead to the endothermal effect (*11*, *12*). It is possible that the endothermal effect is caused by a reversible glass transformation (*29*).

Identification of minerals formed at the exothermal peak temperatures may be accomplished by X-ray diffraction analysis. Schrämali (*13*) found no relationship between the temperature of exothermal peaks and the nature of products of devitrification though he did not rule out this possibility with refined DTA techniques. Mtschedlov-Petrosian, Lewtschuk, and Sstrelkova (*11*) found that for a particular slag the glassy phase shows characteristic devitrification temperatures in steps, akermanite devitrifying at 920-930°C, intermediate melilites at 970-980°C, and gehlenite at 1030-1050°C.

Krüger and co-workers (*15*) have reported that merwinite (C_3MS_2) crystallizes initially but is masked by the formation of melilite with a sharp peak at 870°C. The crystallization of larger quantities of merwinite results in a peak at 870°C, followed by a peak at about 930-940°C representing a solid state reaction in which merwinite forms more melilite, CMS, and β-C_2S.

An interesting relationship between the temperature of exothermic peak and SiO_2/Al_2O_3 ratio in the composition range $CaO \cdot SiO_2$–$CaO \cdot Al_2O_3$ has been worked out by Kondo (*10*). Attempts to formulate a general relationship between the strength potential of various slags and differential thermal effects have not met with success.

Slag Cements. The hydration products of slag cement activated with lime and gypsum or Portland cement would show in thermograms the existence of a low temperature endothermic effect corresponding to a tobermorite-like phase, a peak at about 150°C for the presence of ettringite, a peak at about 200°C for C_4AH_{13}, and an endothermic effect at about 520°C for $Ca(OH)_2$.

Hanada *et al* (*20*) obtained evidence through DTA and X-ray data that in Slag cements only 28% was hydrated after 28 days. Nurse and Midgley (*25*) used DTA to estimate the lime absorbed by slags at various lime concentrations. A plot of lime *absorbed* against lime *added* indicated a change in the rate of absorption at a lime concentration of 15-20%, corresponding to a maximum strength development.

The mechanism of activation of slags with lime and gypsum mixture was studied by Samaddar and Lahiri (*24*) through DTA. They prepared synthetic glasses with a constant SiO_2/Al_2O_3 ratio but with varying proportions of lime. The hexagonal C_4AH_{13} was detected as an initial product on the surface of the grains. It was inferred that during hydration lime was liberated from the hydrate and reacted further with the mass of the slag.

Lommatzsch (*17*) observed thermal effects in the 800-900°C range for sulfate-activated slags and ascribed them to modifications and new formations typical of hydraulic basic slags.

It should be possible to utilize the DTA technique to study the role of activators such as anhydrite. When anhydrite is added, the uncombined form of the activator will form gypsum, which can be identified by two endothermic effects in the thermograms. The presence of excess of gypsum should cause lower strengths in slag cements.

Supersulfate Cements. The DTA and X-ray methods have proved very useful in the examination of set supersulfate cements (*18*, *19*, *21*, *23*). The addition of Portland cement to supersulfate compositions is to supply $Ca(OH)_2$ during the setting and hardening period for stabilizing the calcium sulfoaluminate hydrate. The main products of hydration of supersulfate cements are ettringite and a tobermorite phase.

Chopra and others (*19*, *23*) have shown that supersulfate cements can be prepared from slags containing low CaO/SiO_2 and SiO_2/Al_2O_3 ratios. The best mix (A) contained slag, $CaSO_4$, Portland cement clinker in proportions 70:25:5. The compressive strength of this cement in the 1:3 mortar was 2400, 3500, 6330 and 11,600 lb/in.^{2} at 3, 7, 28, and 90 days respectively. Another mix (B) containing 17% $CaSO_4$ as an activator showed reduced strengths.

Thermograms were obtained at different periods of curing, both for cements A and B and are shown in Figs. VIII.1.3 and VIII.1.4. The supersulfate cement A exhibits an endothermic peak at 150-175°C due to ettringite. A large quantity of ettringite was present at 28 days, which is evident from the intensity of the endothermic peak at 28 days. At longer periods also ettringite was present, but the increase in strength could only be due to increased formation of tobermorite and alumina gel. The endothermic peak at 640°C exhibited by the 7-day old sample could not be explained by Chopra *et al.*

In supersulfate cement B an endothermic peak is seen at 150°C at 7 days of curing. After 28 days the endotherms with peaks at 175 and 240°C correspond to an ettringite-type solid with a $CaSO_4/Al_2O_3$ ratio of 2.5. At 90 days

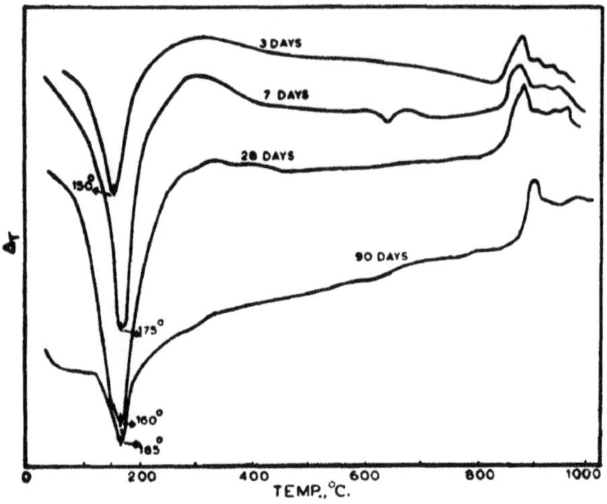

Fig. VIII.1.3: DTA curves of supersulfated cement K.55 (A) (*23*).

ettringite is present together with a low sulfoaluminate hydrate showing a peak at 190°C. The low strengths in sample B are ascribed to the formation of low sulfoaluminate hydrate.

Nurse (*21*) applied the DTA technique to a quantitative evaluation of the

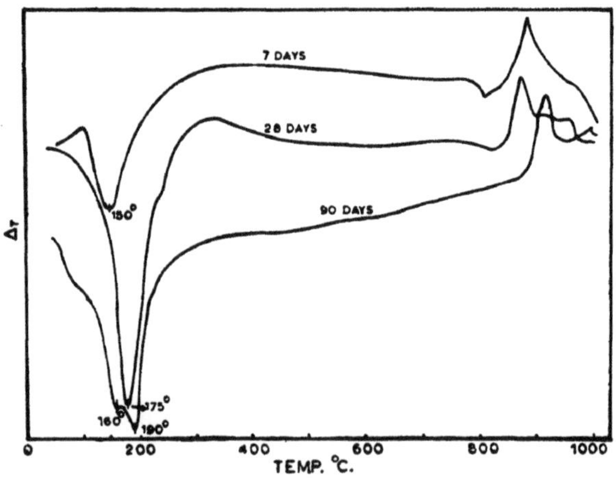

Fig. VIII.1.4: DTA curves of supersulfated cement K.44 (B) (*23*).

hydrated products in a supersulfate cement. He carried out a calibration of the pure complex calcium aluminates according to the method of Midgley and Rosaman (*26*). An examination of the set supersulfate cement showed it to contain a phase corresponding to a member of the solid solution of the series $3CaO \cdot Al_2O_3 \cdot 3CaSO_4 \cdot aq$–$3CaO \cdot Al_2O_3 \cdot 3Ca(OH)_2 \cdot aq$.

Applying this technique the hydroxy compound was estimated at 80% at 1 day, 26% at 7 days, 17% at 14 days, 28 days, 3 months, and 6 months. With slags containing low CaO/SiO_2 very little ettringite was detected at 1 and 3 days. At 28 days tobermorite was present.

Autoclaved Slag Cements. Autoclave treatment of slag with lime results

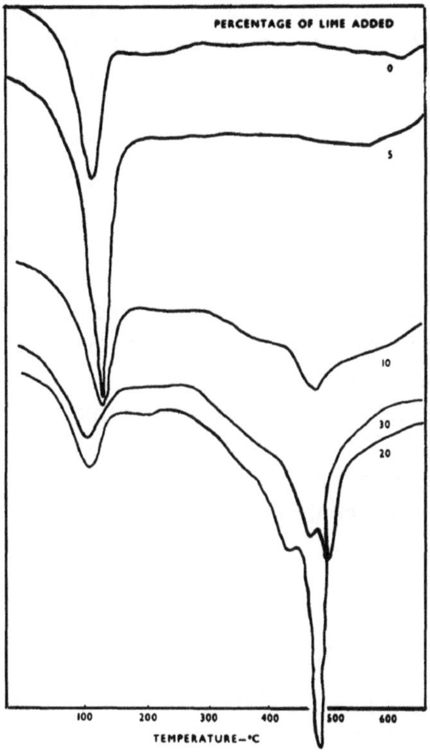

Fig. VIII.1.5: Thermograms of mixes autoclaved for 16 hr and containing slag and various percentages of added lime (*22*) (Crown Copyright, reproduced by permission of Her Majesty's Stationery Office].

in increased strength development. Kalousek (*27*) and Midgley and Chopra (*22*) studied the autoclaved products of slag–lime mixes by DTA and tried to examine the causes leading to strength deterioration in some cases.

An autoclaved sample of foamed or granulated slag showed a compressive strength of 3780 lb/in.^{2} and in admixture with 5% lime the strength increased to 8589 lb/in.^{2} With higher lime contents, strengths decreased, the values for 30 and 40% lime being 3600 and 2434 lb/in.^{2} respectively.

In Fig. VIII.1.5 are shown the thermograms of autoclaved products of a slag containing 0, 5, 10, 20, and 30% lime. The autoclaved slag shows an endothermal peak just above 100°C due to the presence of a poorly crystallized tobermorite gel. The peak intensity of the tobermorite gel increases with 5% lime. In the slag containing 10% lime there is a decrease in the intensity of the tobermorite peak and the development of an endothermic peak at about 500°C. A further decrease in the intensity of the low temperature endothermal peak at 20 and 30% lime contents and an increase in the intensity of the peak at 500°C are noticed.

The increase in the tobermorite content at 5 days corresponds to an enhanced strength. The endothermic effect below 500°C for the mix containing 10% lime is caused by the presence of dicalcium silicate α-hydrate, the formation of which is detrimental to strength development. At 20 and 30% lime contents an increase in the intensity of the endothermic peak at 500°C indicates formation of increased quantities of dicalcium silicate α-hydrate at the expense of tobermorite. Thus there is a decrease in strengths as the lime content increases from 5 to 40%.

The basicity of the calcium hydrosilicates formed in autoclaved blast furnace slags activated with lime and gypsum depends on the specific surface of the slag glass and the duration of curing. This study was extended to 20 slags by adopting DTA technique (*28*).

REFERENCES

1. Kramer, W., *Stahl Eisen*, *73*, 1596, 1953.
2. D'Ans, J., and Eick, H., *Zement-Kalk-Gips*, *7*, 449, 1954.
3. Keil, F., *Proc. 3rd Intern. Symp. Chem. Cements*, London, p. 530, 1952.
4. Leclère, F., *Rev. Mater. Construct. No. de Commémoration*, 29, 1950.
5. Nicol, A., *Rev. Mater. Construct. No. de Commémoration*, 34, 1950.
6. Sersale, R., and Gregorio, E., *Ric. Sci.*, *21*, 2152, 1951.
7. Ota, C., *Semento Gijutsu Nenpo*, *10*, 85, 1956.
8. Kurdowski, W., and Stoch, L., *Cement-Wapno-Gips*, *13*, 93, 1957.
9. Krüger, J.E., *Cement and Lime Manufacture*, *35*, 1, 1962.
10. Kondo, R., *Proc. 4th Intern. Symp. Chem. Cements*, Washington, p. 973, 1960.
11. Mtschedlov-Petrosian, O.P., Lewtschuk, N.A., and Strelkova, I.S., *Silikat*

Tech., *13*, 153, 1962.
12. Zavgorodnii, N.S., Mtschedlov-Petrosian, O.P., Sidochenko, I.M., and Sstrelkova, I.S., *Tsement*, *28*, 13, 1962.
13. Schrämali, W., *Zement-Kalk-Gips*, *16*, 140, 1963.
14. Webb, T.L., and Krüger, J.E., *Symp. Analysis Calcareous Materials*, London, pp. 419-439, 1963.
15. Krüger, J.E., Sehlke, K.H.L., and Van Aardt, J.H.P., *Cement and Lime Manufacture*, *37*, 2, 1964.
16. Karsch, K.H., and Schwiete, H.E., *Arch. Eisen.*, *35*, 21, 1964.
17. Lommatzsch, A., *Silikat Tech.*, *7*, 468, 1956.
18. Chatterji, S., *Studies of the Hydraulic and Related Properties of Blastfurnace Slags Available in India*, D. Phil. Thesis. Calcutta University, India, 1957.
19. Williams, H., and Chopra, S.K., *Proc. 4th Intern. Symp. Chem. Cements*, Washington, p. 979, 1960.
20. Hanada, M., Tanaka, H., Sakurai, S., Chikano, T., and Murakami, K., *Yogyo Koyaki Shi*, *68*, 307, 1960.
21. Nurse, R.W., *Atti del Convegno sulla Produzione e le applicazione dei cementi siderurgici*, Napoli, 187-195, 1960.
22. Midgley, H.G., and Chopra, S.K., *Mag. Concrete Res.*, *12*, 73, 1960.
23. Chopra, S.K., and Kishan Lal, *J. Sci. Ind. Res.* (*India*), *20D*, 218, 1961.
24. Samaddar, B., and Lahiri, D., *Trans. Indian. Ceram. Soc.*, *21*, 75, 1962.
25. Nurse, R.W., and Midgley, H.G. (Cited by Taylor, H.F.W.), *The Chemistry of Cements*, Vol. 2, Academic Press, New York, 1964.
26. Midgley, H.G., and Rosaman, D., *Proc. 4th Intern. Symp. Chem. Cements*, Washington, p. 259, 1960.
27. Kalousek, G.L., *J. Am. Concrete Inst.*, *25*, 365, 1954.
28. Vinogradov, B.N., Grenik, E.A., and Gladkikh, K.V., *Stroit Materialy*, *9*, 12, 1963.
29. Webb, T.L. (personal communication).

VIII.2. HIGH ALUMINA CEMENT

High alumina cement, also known as aluminous cement or calcium aluminate cement, is formed by firing a mixture of limestone or chalk and bauxite to a fusion temperature of 1500-1600°C. This cement is manufactured in the United Kingdom and France under the name *Ciment Fondu* and in the U.S.A., *Lumnite*. The color of the alumina cement is generally dark due to the presence of iron in the ferrous state.

The high alumina cement develops very high strengths within 24 hr of curing in water. A mortar containing 1 part alumina cement and 3 parts of aggregate may attain a compressive strength higher than 10,000 $lb/in.^2$ within 24 hr, whereas the corresponding strength even for a rapid hardening Portland cement would be less than 2000 $lb/in.^2$

High alumina cement has a very good resistance to sulfate attack and can be used in refractory concrete at a temperature of 1600°C. In addition, this cement does not contain free lime, and hence unsoundness due to lime is completely avoided. This cement is also used as an expanding or stressing cement in combination with gypsum and other constituents (*1*, *2*).

Clinker Composition. The most important constituent of alumina cement is calcium monoaluminate, CA. In cements with a high Al_2O_3/CaO ratio, CA_2 (previously designated by the formula C_3A_5) is also formed. Another aluminate—$C_{12}A_7$—is also reported to be present in commercial cements, and earlier literature refers to this compound by the formula C_5A_3. The other aluminate, C_3A (present in Portland cement), does not appear in alumina cement.

Two siliceous compounds have also been detected in commercial samples. In cements containing less than 5% SiO_2, β-C_2S is present, and in those having higher SiO_2 contents gehlenite (C_2AS) or melilites predominate. Alumina cements also contain iron oxides in ferrous and ferric states to an extent of 18%. In the ferric state, iron forms compounds lying between C_4AF and C_6AF_2. The ferrous iron occurs in the fibrous constituent of alumina cements and also as FeO. Tavasci (*3*) reported the formation of a compound with the formula CT (Pervoskite) in cements containing TiO_2.

Hydration Reactions. The principal cementing constituent in hydrated alumina cement is CAH_{10}—a metastable product of hydration of CA—which was reported by Assarsson (*4*) in 1936. Other metastable hydrates such as hexagonal C_2AH_8 and C_4AH_{13} (or C_4AH_{19}) are also present in hydrated cements, and these transform ultimately to the stable cubic C_3AH_6. The transformation of the metastable hydrates to the stable hydrate is dependent on the temperature of hydration. Some references to the existence of a compound of the formula $C_3AH_{10\text{-}12}$ are also found in literature, but its presence in alumina cement is generally discarded.

Alumina gel is always formed in the hydrated products and is converted to a stable form known as gibbsite (AH_3). Under normal conditions of exposure the alumina cement may react with carbon dioxide and form calcium aluminate carbonates of the formula $C_3A \cdot CaCO_3 \cdot 11H_2O$ and $C_3A \cdot 3CaCO_3 \cdot 32H_2O$.

The $C_{12}A_7$ component reacts very rapidly, is quick setting, and gives hydrated aluminates and hydrated alumina. The other aluminate—CA_2—is practically unreactive towards water, but in the presence of CA reacts at a moderate rate.

The dicalcium silicate, β-C_2S, hydrates very slowly as in Portland cement and its contribution to strength development at earlier stages is not significant. Calcium hydroxide formed in this reaction is consumed by alumina gel.

Pure gehlenite, C_2AS, is known to be very unreactive to water but in alumina cements it may be present as a glass and hence may attain hydraulicity. The hydration product of gehlenite is C_2ASH_8 (*5*).

The hydration products of iron-bearing constituents may include a series of solid solutions of composition between C_3AH_6 and C_3FH_6.

Differential Thermal Analysis

The method of DTA has been applied in conjunction with X-ray diffraction technique for the study of the hydration products of commercial as well as synthetic calcium aluminates (*6-16*).

A study of the hydration of CA is of great interest in high alumina cement chemistry. Hence this is described in some detail below; the thermal investigations of hydration products of other aluminates have already been described.

A systematic study was carried out applying DTA technique to the hydration of CA by Tulovskaya and Segalova (*16*). They obtained a synthetically prepared sample of CA by calcining CaO and Al_2O_3 in equimolar proportions to 1400°C. The resulting CA was hydrated at a water/solid ratio of 5 at temperatures of 5, 20, 30, and 60°C. Suspensions were removed at known intervals and washed with alcohol and ether and then subjected to DTA.

Figs. VIII.2.1-2.4 refer to the thermograms obtained by these workers at different temperatures and periods of hydration. Thermal effects have been discussed by obtaining thermograms of pure $CA \cdot aq.$, $C_2A \cdot aq.$, $C_4A \cdot aq.$ and C_3AH_6.

Fig. VIII.2.1 pertains to the hydration behaviour of CA at 5°C. It is evident that the peak at 140-170°C represents $CA \cdot aq$ and it persists even after a

Fig. VIII.2.1: Hydration of monocalcium aluminate at 5°C (*16*).

Fig. VIII.2.2: Hydration of monocalcium aluminate at 20°C (*16*).

period of 3 months. The hexagonal $C_2A \cdot aq$ appears after 1½ months and its presence is indicated by the endothermic peak at 290°C. At 3 months, only $CA \cdot aq$ and $C_2A \cdot aq$ are present and conversion to C_3AH_6 has not occurred. The increase in peak temperature for $CA \cdot aq$ from 140°C at 24 hr to 170°C at longer periods is attributed to an increase in the size of crystallites.

At 20°C there is a slight increase in the rate of conversion of $CA \cdot aq$ to other products (Fig. VIII.2.2). At 5°C it takes 1½ months for $C_2A \cdot aq$ to appear whereas, at 20°C it appears at 9 hr after hydration. At 5 days the peak due to $C_2A \cdot aq$ increases at the expense of the peak due to $CA \cdot aq$. After 2 months C_3AH_6 predominates and at 3 months all the three hydrates, *viz.* $CA \cdot aq$, $C_2A \cdot aq$, and C_3AH_6 co-exist.

At 30°C there is a further enhancement in the rate of conversion of $CA \cdot aq$ (Fig. VIII.2.3). At 8 hr both $CA \cdot aq$ and $C_2A \cdot aq$ are present and after 3 days the peaks due to C_3AH_6 appear. At 5 days there is no indication of the presence of $CA \cdot aq$ and at 3 months only $C_2A \cdot aq$ and stable C_3AH_6 are detected.

In Fig. VIII.2.4 the DTA of hydrated products obtained at 60°C is shown. Even at 10 min the product contains all the three hydrates and $CA \cdot aq$ is completely absent after 24 hr. The main product of hydration at 24 hr is C_3AH_6. It is evident from these studies that if the hydration had been carried out at longer periods at lower temperatures or at higher temperatures, only C_3AH_6

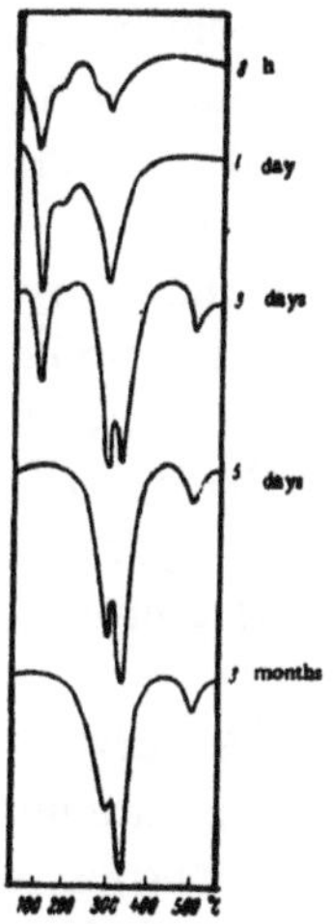

Fig. VIII.2.3: Hydration of monocalcium aluminate at 30°C (*16*).

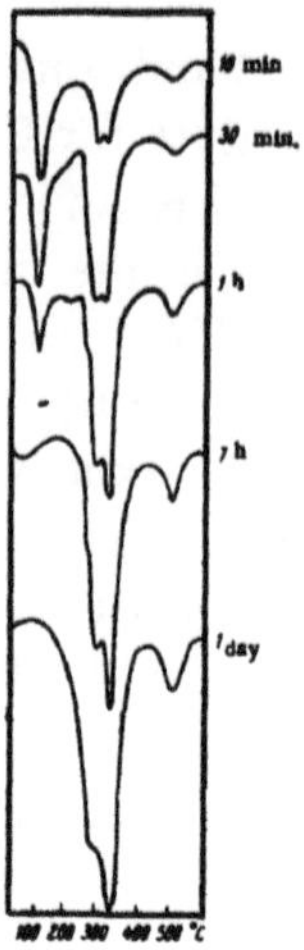

Fig. VIII.2.4: Hydration of monocalcium aluminate at 60°C (*16*).

would be the final product.

Thermographic studies, in addition to providing a clear picture of the stability of the metastable hydration products, indicate that only $C_2A \cdot aq$ is the intermediate product in the conversion of $CA \cdot aq$ to C_3AH_6. There is no evidence of the presence of either the hexagonal $C_3A \cdot aq$ or $C_4A \cdot aq$ in any of the samples. The hexagonal $C_4A \cdot aq$ would form when CA or C_3A hydrates in excess of $Ca(OH)_2$ solution.

Budnikov and Kravchenko (*14*) studied by DTA the hydration of CA at 15-18°C and at 45°C. The rate of conversion of $C_2A \cdot aq$ to C_3AH_6 was higher at 45°C. In addition they found that hardening at normal or elevated temperatures in air reduced the rate of hydration of CA. Hexagonal aluminates have also been reported to form by hydration of synthetic aluminates and commercial alumina cements, these converting to C_3AH_6 under autoclave conditions (*10*).

Schneider (*13*) examined the DTA curves of CAH_{10} and four hydrated alumina cements (Fig. VIII.2.5). An attempt was made to find a correlation between the mechanical properties and compound compositions. Two endothermic peaks at 155 and 285°C in CAH_{10} were thought to represent stepwise dehydration. However, the results of Tulovskaya and Segalova (*16*) show that the sample studied by Schneider must have been contaminated with $C_2A \cdot aq$, which should explain an endothermic peak at

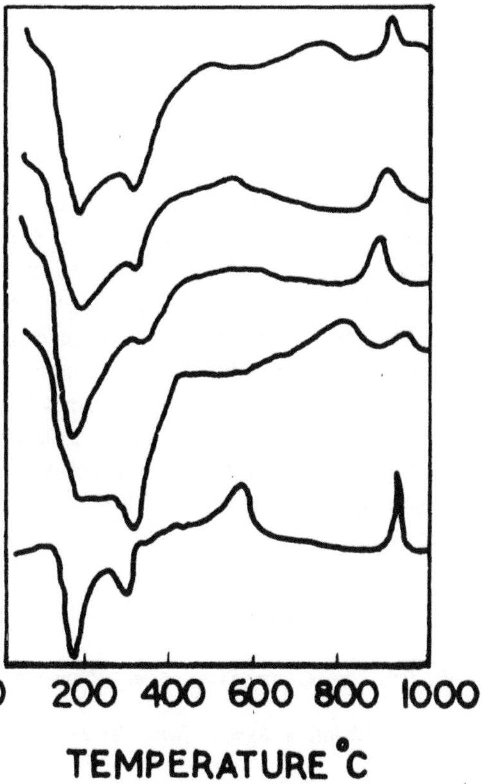

Fig. VIII.2.5: Differential thermal analysis curves of four hydrated aluminous cements and CAH_{10} (*13*).

about 285°C.

Two exothermic peaks at about 545 and 930°C were attributed to the crystallization of CA in two steps. But Heindl and Post (*8*) in their work on hydrated alumina cements attribute an exothermic peak at 900°C to the crystallization of Al_2O_3, although according to Schneider DTA gives no evidence of conversion of CAH_{10} to C_3AH_6 and AH_3. Conversion occurred by heating CAH_{10} under slight pressures in presence of water. Apparently the thermograms of the hydration products of the four commercial alumina cements are similar to that of CAH_{10}. The X-ray results, however, showed that the cements formed C_3AH_6 and AH_3 on heating.

Schneider (*13*) also determined the mechanical properties such as transverse strength and modulus of elasticity after heat treatment and tried to relate the results with the dehydration behavior in alumina cements. A fall in transverse strength from 50 to 75°C corresponded to a partial dehydration of CAH_{10} or to the formation of cubic C_3AH_6 and AH_3. An increase in strength was noticed from 75 to 110°C and this was caused by an increased crystallinity of C_3AH_6 and AH_3. There was also a progressive fall in strength from 110 to 500°C, the range at which CAH_{10}, C_3AH_6 and AH_3 completely dehydrate.

Increased strengths up to 700°C could be correlated with increase in the crystallinity of the dehydrated products, but thermal expansion data were not consistent with this explanation. A slow decrease in strengths from 700 to 1100°C may be governed by the formation of new compounds.

Ando, Hiraska, and Matsuno (*12*) studied the nature of thermal changes occurring in refractory castables containing alumina cement and a refractory aggregate. Solid state reactions result at temperatures from 600 to 1000°C between calcium aluminates (CA, $C_{12}A_7$), CaO, Al_2O_3 and the aggregate. Impurities such as iron oxide promote these reactions. An increase in strength with alumino-silicate aggregates at 700-800°C is shown as endothermal effects in thermograms.

REFERENCES

1. Mikhailov, V.V., *Proc. 4th Intern. Symp. Chem. Cements*, Washington, p. 927, 1962.
2. Budnikov, P.P., *Proc. 4th Intern. Symp. Chem. Cements*, Washington, p. 469, 1962.
3. Lea, F.M., *The Chemistry of Cement and Concrete*, Edward Arnold (Publishers) Ltd., London, p. 435, 1956.
4. Assarsson, G., *Sveriges Geol. Undersokn*, *30*, 5, 1936.
5. Zur Strassen, H., *Zement-Kalk-Gips*, *11*, 137, 1958.
6. Pole, G.R., and Moore, D.G., *J. Am. Ceram. Soc.*, *29*, 20, 1946.
7. Nagai, S., and Harada, T., *J. Japan. Ceram. Soc.*, *60*, 386, 1952; *62*, 618, 1954; *63*, 189, 1955.
8. Heindl, R.A., and Post, Z.A., *J. Am. Ceram. Soc.*, *37*, 206, 1954.
9. Turriziani, R., and Schippa, G., *Ind. Ital. Cemento*, *26*, 295, 1956.
10. Talabér, T., *Epitoanyag*, *8*, 161, 202, 295, 349, 1956.
11. Rey, M., *Silicates Ind.*, *22*, 533, 1957.
12. Ando, J., Hiraska, G., and Matsuno, S., *Yogyo Kyokai Shi*, *66*, 253, 1958.
13. Schneider, S.J., *J. Am. Ceram. Soc.*, *42*, 184, 1959.
14. Budnikov, P.P., and Kravchenko, I.V., *Kollodn. Zh.*, *21*, 9, 1959.
15. Lehmann, H., and Mitusch, H., *Feuerfester Beton aus Tonerde-Schmelzzement*, Hübener, H., Verlag, Goslar, Germany, 1959.
16. Tulovskaya, Z.D., and Segalova, E.E., *Zh. Prikl. Khim.*, *37*, 267, 1964.

VIII.3. POZZOLANAS AND POZZOLANIC CEMENTS

Pozzolanas are defined by Lea (*1*) as

> Siliceous materials which, though not cementitious in themselves, contain constituents which at ordinary temperatures will combine with lime in presence of water to form compounds which have low solubility and possess cementing properties.

A pozzolanic cement is obtained by grinding together a mixture of pozzolana and Portland cement clinker or by intermixing a pozzolana with calcium hydroxide.

A pozzolana may be either a material of natural origin or an artificial preparation. The well known examples of natural pozzolanas are volcanic glasses, pozzolanic tuffs, and siliceous types such as diatomite and rocks. Artificial pozzolanas include pulverized fuel ash (fly ash) formed as a residue of combustion of pulverized coal and calcined clays long known in India by the name *Surkhi* and in Egypt by *Homra*. Granulated blast-furnace slag is also considered as a pozzolana by some cement chemists.

Pozzolanas are used in mortars in combination with lime or Portland cement and in concretes. It is recognized that concretes containing a pozzolana possess a very good resistance to sulfate attack. Addition of a pozzolana reduces the heat evolved from the Portland cement products. The pozzolana also reduces the expansion due to alkali-aggregate reaction and serves as a part replacement of Portland cement.

In general, pozzolanic materials develop strength very gradually, but in certain proportions with pozzolanic cement and slags they may not only compare with Portland cements but may have higher ultimate strengths and greater resistance to aggressive waters. Table VIII.3.1 illustrates this fact.

TABLE VIII.3.1
Compressive Strengths of 1:3 Mortars of Some Cements
[Central Road Research Institute (2)]

Cement	*Compressive strength, lb/in.^{2}* 2 days	7 days	28 days	90 days
C.P.-M.F.2	2120	3870	6200	8250
Portland cement	—	2270	3550	—
High early strength Portland cement	—	3550	4450	—
Super cement	—	5050	7100	—
Fly ash pozzolanic cement	1450	2670	4700	—
Slag-fly ash cement	850	1390	2900	—

Compressive strengths of 1:3 mortar of a patented special pozzolana-based cement C.P.-M.F.2 used by *La Societe des Materiaux de Construction de la Loisne* of France are compared with those of other known cements.

Chemistry. The strength development in a pozzolanic cement is ascribed to the direct combination of lime with the pozzolana. Some investigators consider the phenomenon of cation exchange as being the main factor contributing to the strength. Alkalies are known to be liberated when a pozzolana is mixed with lime, but it is doubtful if this contributes to the strength because the total exchange capacity of a pozzolana is too low to explain the combination of a substantial amount of Ca^{++} and the development of strength. Further, in the cation exchange reaction, the structure of the pozzolana is not expected to undergo any alteration.

The direct combination of $Ca(OH)_2$ with the active silica and alumina constituents of a pozzolana results in the formation of various hydrated compounds. The hydrated calcium silicate of the tobermorite group of variable CaO/SiO_2 ratio invariably forms, along with tetracalcium aluminate hydrate, C_4AH_x and cubic tricalcium aluminate hydrate, C_3AH_6. Strätling (*3, 4*) demonstrated the presence of a calcium aluminosilicate hydrate of the formula C_2ASH_x in the reaction between calcined kaolin and lime. The reaction could be formulated as follows:

$$\underset{\textit{calcined kaolin}}{2(Al_2O_3{\cdot}2SiO_2)}+\underset{\textit{lime}}{7Ca(OH)_2}+\underset{\textit{water}}{aq} \rightarrow \underset{\textit{calcium silicate hydrate}}{3CaO{\cdot}2SiO_2{\cdot}aq}+\underset{\textit{gehlenite}}{2(2CaO{\cdot}Al_2O_3{\cdot}SiO_2{\cdot}aq)}$$

It is doubtful whether C_2ASH_x is a hydration product of all pozzolanic cements. A pozzolana in combination with lime, gypsum and water forms the same compounds as a hydrated Portland cement, *viz.* ettringite ($3CaO{\cdot}Al_2O_3{\cdot}3CaSO_4{\cdot}32H_2O$), and under certain conditions the low sulfate form ($3CaO{\cdot}Al_2O_3{\cdot}CaSO_4{\cdot}12H_2O$), calcium silicate hydrate, and C_4AH_{13}. In hydrated pozzolana–Portland cement mixes the same type of compounds are expected but containing lower quantities of free $Ca(OH)_2$.

Differential Thermal Analysis

Amongst various investigators, Turriziani *et al* seem to have applied the method of DTA to the greatest extent for investigating pozzolanic reactions (*5-10*). Others who have investigated the pozzolanic reactions by this technique are Rey (*11*), Leonard (*12*), Ariizumi (*13*), Ludwig and Schwiete (*14*), Benton (*15*), Fernand (*16*), Wessley (*17*), Surovkin and Krylov (*18*), Jambor (*19*), and Chebukov and Egorova (*20*).

Turriziani (*10*) investigated the hydrated products in mortars made from a natural Segni pozzolana–lime mixture having a lime/pozzolana ratio of 0.15. The mortars were cured at room temperature or at 40°C. DTA revealed endothermic peaks that could be ascribed to the presence of C_4AH_x, calcium silicate hydrate, and unreacted $Ca(OH)_2$. In mortars cured at 40°C alumina was present mainly as C_3AH_6. In none of the mortars C_2ASH_8 was present.

In a saturated lime suspension the Segni pozzolana exhibited an endothermic peak at 140°C showing the presence of Taylor's silicate of the formula $C_{0.8\text{-}1.5}$ SH(I), and an endothermic effect at 220°C signifying the presence of C_4AH_{13}. However, when the reaction was carried out with a low water/pozzolana ratio or in a percolating bed, C_2ASH_8 was detected.

In their work on the hydrated products of calcined kaolin and $Ca(OH)_2$, Turriziani and Schippa (*5*) assign the endothermic peak at 220°C to the dehydration of the Stratling compound C_2ASH_8. This observation is in consonance with that of Stratling (*3*) and Strätling and Zur Strassen (*4*). DTA was also successfully employed by Turriziani and Rio (*8*) to estimate $Ca(OH)_2$ in pozzolanic cements at different time intervals. The estimation by the DTA method was within 0.5% of the Franke's standard method.

Jambor (*19*) presented differential thermograms of 10 pozzolanas and 22 hydrated pozzolanic cements. The pozzolanic cements contained pozzolana and $Ca(OH)_2$ in proportions of 2:1, 1:1, or 1:2 by weight. Mixes were pre-

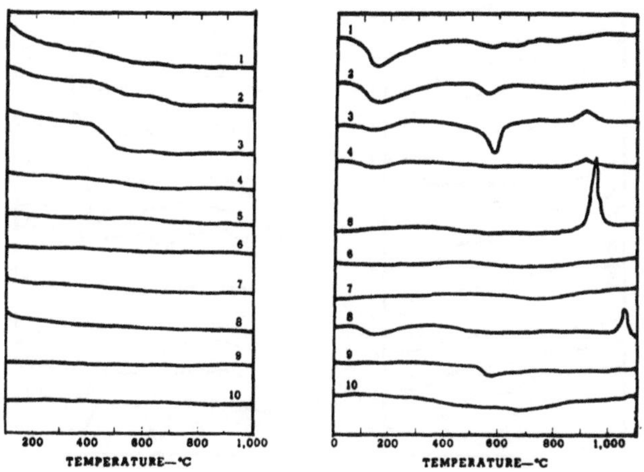

Fig. VIII.3.1: Gravimetric and differential thermograms of pozzolanas using (1) dacite tuff, (2) basalt tuff, (3) diatomite I, (4) diatomite II, (5) activated kaolin, (6) fly ash I, (7) fly ash II, (8) silica gel, (9) quartz, (10) sial (*19*).

pared to a plastic consistency and cured in moist air for 24-48 hr and subsequently in water for 400 days at 19-21°C.

Fig. VIII.3.1 gives the thermal behavior of dalcite tuff, basalt tuff, two diatomites, an activated kaolin (heated at 650°C for 2 hr.), silica gel, two fly ash samples, β-quartz, and sial (chemical glass). The thermal effects in 22 hydrated products of these pozzolanas are shown in Fig. VIII.3.2.

Thermograms of samples having the same mix proportions are similar in nature. The endothermic effect at 140-170°C is caused by the dehydration of bound water from CSH(I). The endothermic peak at 560°C is due to the decomposition of $Ca(OH)_2$ and the endotherm at 800-900°C represents the

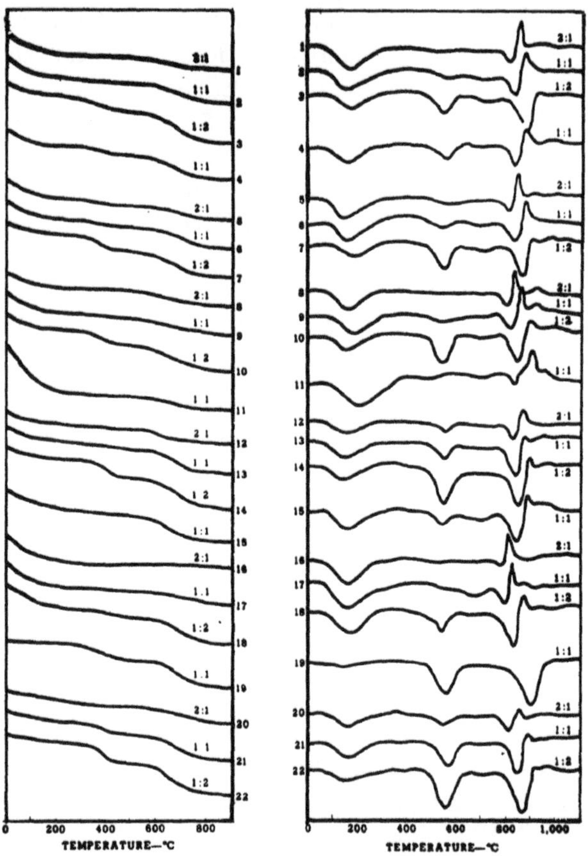

Fig. VIII.3.2: Gravimetric and differential thermograms of lime-pozzolana pastes after 400 days' hardening at 20°C. Pozzolanas studied (1-3) dalcite tuff, (4) basalt tuff, (5-7) diatomite I, (8-10) diatomite II, (11) activated kaolin, (12-14) fly ash I, (15) fly ash II, (16-18) silica gel, (19) quartz, (20-22) sial (*19*).

decarbonation of $CaCO_3$. The high temperature endothermic effect indicates the development of wollastonite or β-C_2S.

Some samples of mix proportion 1:2 failed to exhibit an exothermic effect at higher temperatures though they contained CSH(I). The peak areas of the low temperature endothermic effect are larger in mixes with dehydrated kaolin. Actual estimation shows that a kaolin-based product contains more than 75% of the material in the hydrated state. Evidence is also obtained for the presence of C_2ASH_8 in mixes containing calcined kaolin. The calculations show that in a lime–quartz mixture only 5.2% is hydrated in the product. DTA also confirms this by showing practically no endothermic effect at low temperatures.

Benton (*15*) also studied the DTA of $Ca(OH)_2$–pozzolana reactions. His study included some pozzolanas not investigated by Jambor (*19*). The eight samples in his study were pumicite, diatomite, kaolinite, bentonite, illite, gibbsite, quartz, and fly ash. The clay samples were calcined at 760°C for 2 hr before use. Mixtures containing $Ca(OH)_2$/pozzolana/water in the ratio 1:1:1.6 by weight were prepared and the DTA of the products was examined after 1,3, and 11 months. Thermogravimetric analysis estimated the combined $Ca(OH)_2$ in these products.

In mixtures with about 80% kaolinite, $Ca(OH)_2$ was in the combined form as was evident by a small endothermal effect. These results are in agreement with those of Jambor. DTA of quartz-based products showed a large endothermic peak for $Ca(OH)_2$ decomposition, and in this sample only 15% $Ca(OH)_2$ was in the combined form.

Endothermic effects starting at about 100°C indicated the presence of calcium silicate hydrate. The cubic C_3AH_6 with an endothermic effect in the range 300-400°C was evident only in mixtures containing gibbsite. Though DTA indicated a peak at 700-830°C for calcite, X-rays failed to show lines corresponding to calcite. This led Benton to infer that CO_2 was in a combined form in tetracalcium aluminate hydrate. The endothermic peak at 200°C was thus attributed to tetracalcium aluminate carbonate hydrate. The peak at the same temperature in kaolinite mixes, however, represented C_2ASH_x.

Fernand (*16*), Leonard (*12*), Ariizumi (*13*), and Wessley (*17*) have also used DTA to study pozzolanas. Ariizumi found C_2ASH_x as the reaction product of allophane and $Ca(OH)_2$.

The artificial pozzolanas such as clays require an optimum temperature of calcination for the development of maximum pozzolanic activity and this will depend on the type of clay mineral. Mielenz *et al* (*21*) and Mackenzie *et al* (*22*) studied the effect of temperature of calcination on the pozzolanic activity of clay minerals such as kaolinite, montmorillonite, illite, palygorskite, and mixed clays containing vermiculite and chlorite.

The kaolinitic clay mineral was found to exhibit maximum activity just above the temperature of dehydroxylation. On theoretical grounds also enhanced activity is expected at this temperature due to the modifications in the lattice.

At temperatures of 870-982°C nucleation to mullite should result in decreased reactivity. The montmorillonitic clay mineral develops strength at a temperature after dehydroxylation but before the temperature at which crystallization occurs. Illites do not produce as reactive pozzolanas as kaolinites because potassium in them tends to produce a glass after dehydroxylation. Illites, as also vermiculite and chlorite, develop maximum strength at 982°C. Attapulgite is a better pozzolana than montmorillonite.

The technique of DTA should prove a valuable tool in the identification of clay minerals and should also be able to provide information on the temperatures of dehydration and development of various phases. The application of DTA to the study of clay minerals is discussed in Chapter III.

The DTA technique was also employed by Turriziani and Schippa (*5*, *9*) and Turriziani (*10*) to study the reactions in mixtures containing a pozzolana, calcium hydroxide, gypsum, and water. Evidence was obtained for the existence of ettringite and under certain conditions for the low sulfoaluminate hydrate. No indication was available for the presence of C_2ASH_x. Mortars containing 6% gypsum showed increased strengths but an expansion and disintegration was noticed in mortars having 10% gypsum.

The rapid formation of sulfoaluminate at the expense of $Ca(OH)_2$ fixation was given as the cause for disintegration. Ludwig and Schwiete (*14*) investigated the thermal behaviour of the system trass–lime–gypsum–water and found that the hydrates were the same as those in Portland cements, *viz.* $Ca(OH)_2$, $3CaO \cdot 2SiO_2 \cdot aq$, $4CaO \cdot Al_2O_3 \cdot 13H_2O$, $3CaO \cdot Al_2O_3 \cdot 3CaSO_4 \cdot 32H_2O$, and $3CaO \cdot Al_2O_3 \cdot CaSO_4 \cdot 12H_2O$.

In mixtures consisting of pozzolana, Portland cement, and water the Portland cement hydrates to form $Ca(OH)_2$ which combines with the pozzolana. DTA results (*10*) of mixes with Segni pozzolana and Portland cement showed the hydrated products to consist of calcium silicate hydrate of the tobermorite type, tetracalcium aluminate hydrate, the high sulfate form of sulfoaluminate, and a small quantity of calcium hydroxide. Neither C_2ASH_x nor the low sulfoaluminate hydrate was identified in these cements.

Benton's study (*15*) on pozzolanic cements consisted of obtaining thermograms of the reaction products of 20% pozzolana in Portland cement (type III or V). The results are shown in Figs. VIII.3.3 and 3.4. The thermograms

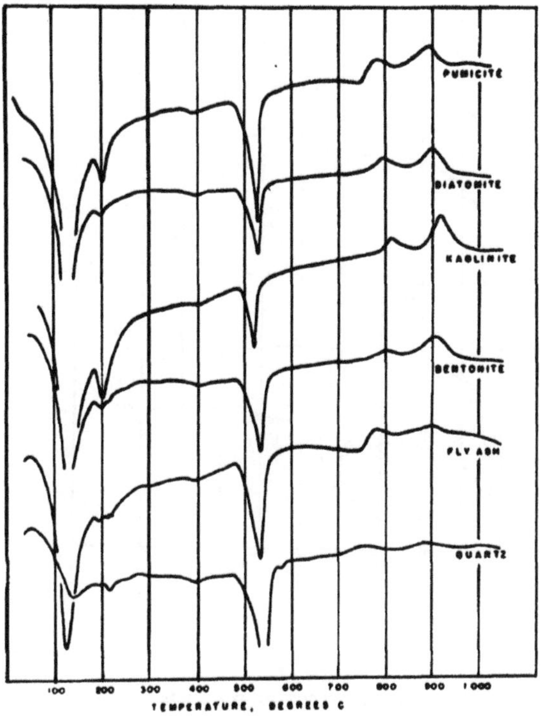

Fig. VIII.3.3: DTA curves of reaction products of mixtures containing 4 g pozzolana, 16 g high alkali type III cement, 15 ml water, and reacted for 11 months at 38°C (*15*).

of mixtures with type III cement have an endothermal effect at 200°C, with a doublet in some due to the presence of C_4AH_x and a related sulfate or a carbonate complex. The reactive pozzolanas always exhibit smaller endothermal peaks for $Ca(OH)_2$ decomposition.

The endothermic effects at 200°C for mixes with type V cement are generally smaller in magnitude than the corresponding effects with type III cement. It could be argued that with type V cement, C_4AH_x or a related sulfate or carbonate is not present as a reaction product. It is also noticed that the intensity of high temperature exothermic peaks is greater in mixes containing more active pozzolanas.

Chebukov and Egorova (*20*) used the DTA technique to examine the reaction products in a mixture consisting of 25% Portland cement, 30% lime-fly ash mixture, and 45% unground fly ash at a water/cement ratio of 0.5. Cements were either cured under normal conditions of exposure or in steam. Thermograms of the products cured at intervals of 2, 7, and 28 days

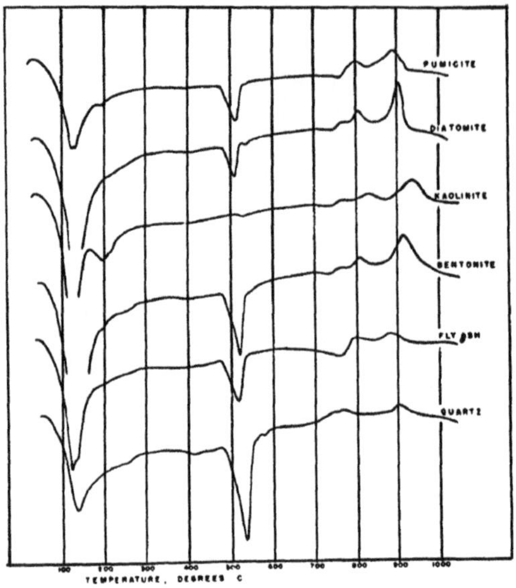

Fig. VIII.3.4: DTA curves of reaction products of mixture containing 4 g pozzolana, 16 g low alkali type V cement, 15 ml water, and reacted for 11 months at 38°C (*15*).

demonstrated the presence of calcium hydroaluminate by an endothermic effect at 190-200°C, of $Ca(OH)_2$ by an endothermic effect at 475-500°C, and of calcium silicate hydrate by an endotherm at 780-800°C.

REFERENCES

1. Lea, F.M., *Proc. 2nd Symp. Chem. Cements*, Stockholm, p. 460, 1938.
2. *Pozzolanic Clays of India: Their Industrial Exploitation and Use in Engineering Works*, Special Report No. 1, Central Road Research Institute, India, p. 152, 1962.
3. Strätling, W., *Zement*, *29*, 311, 1940.
4. Strätling, W., and Zur Strassen, H., *Z. Anorg. Allgem. Chem.*, *245*, 257, 1940.
5. Turriziani, R., and Schippa, G., *Ric. Sci.*, *24*, 366, and 1895, 1954.
6. Fratini, N., and Turriziani, R., *Ibid*, *24*, 1654, 1954.
7. Turriziani, R., *Ibid*, *24*, 1709, 1954.
8. Turriziani, R., and Rio, A., *Ann. Chim.*, *44*, 787, 1954.
9. Turriziani, R., and Schippa, G., *Ric. Sci.*, *26*, 3387, 1956.
10. Turriziani, R., *Silicates Ind.*, *23*, 181 and 265, 1958.
11. Rey, M., *Silicates Ind.*, *22*, 533, 1957.

12. Leonard, R.J., *The Pozzolanic Activity of Certain Fly Ashes and Soil Minerals*, Ph.D. Thesis, Iowa State College, 1958.
13. Ariizumi, A., *Semento Gijutsu Nenpo, 13*, 339, 1959.
14. Ludwig, U., and Schwiete, H.E., *Proc. 4th Intern. Symp. Chem. Cements*, Washington, pp. 1093-1098, 1960.
15. Benton, E.J., *Highway Res. Board Bull. 239*, 56, 1960.
16. Fernand, G., *Intern. Congr. Cements*, Lisbon, pp. 193-231, 1960.
17. Wessley, von I., *Zement-Kalk-Gips, 14*, 284, 1961.
18. Surovkin, V.M., and Krylov, G.M., *Uzbeksk Khim. Zh., 6*, 68, 1962.
19. Jambor, J., *Mag. Concrete Res., 15*, 131, 1963.
20. Chebukov, M.F., and Egorova, A.M., *Zh. Prikl. Khim., 37*, 255, 1964.
21. Mielenz, R.C., White, L.P., and Glantz, O.J., *Am. Soc. Test. Mat., Special Publ. 99*, pp. 43-91, 1950.
22. Mackenzie, R.C., Greene, K.T., and Schieltz, N.C., *Econ. Geol., 46*, 311, 1951.

VIII.4. OIL WELL CEMENT

Cements with special properties are used to seal oil and gas wells. These cements are particularly used in the petroleum industry in which the steel casing of the wells is cemented to the rock. An oil well cement under these conditions has to withstand temperatures up to about 175°C and pressures of about 18,000 $lb/in.^2$ In addition, it has to remain fluid for a few hours and then rapidly harden and also should be resistant to corrosive conditions. Cements in which C_3A content is negligibly small, pozzolanic cements, or cements with a suitable retarder are used as oil well cements.

Budnikov *et al* (*1*) were interested in the high temperature behaviour of cements and investigated the stability and mineralogical characteristics of synthetic mixtures containing $CaO/SiO_2=0.8$, 1, or 2 by subjecting them to steam pressures of 700 atm at 200°C or to 16 atm at 200°C. The techniques used for this study included X-ray diffraction, microscopy, and DTA.

Differential thermal curves showed the formation of crystalline C_2SH with endothermic effects at 680 and 730°C in mixtures containing $CaO/SiO_2=2$. Mixtures containing $CaO/SiO_2=1$ showed xonotlite, mixtures with a ratio 0.8 showed xonotlite, tobermorite, and fibrous $CaO \cdot SiO_2 \cdot H_2O$. The presence of tobermorite was indicated by an endothermic effect at 230°C and of the fibrous product, by an exothermic effect at 845°C. In general, the products formed at 700 atm were more homogeneous than those obtained at 16 atm.

REFERENCES

1. Budnikov, P.P., Lopatnikova, L.Y., and Dmitriev, A.M., *Dokl. Akad. Nauk, SSSR, 134*, 591, 1960.

VIII.5. HYDRAULIC LIME

Certain limestones containing an argillaceous material, when calcined at temperatures 950-1200°C, give a cementitious product. The calcined material hardens under water and is known as a hydraulic lime.

The cementing action in hydraulic limes is attributed to the formation of hydrated calcium silicates and aluminates. Standard specifications define the hydraulicity of these limes on the basis of chemical composition and strengths. The classification of hydraulic limes according to the Indian Standard Specification is given in Table VIII.5.1.

The nature and amount of cementing materials in hydraulic limes are not exactly the same as in Portland cement. Azbe, Bradley, and Clark (*1*) reported the presence of CaO, MgO, β-C_2S, and C_3A in hydraulic limes. Roberts (*2*) identified the presence of β-C_2S, γ-C_2S, $Ca(OH)_2$, γ and β-CS, C_4AF, C_2AS, and C_3S_2 in some of the samples he studied.

DTA could have been applied to obtain information on the polymorphism of C_2S, and obviously this method was not used. Some work was done on the products formed by reacting CaO with pure clay minerals by Ludwig, Muller-Hesse, and Schwiete (*3*).

Differential Thermal Analysis

An extensive application of DTA to a study of the hydraulic limes is to be found in the work of Mohan Rai (*4*). He investigated 7 argillaceous limestones and 23 artificially prepared mixtures containing pure $CaCO_3$ and clay minerals such as kaolinite, illite, montmorillonite, and nontronite. The X-ray diffraction results were obtained for 4 artificial hydraulic limes formed with 4 parts of $CaCO_3$ and 1 part of kaolinite, illite, montmorillonite, or nontronite and calcined at 1150°C.

The results show that kaolinite-based lime contains γ-C_2S and C_3A, illite-based lime contains β-C_2S, C_4AF, and C_3A, montmorillonite-based lime contains γ-C_2S and C_3A, and nontronite-based lime contains β-C_2S, C_4AF, and C_3A as the principal constituents. An attempt was made to obtain a quantitative estimation of C_2S, C_3A, and C_4AF of the above mixes by the method of Jander and Hoffman (*5*). These results are compared with strength data (Table VIII.5.2).

Highest strength is observed in products with illite as the raw material and this is presumably due to the presence of larger amounts of C_2S. Lowest strengths in nontronite correspond to the presence of lowest amounts of C_2S. Mohan Rai concluded that under comparable conditions of calcination the compressive strengths of hydraulic lime with various clay minerals were in the

TABLE VIII.5.1

Indian Standard Specification (IS: 712-1964) for Building Limes [by permission of I.S.I. New Delhi]

Class and purpose	*CaO+MgO %**	*Sol. SiO_2 + Al_2O_3 + Fe_2O_3, %*	*Insol, %*	*CO_2, %*	*Compressive strength 1:3 mortar, lb/in.²*	*Slaking method*
A. Eminently hydraulic	70 min.	25 min.	—	5 max.	250 (14 days)	1 hr at 50
	70 max.	—	2 max.		400 (28 days)	or 100°C
B. Semi-hydraulic	70 min.	15 min.	3 max.	5 max.	—	Total time
C. Non-hydraulic	85 min.	—	—	5 max.	—	16 or 24 hr

* No building lime should contain more than 5% MgO.

TABLE VIII.5.2
Composition of Artificial Hydraulic Limes and Their Strengths [Mohan Rai (4)]

Starting material (1:4)	C_2S, %	C_3A, %	C_4AF, %	*Compressive strength of lime: sand (1:3) mortar at 180 days (lb/in.²)*
Kaolinite: $CaCO_3$	30.69	12.42	0.66	1523
Illite: $CaCO_3$	47.72	6.08	1.87	2330
Montmorillonite: $CaCO_3$	35.32	5.19	1.59	1523
Nontronite: $CaCO_3$	29.63	1.20	4.56	1478

order illite>montmorillonite>kaolinite>nontronite. The hydraulic limes were formed at 1150°C at a retention time of 3 hr and hence conclusions cannot be applied to these mixes fired to near-equilibrium conditions.

Thermograms of hydrated hydraulic limes were obtained after a hydration period of 7 days. The interpretation of various peaks is given in Table VIII.5.3.

An explanation of the DTA curves in terms of the amounts of products formed would facilitate interpretation of strength data, especially if the curves had been obtained at longer periods of hydration, but Mohan Rai made no attempts to estimate the products of hydration quantitatively.

REFERENCES

1. Azbe, V.J., Bradley, W.F., and Clark, G.L., *Ind. & Eng. Chem.*, *32*, 972, 1940.
2. Roberts, M.H., *Cement and Lime Manufacture (London)*, *29*, 27, 1956.
3. Ludwig, U., Müller-Hesse, H., and Schwiete, H.E., *Zement-Kalk-Gips*, *13*, 449, 1960.
4. Mohan Rai, *Physico-Chemical Aspects of Hydraulic Lime*, Ph.D. Thesis. Banaras Hindu University, India, 1963.
5. Jander, W., and Hoffman, E., *Angew. Chem.*, *46*, 76, 1933.

TABLE VIII.5.3
Hydration Products of Artificially Prepared Hydraulic Limes According to DTA [Mohan Rai (4)]

Kaolinite		*Nontronite*		*Illite*		*Montmorillonite*	
Peak temp., °C	*Identification*	*Peak temp., °C*	*Identification*	*Peak temp., °C*	*Identification*	*Peak temp., °C*	*Identification*
100 (En)	H_2O	90 (En)	H_2O	100 (En)	H_2O	100 (En)	H_2O
160 (En)	CSH	160 (En)	CSH; C_4FH_{14}	160 (En)	CSH; C_4FH_{14}	160 (En)	CSH
300 (En)	C_3AH_6	300 (En)	C_3AH_6; C_2FH_5	300 (En)	C_3AH_6; C_2FH_5	300 (En)	C_3AH_6
—	—	—	—	—	—	400 (En)	$Mg(OH)_2$ (?)
500 (En)	$Ca(OH)_2$	500 (En)	$Ca(OH)_2$	500 (En)	$Ca(OH)_2$	500 (En)	$Ca(OH)_2$
—	—	620 (Ex)	?	630 (En)	KCS.aq.	620 (En)	(N,K) CS.aq.
				650 (Ex)	?		
740 (En)	C_2S (trans)	730 (En)	C_2S (trans)	740 (En)	C_2S (trans)	700 (En)	C_2S (trans)
850 (En)	CSH	850 (En)	CSH	—	—	850 (En)	CSH
980 (Ex)	?	980 (Ex)	?	1040]Ex)	?	—	—

Note: En = Endothermic; Ex = Exothermic; trans = transition.

VIII.6. EXPANDING CEMENT

Concretes containing Portland cement and aggregates exhibit a linear shrinkage of about 0.05-0.1% under normal conditions of exposure due to evaporation of water and carbonation. The so-called expanding cement is obtained by adding certain components to normal Portland cement to cause an expansion equal to or greater than the shrinkage.

Lossier (*1*) produced an expanding cement by burning a mixture of bauxite, calcium carbonate, and gypsum and mixing it with Portland cement. Budnikov and Kosuireva (*2*) prepared an expanding cement by adding a mixture of gypsum and a hydration product of lime–pozzolana reaction to Portland cement. The cement produced by Klein and co-workers (*3, 4*) was formed by firing at 1350°C a mixture of calcium hydroxide or calcium carbonate with gypsum and bauxite or aluminum sulfate. Recently, Slatanoff and Djabaroff (*5*) have reported that expansion can be brought about by adding periclase (dead burnt MgO) to cement.

It is generally believed that the formation of sulfoaluminate of the formula $3CaO \cdot Al_2O_3 \cdot 3CaSO_4 \cdot 30\text{–}32H_2O$ (ettringite) is the basis for expansion in cements produced by Lossier, Budnikov and Klein.

Differential Thermal Analysis

According to Lafuma (*6*), the expanding agent described by Lossier (*1*) contained a mixture of calcium sulfate, $5CaO \cdot 3Al_2O_3$, and β-$2CaO \cdot SiO_2$. Klein and Troxell (*3*) examined the products by gradually heating a mixture of gypsum, alumina, and calcium hydroxide in a DTA furnace up to a temperature of 1300°C. The DTA of raw materials and two mixtures are shown in Fig. VIII.6.1. Gypsum shows two endothermal peaks at 195 and 230°C, indicating a characteristic stepwise dehydration of water molecules. An endothermic peak at 1240°C is caused by an inversion of monoclinic $CaSO_4$ to rhombohedral form. The endothermic peak at 560-570°C is due to the dehydration of $Ca(OH)_2$. The endothermic peak at 840°C in mixtures indicates the formation of nuclei of a compound not detected by X-ray. This may be a nucleation of the ultimate calcium sulfoaluminate component.

Chopra and Mohan Rai (*7, 8*) used DTA technique to investigate the expanding agent formulated by Budnikov and Kosuireva (*2*). According to Budnikov and Kosuireva the following reactions occur:

$$Al_2O_3 + 2SiO_2 + 4Ca(OH)_2 + aq \longrightarrow 2(CaO \cdot SiO_2 \cdot aq) + 2CaO \cdot Al_2O_3 \cdot 7H_2O$$

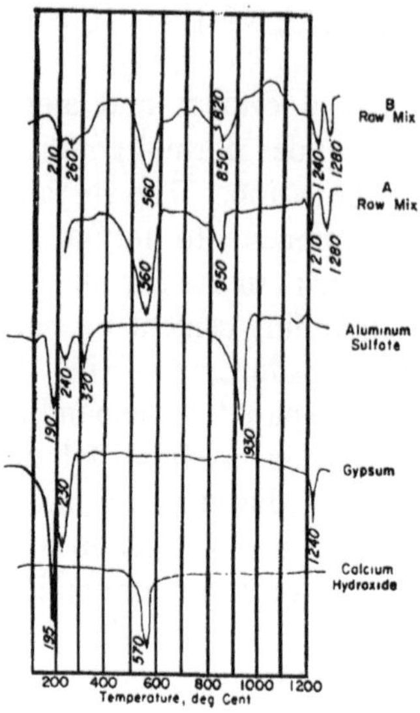

Fig. VIII.6.1: DTA of raw materials and raw mixes A and B (*3*).

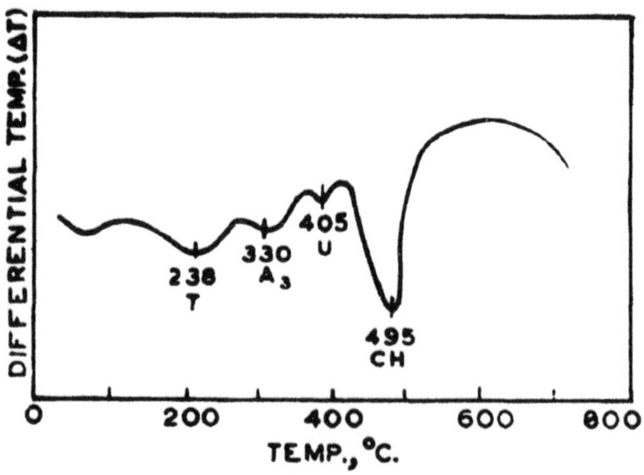

Fig. VIII.6.2: Thermograms of reaction product of lime and kaolin (*7*).

$$3(2CaO \cdot Al_2O_3 \cdot 7H_2O) + 6CaSO_4 \cdot 2H_2O + aq \longrightarrow$$
$$2(3CaO \cdot Al_2O_3 \cdot 3CaSO_4 \cdot aq) + Al_2O_3 \cdot aq$$

According to Chopra and Mohan Rai (Fig. VIII.6.2), in addition to $2CaO \cdot Al_2O_3 \cdot 7H_2O$, large quantities of $4CaO \cdot Al_2O_3 \cdot aq$ are formed in the products of hydration of lime and kaolin. The endothermic effect at 238°C is attributed to the formation of tetracalcium aluminate hydrate. The unreacted $Ca(OH)_2$ at 10 days was indicated by an endothermic peak at about 495°C. The results were confirmed by X-ray diffraction studies.

The expanding cement developed by Klein and Troxell (*3*) consisted of a mixture of Portland cement with an appropriate mixture of $4CaO \cdot 3Al_2O_3 \cdot SO_3$, $CaSO_4$, and $Ca(OH)_2$. The hydration products of the system $C_4A_3\bar{S}$–$C\bar{S}$–C–H was studied by DTA at periods of 6, 12, 24, 48, and 72 hr (*9*). The formation of high sulfoaluminate was distinguished by a strong peak at 190°C and a weaker peak at 270°C. In mixtures containing a low proportion of the high sulfoaluminate two peaks at 150 and 250°C were observed. The monosulfoaluminate could be identified in presence of the high sulfoaluminate by two peaks at 230 and 300°C. Except in pastes with $CaO/SO_3 > 4$ and $Al_2O_3/SO_3 > 1$, the high sulfoaluminate was the stable phase at other concentrations.

REFERENCES

1. Lossier, H., *Trav.*, *27*, 203, 1943.
2. Budnikov, P.P., and Kosuireva, Z.Z., *Dokl. Akad. Nauk SSSR*, *61*, 681, 1948.
3. Klein, A., and Troxell, G.E., *Proc. Am. Soc. Test. Mater.*, *58*, 986, 1958.
4. Klein, A., Karly, T., and Polivka, M., *J. Am. Concrete Inst.*, *58*, 59, 1961.
5. Slatanoff, V., and Djabaroff, N., *Rev. Matér. Constr.*, *544*, 30, 1961.
6. Lafuma, H., *Proc. 3rd Intern. Symp. Chem. Cements*, London, p. 574, 1952.
7. Chopra, S.K., and Mohan Rai, *J. Sci. Ind. Res. (India)*, *21D*, 235, 1962.
8. Chopra, S.K., and Mohan Rai, *Cement and Lime Manufacture*, *35*, 17, 1962.
9. Mehta, P.K., and Klein, A., *Symposium on Structure of Portland Cement and Concrete*, Highway Res. Board, Special Report No. 90, Washington, pp. 328-352, 1966.

VIII.7. ACID-PROOF CEMENT

Acid-proof cements are used for the lining of equipments that come into contact with corrosive liquids and vapours. Acid-proof cement contains water glass, sodium fluosilicate, and a filler such as quartz sand, andesite, or diabase.

Dombrovskaya and Mitelman (*1*) examined the DTA of six compositions of acid-proof cements containing water glass of modulus $(SiO_2/Na_2O)=1.2$ to 3.3. In all the samples an irreversible endothermal effect at 100°C was observed, corresponding to the loss of water. A reversible endothermal effect at 575°C signified the $\alpha \rightleftharpoons \beta$ transition in quartz and an endothermic effect at about 550°C was attended by a violent decomposition of Na_2SiF_6 to NaF and SiF_6. Repeated heating of the cements resulted in only a single endothermic effect at 575°C.

It was concluded from DTA that the cement can be satisfactorily used at temperatures up to 575°C, provided it is initially heated to 100°C. In addition, excess of Na_2SiF_6 is to be avoided in the mix as it has a tendency to decompose violently at 542-556°C. Cements with a water glass of high modulus differ from those with a low modulus in exhibiting an endothermal effect at 542°C.

REFERENCES

1. Dombrovskaya, N.S., and Mitelman, M.R., *Zh. Prikl. Khim.*, *26*, 18, 1953.

IX. Other Non-Portland Cements

IX.1. MAGNESIUM OXYCHLORIDE AND OXYSULFATE CEMENT

Sorel (*1*) in 1867 announced the discovery of an excellent cement formed by the addition of magnesium chloride solution to magnesium oxide. The cement is variously known as Sorel cement, magnesite cement, plastic magnesia cement, and magnesium oxychloride cement. The material is now popularly known as magnesium oxychloride cement. This cement has many properties superior to those of Portland cement. It does not need wet curing, has high fire resistance, low thermal conductivity, very good resistance to abrasion, warmth, and high transverse and crushing strengths. In addition, this cement forms a good bond with various types of organic and inorganic aggregates, gives high early strength, is insecticidal, resilient, conducting, light, and unaffected by grease, oils, and paints.

Different compositions of oxychloride cement are known and the mixes are prepared to suit a particular need. The usual formulations may contain magnesium oxide and magnesium chloride mixed with saw dust, wood flour, chalk, marble flour, soapstone flour, asbestos powder, sand, ground quartz,

TABLE IX.1.1
Composition and Strength Characteristics of Magnesium Oxychloride Cement [Mohan Rai (2)]

	Composition (Wt. %)					
No.	*MgO*	*Saw dust*	*Pig-ment*	*Coarse sand*	*Fine sand*	*Stone aggregates*
1	60	15	5	20	Nil	Nil
2	50	15	5	20	10	Nil
3	40	15	5	20	20	Nil
4	30	15	5	20	30	Nil
5	20	15	5	20	40	Nil
6	20	15	5	20	40	200

stone chips and gravel, bitumen emulsion, magnesium sulfate, and pigments. The strength data of some of the oxychloride compositions are given in Table IX.1.1.

The mix containing cement and aggregate in the ratio 1:14 gives a crushing strength of 2360 lb/in.^{2} and a corresponding mix with Portland cement may yield a strength of only 600 lb/in.^{2}

The magnesium oxychloride cements are used in making adhesive and insulating compositions, artificial stones, dental cements, tiles, refractory bricks, paints for plaster or foam concrete walls, and jointless flooring.

One of the main reasons why the use of oxychloride cement is not very popular in building industry is that it disintegrates on prolonged contact with water.

Chemistry

The hardening of magnesium oxychloride cement is due to the chemical reaction between MgO and $MgCl_2$ solution. Magnesium oxide is obtained by the calcination of magnesite at 800-1000°C, while magnesium chloride hexahydrate is obtained as a by-product in the sea salt industry.

The cement was denoted until 1932 by the general formula $x\text{MgO}\cdot\text{MgCl}_2\cdot y\text{H}_2\text{O}$, with x varying from 1 to 10 and y varying from 10 to 21. Lukens (*3*), Bury, and Davies (*4*) and Chassevent (*5*) showed by X-ray technique that x could have a value of 3 or 5. The compound having $x=3$ had a value of 11 for y and for that having $x=5$, the value of y was not known.

Feitknecht and Held (*6*) showed the existence of two compounds with the formula $3\text{Mg(OH)}_2\cdot\text{MgCl}_2\cdot7\text{H}_2\text{O}$ and $5\text{Mg(OH)}_2\cdot\text{MgCl}_2\cdot7\text{H}_2\text{O}$. Later workers (*7-12*) concluded that both the complexes reported by Feitknecht and Held contain 8 molecules of water. In the system $\text{MgO–MgCl}_2\text{–H}_2\text{O}$, four complexes are known to exist. At lower temperatures the compounds

TABLE IX.1.1 *(Continued)*

Tensile strength (lb/in.^{2})			*Transverse strength (lb/in.^{2})*			*Compressive strength (lb/in.^{2})*		
3 days	*7 days*	*28 days*	*3 days*	*7 days*	*28 days*	*3 days*	*7 days*	*28 days*
680	900	1050	1050	1350	2350	4380	5900	7570
570	640	880	930	1270	1890	4190	5180	6990
470	560	800	780	1150	1780	3840	4600	6930
410	550	740	700	1070	1550	3670	4250	5870
300	460	630	580	850	1220	2420	3270	4270
—	—	—	300	420	560	1520	1800	2360

formed are $3Mg(OH)_2 \cdot MgCl_2 \cdot 8H_2O$ (3-form) and $5Mg(OH)_2 \cdot MgCl_2 \cdot 8H_2O$ (5-form) and at temperatures higher than 100°C the two compounds formed are $9Mg(OH)_2 \cdot MgCl_2 \cdot 5H_2O$ (9-form) and $2Mg(OH)_2 \cdot MgCl_2 \cdot 4H_2O$ (2-form). It has also been inferred that the 3-form is always preceded by the formation of 5-form. Higher initial concentrations of magnesium chloride promote the formation of the 3-form.

Differential Thermal Analysis

Magnesium Chloride Hexahydrate. Differential thermal studies of magnesium oxychloride complexes have been reported by several workers (*10-16*). The solids formed in the system $MgO–MgCl_2–H_2O$ exhibit complex thermal effects. It has been reported that MgO on treatment with magnesium chloride solution contains, in addition to oxychloride complexes, magnesium hydroxide and magnesium chloride. Mutual interference in the thermal effects makes identification difficult. For instance, the thermal curve of $MgCl_2 \cdot 6H_2O$ alone is so complex that all the effects have not been accounted for with certainty (*17-20*). Mitchell [cited by Mackenzie (*19*)] reported four endothermal peaks below 250°C and three at 297, 488, and 700°C. Berg, Nikolaev, and Rode (*17*) obtained three peaks below 250°C and two at 300 and 550°C. Berg and Burmistrova (*18*) found three peaks below 250°C and three more at 295, 545, and 665°C. A careful examination of $MgCl_2 \cdot 6H_2O$ indicated six endothermal peaks in the thermogram (*20*). Details of these results are given in Table IX.1.2.

Magnesium Oxychloride Complexes. The DTA and DTG (differential thermogravimetric analysis) results of four oxychloride complexes are shown in Fig. IX.1.1. A good correlation between DTA and DTG results is evident. In general the peak of any reaction occurs at a higher tempera-

TABLE IX.1.2
DTA Results on $MgCl_2 \cdot 6H_2O$* [Behl and Gaur (20)]

Peak No.	*Reaction temp., °C*	*Phase change*
1	70	$MgCl_2 \cdot 6H_2O \longrightarrow MgCl_2 \cdot 4H_2O + 2H_2O(g)$
2	100	$MgCl_2 \cdot 4H_2O \longrightarrow MgCl_2 \cdot 2H_2O + 2H_2O(g)$
3	116	$MgCl_2 \cdot 2H_2O \longrightarrow MgCl_2 \cdot 2H_2O(l)$
4	150	$MgCl_2 \cdot 2H_2O(l) \longrightarrow MgCl_2 \cdot H_2O(s) + H_2O(l)$
5	196	$H_2O(l) \longrightarrow H_2O(g)$
6	236	$MgCl_2 \cdot H_2O \longrightarrow MgCl_2 + H_2O(g)$

* Reaction temperature corresponds to the commencement of reaction peak. All effects are endothermal.
At the endoethermal effects corresponding to peak Nos. 3, 4 and 6, water was condensing on cooler parts.
At peak No. 5 the reaction was accompanied by partial hydrolysis.

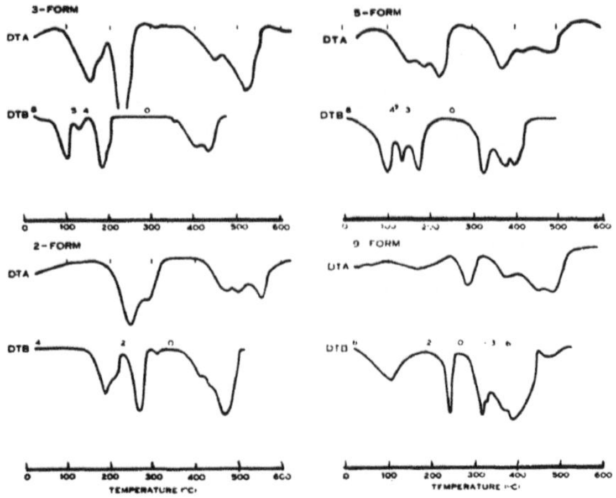

Fig. IX.1.1: Differential thermal analysis and differential thermobalance curves of four forms of oxychloride cement (*11*).

ture in DTA. In all the four complexes the water of hydration is lost in a stepwise manner below 360°C. The 3-form shows three endothermal peaks corresponding to loss of water.

The DTG also shows three effects but the second peak is not well defined. The first effect at 145°C is believed to be due to loss of 3 moles of H_2O, the second at 170°C due to loss of 1 mole of water, and the third at 220°C due to loss of 4 moles of remaining water. Similar results were also obtained by Newman (*12*).

Both DTA and DTG of the 5-form indicate three low temperature inflections. The first effect corresponds to loss of 4 moles of water, the second to loss of 1 mole of water, and the third to loss of 3 moles of water. The 2-form of complex has two low-temperature endothermal peaks, each effect resulting from the expulsion of 2 moles of water. A broad hump and a sharp endothermal effect are shown by the 9-form. The endothermal hump is due to a slow dehydration of about 4 moles of water and the sharp endothermal effect is due to the dehydration of 2 moles of water.

Several endothermal effects observed in all the complexes beyond 360°C represent the overall effect of a number of reactions. The complexes dehydrate to form a compound $n\mathrm{Mg(OH)_2 \cdot MgCl_2}$ at about 360°C and this decomposes to the end product MgO. The 5-, 2- and 9-forms exhibit three endothermic effects each owing to the decomposition of the anhydrous form, whereas the 3-form shows only two endothermal effects. The course of these reactions

TABLE IX.1.3
Differential Thermal Effects of Magnesium Oxychloride Complexes [Mackenzie (19)]

No.	*Oxychloride complex*	*Endothermal effects due to dehydration, °C*	*Decomposition of anhydrous phases, °C*
1	$2Mg(OH)_2 \cdot MgCl_2 \cdot 4.14H_2O$	245 ml	462 ml
		284 ml	486 ml
			546 ml
2	$3Mg(OH)_2 \cdot MgCl_2 \cdot 7.7H_2O$	145 m	380 l
		170 m	430 l
		220 ml	505 l
3	$5Mg(OH)_2 \cdot MgCl_2 \cdot 7.9H_2O$	160 ml	365 l
		185 ml	410 l
		220 ml	480 l
4	$9Mg(OH)_2 \cdot MgCl_2 \cdot 5.6H_2O$	173 s	368 l
		292 ms	452 l
			481 l

Note: l=large, ml=medium large, m=medium, ms=medium small, s=small.

is formulated as follows (*11*):

$$nMg(OH)_2 \cdot MgCl_2 \xrightarrow{\text{endothermic}} nMgO + (n\text{-}1)H_2O + H_2O + MgCl_2$$

$$\xrightarrow[\text{exothermic}]{\text{endothermic followed by}} MgO + 2HCl \qquad (\text{where } n = 2, 3, 5 \text{ or } 9)$$

Table IX.1.3 summarizes the thermal effects reported by Mackenzie (*19*) based on the results of Cole and Demediuk (*11*).

MgO–$MgCl_2$–H_2O System. Admoavichyute and others (*15*) studied the nature of compounds formed after 3 months in a mixture containing $MgCl_2$ and MgO in the ratio of 0.6, 0.4, and 0.2. With $MgCl_2/MgO > 0.4$, the 3-form predominated with the formula $3Mg(OH)_2 \cdot MgCl_2 \cdot 5H_2O$, and with $MgCl_2/MgO$ between 0.4 and 0.2, both oxychloride complex and magnesium hydroxide were observed. If the ratio was kept less than 0.2 the product was $Mg(OH)_2$.

A detailed work on the thermal analysis in the system MgO–$MgCl_2$–H_2O carried out by Vyrodov (*13*) contains 120 curves. Mixes with different concentrations of $MgCl_2$ up to 40% were prepared and the effect of drying conditions on the thermal curves noted. Thermal effects of some of the compositions obtained by Vyrodov are summarized in Table IX.1.4. A comparison of the thermal effects of pure $MgCl_2 \cdot 6H_2O$ and $Mg(OH)_2$ with those of the mixtures may apparently indicate mutual interference. However, conclusions can be drawn by comparing the intensity of the effects with

TABLE IX.1.4

Differential Thermal Analysis Data of the System $MgO–MgCl_2–H_2O$ (Peak Temperatures, °C) [Vyrodov (13)]

Composition	*20-100°*	*100-200°*	*200-300°*	*300-400°*	*400-500°*	*500-600°*	*600-700°*	*700-800°*
1	*2*	*3*	*4*	*5*	*6*	*7*	*8*	*9*
$MgCl_2 \cdot 6H_2O$	110	140	200 s	305 m	—	530 w	690 w	720 w
$Mg(OH)_2$	—	—	—	—	410 s	—	—	740
$MgO:MgCl_2:H_2O$								
1.0:1:8	100 +	190 s	—	—	450 +	535 m	—	720 w
2.0:1:8	—	180 s	215 m 270 +	—	455 w	520 m	—	720
2.0:1:26	—	200 s	—	380 +	450 m	520	620	715
2.0:1:30	100	150 s 190 +	—	385 +	450 m	515 w	690	760
3.0:1:8	—	—	220 s	—	500 m	565 m	—	720 w
3.0:1:26	—	150 s	—	380 w	445 w	510 w	—	705 w
5.0:1:14	—	190 s	230 s	400 m	445 +	530 m	—	720
5.0:1:18	—		210 s	390 +	460 m	525 m	—	720
5.0:1:26	100	150 s 170 s	—	380 +	450 m	515 m	—	705

6.0:1:35	—	185 s	—	385 +	460 m	525 +	600 +	730
6.2:1:10	—	—	240 m	—	480 s	560 +	—	725
7.0:1:26	—	190 s	—	380 +	450 m	515 w	—	—
7.0:1:35	—	—	—	380 m	450 m	525 w	—	—
7.2:1:10	—	—	225 m	—	470 s	540 +	—	—
7.2:1:12	—	—	220 m	—	410 s	520 +	—	—
9.0:1:24	—	—	230 s	—	475 s	—	—	—
9.0:1:26	—	—	220 s	380 +	460 m	545 m 565 +	—	720
9.0:1:35	100	150 + 175 s	—	385 +	455 m	515 w	—	—
9.0:1:259	—	140 s	—	—	415 m	—	—	—
10.0:1:22	—	—	230 s	—	430 s	—	—	—
12.0:1:35	—	150 m 170 s	—	380 m	450 m 495 +	—	—	—
15.0:1:35	—	150 + 170 s	—	380 +	455 m	515 +	—	—
15.0:1:40	—	160 + 175 s	—	—	405 s 465 +	525 +	—	—
18.0:1:40	—	150 m	—	380 +	—	515	—	—
21.0:1:50	—	190 m	—	—	420 s	—	—	—
24.0:1:50	—	—	210 w	—	440 s	—	—	—
48.0:1:259	—	185 s	—	400 m	—	—	—	—
64.0:1:206	—	120 s	—	—	—	465 m	—	—

Designation of the peaks: no designation=very weak; w=weak; m=moderate; s=strong; +=unresolved, superposition effect.

changing ratios of $MgO/MgCl_2$ in the starting material.

Oxychloride Cements from Partially Calcined Dolomites. Ramachandran, Kacker, and Mohan Rai (*21*) presented data on magnesium oxychloride cement made from magnesium chloride solution and partially calcined dolomite obtained at different temperatures and retention times. Mixes were prepared at standard consistency, and thermal behavior and compressive strengths were determined at different periods of curing. Attempts were made to correlate the thermal behavior with strength results. DTA results could also be used to determine optimum conditions of temperature and retention time of calcination.

The dolomite used for the study contained MgO=18.06%. The material was crushed to 20-30 mm and calcined at 700, 750, and 800°C at retention times of 1, 2, or 3 hr. The calcined material was ball-milled to pass through 100 BSS sieve and mixed with magnesium chloride solution of 22 Bé to a standard consistency according to the Indian Standard Specification 657: 1956. For calcined dolomite mixes the weight ratio $MgO/MgCl_2 \cdot 6H_2O$ varied from 1.5 to 2.1 and for calcined magnesite with precipitated calcium carbonate as the filler the ratio was only 0.58. Compressive strengths were determined on cubes $2 \times 2 \times 2$ in.^{2} Differential thermal and thermogravimetric analyses were carried out on samples cured for 1, 2, 3, 7, and 28 days. A constant quantity of the sample ground to pass through 100 BSS sieve was used for DTA work.

The technique of DTA was applied by these workers as a prognostic tool for calcination of dolomite. The DTA of uncalcined and calcined dolomite is present in Fig. IX.1.2. The dolomite sample shows two large endothermal effects with peaks at 810 and 900°C (Fig. IX.1.2A).

The first endotherm represents the decomposition of $MgCO_3$ and the second, the decomposition of $CaCO_3$. The characteristic temperature at which $MgCO_3$ starts to decompose is about 700°C, and this is completed between 810 and 840°C. The onset of decomposition of $CaCO_3$ occurs at about 840°C. Hence, temperatures of 700, 750, and 800°C were chosen to arrive at an optimum retention time and temperature of calcination.

Typical curves of a few calcined samples are given in Fig. IX.1.2. The heat treatment at 700°C for 2 hr results only in a partial decarbonation of $MgCO_3$, as is evidenced by the persistence of a large endothermic peak at 810°C. A very small endothermal effect at 810°C for the sample calcined at 750°C for 2 hr indicates that a major portion of $MgCO_3$ is decomposed. The specimen fired to 800°C for 1 hr fails to exhibit an endothermal effect characteristic of $MgCO_3$ decomposition, with the possibility that partial decomposition of $CaCO_3$ has occurred.

Thermograms reveal that for oxychloride cements, the dolomite calcined

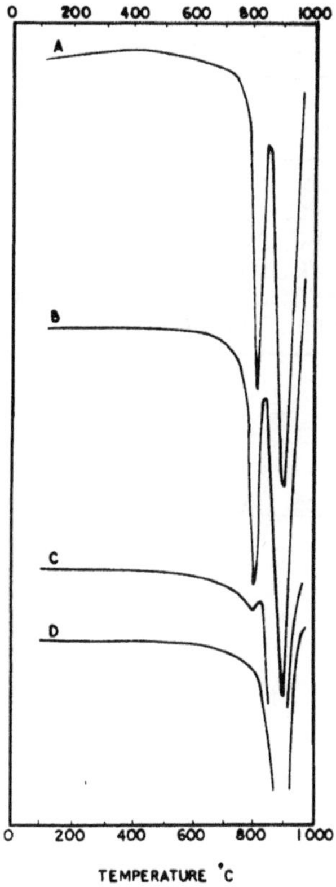

Fig. IX.1.2: DTA of calcined dolomite (*21*). (A) Dolomite; (B) 2 hr, 700°C; (C) 2 hr, 750°C; (D) 800°C, 1 hr.

to temperatures of 700 and 800°C tends to give lower strengths than that fired to 750°.C The persistence of a small endothermal effect due to $MgCO_3$ for the dolomite fired to 750°C should ensure that $CaCO_3$ is practically undecomposed. The strength data (Table IX.1.5) confirm the observations based on DTA.

In oxychloride cements the strength should depend on the type and the amount of the complex formed. Four compounds are known to be formed in cement pastes. At lower temperatures the compounds formed are $3Mg(OH)_2 \cdot MgCl_2 \cdot 8H_2O$ (referred to as 3-form) and $5Mg(OH)_2 \cdot MgCl_2 \cdot 8H_2O$ (5-form), the 3-form preceded always by the formation of 5-form. In

TABLE IX.1.5
Compressive Strength Data of Magnesium Oxychloride Cements [Ramachandran et al (21)]

Sample No.	*Source of MgO*	*Compressive strength* (kg/cm^2) *3 days*	*7 days*	*28 days*
1	1 MgO (from magnesite): 3 sand	211	323	338
2	Calcined dolomite (700°C, 2 hr)	42	47	59
3	Calcined dolomite (750°C, 2 hr)	264	539	591
4	Calcined dolomite (800°C, 1 hr)	48	139	146
5	Calcined dolomite (800°C, 2 hr)	98	106	79
6	1 MgO (from magnesite): 3 pptd.$CaCO_3$	42	71	106

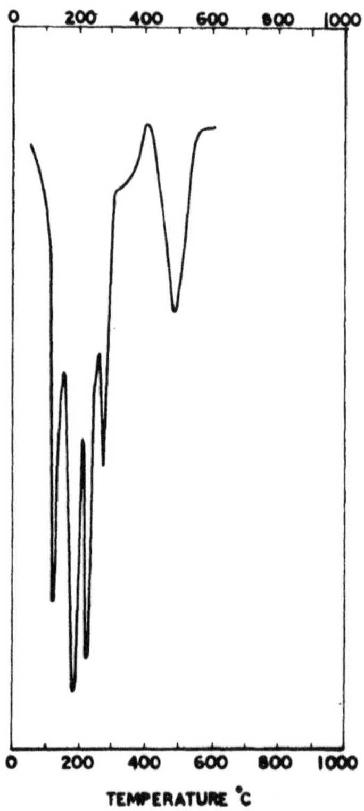

Fig. IX.1.3: DTA of $MgCl_2 \cdot 6H_2O$ (*21*).

the cement paste the complexes are formed at the expense of magnesium chloride and magnesium oxide.

Interpretation of DTA requires also a knowledge of the thermogram of pure $MgCl_2 \cdot 6H_2O$. In Fig. IX.1.3 the DTA of pure $MgCl_2 \cdot 6H_2O$ mixed with α-Al_2O_3 shows five endothermal peaks at 120, 185, 230, 275, and 480°C. These results are in accordance with the observations of Berg and co-workers (*17*, *18*). The effects below 275°C are attended by a stepwise dehydration of 6 molecules of water (*20*, *22*). Pure magnesium oxide does not show any inflection in DTA and is not shown.

In Fig. IX.1.4 is shown the DTA of oxychloride cement prepared from one

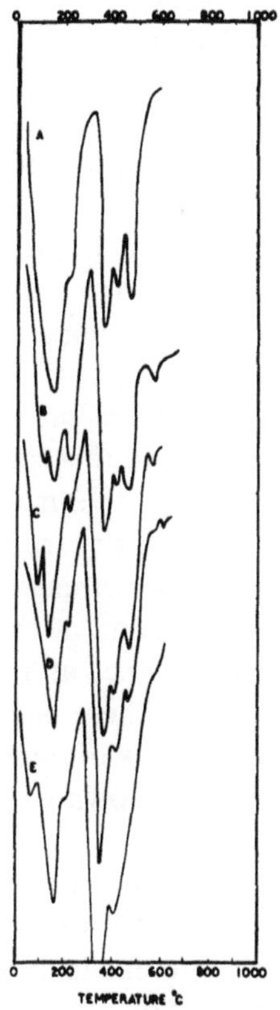

Fig. IX.1.4: Thermograms of oxychloride prepared from calcined magnesite. (A) 1 day, (B) 2 days, (C) 3 days, (D) 7 days, (E) 28 days (*21*).

part of calcined magnesite and three parts of sand, and at 1, 2, 3, 7, and 28 days after hydration. Earlier workers (*10-16*) have reported complex thermal effects of the pure system $MgO–MgCl_2–H_2O$ and these have not been fully explained. All the samples exhibit an intense endothermic effect in the range of about 50-200°C.

Interpretation of the endothermal effect in the low temperature range is not easy because the adsorbed water, $MgCl_2 \cdot 6H_2O$ and the 3- and 5-forms of oxychloride give endothermal effects within this temperature range. A single effect is observed on the 1-day sample, with a resolution on the 2-day sample with the intensity decreasing from the 2nd day onwards. A sharp decrease from the 1st to 2nd day can be attributed to a fall in $MgCl_2 \cdot 6H_2O$ content with the formation of a complex and to an increase in strength from 131 kg/cm^2 on the 1st day to 211 kg/cm^2 on the 3rd.

A small endothermic effect is noticed at 220-230°C. Both the 3- and 5-forms of the complex and magnesium chloride exhibit a peak at this temperature. The peak persists up to 28 days and may represent the presence of $MgCl_2 \cdot 6H_2O$, 5-form, and 3-form, the last predominating at longer periods of curing.

The peak at 370°C is significant and is present in all the thermograms shown in Fig. IX.1.4 and its intensity increases with time. Pure $MgCl_2 \cdot 6H_2O$ does not exhibit a peak at this temperature. At this temperature, $Mg(OH)_2$ and the 3- and 5-forms of oxychloride may exhibit endothermal changes. The fact that the intensity increases with time shows a consumption of $Mg(OH)_2$ to form the 5-form and this converting further to 3-form at longer periods of curing. At 28 days this may be mainly due to 3-form.

There is a sharp enhancement of intensity from the 3rd to the 7th day and this is expected to correspond to higher strengths. The strength at the 7th day shows a 50% increase over that found at the 3rd day. The intensity of 370°C effect does not show much variance at 7 and 28 days. The strength results show less than 5% increase at these periods.

An endothermic effect at 410-420°C is of equal intensity at all periods of curing. Both the 3- and 5-forms are known to exhibit a small endothermal effect in this range.

An endothermic peak at 470-480°C decreases in intensity with time and disappears altogether at 28 days. The 5-form and $MgCl_2 \cdot 6H_2O$ give a peak at this temperature and these are present in large quantity at one day but progressively decrease with curing period. The diminution is marked on the 7th day.

Fig. IX.1.5 refers to the DTA behaviour of oxychloride using calcined dolomite at 700°C for 2 hr. As already shown (Fig. IX.1.2B) this treatment only partly decomposes $MgCO_3$. Hence the strength and the amount of

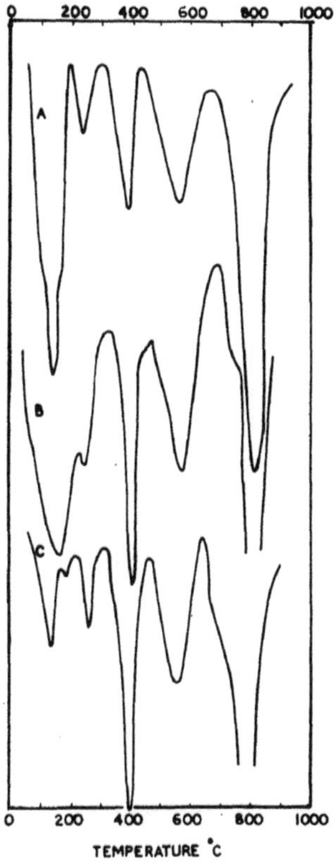

Fig. IX.1.5: DTA of oxychloride by calcination of dolomite at 700°C for 2 hr. (A) 3 days, (B) 7 days, (C) 28 days (*21*).

cementitious material formed should be lower than that obtained from dolomite fired to 750°C for 2 hr. The notable feature of the curves is an intense endothermic effect at 810-820°C, representing decarbonation of $CaCO_3$, and a moderately strong endothermic peak at 560-570°C, suggestive of the presence of $MgCl_2{\cdot}6H_2O$ with the peak shifted to a higher temperature.

The intensity of the peak at 400°C increases only gradually and the strength results also show the same trend, the values being 42, 47, and 59 kg/cm^2 at 3, 7, and 28 days respectively. In general, the peaks at 400°C for these samples are less intense than the corresponding peaks for the samples obtained at 750°C (Fig. IX.1.6).

DTA of the sample heated to 750°C for 2 hr shows that a major portion

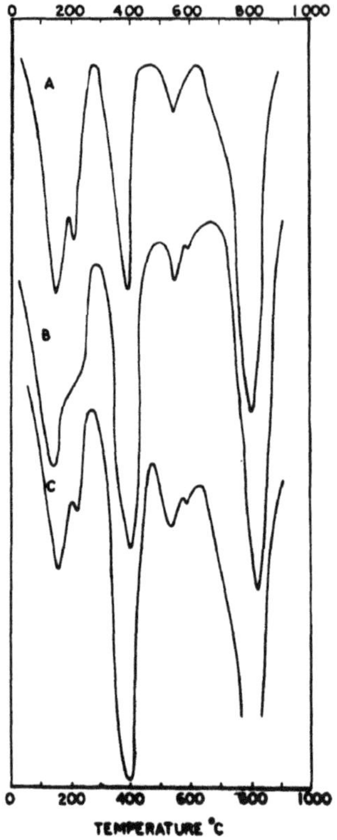

Fig. IX.1.6: DTA of oxychloride by calcination of dolomite at 750°C for 2 hr. (A) 3 days, (B) 7 days, (C) 28 days (*21*).

of $MgCO_3$ has decomposed to MgO (Fig. IX.1.2C). Hence the strength and the amount of oxychloride formation in these samples must be more than what is expected at 700°C for corresponding periods of curing. All the thermograms (Fig. IX.1.6A, B, and C) show large endothermic peaks above 800°C for the decarbonation of $CaCO_3$. The endothermic peak representing magnesium oxychloride complex at 390-400°C increases with time of curing. A substantial increase in intensity is observed between the 3rd and 7th day and the strengths at corresponding periods are 264 and 539 kg/cm². A small increase in intensity of the peak between the 7th and 28th day is comparable to a corresponding small increase in strength from 539 to 591 kg/cm.²

Fig. IX.1.7: DTA of oxychloride by calcination of dolomite at 800°C for 1 hr or 2 hr. (A) 3 days; (B) 7 days; (C) 28 days, 1 hr, 800°C; (D) 3 days; (E) 28 days, 2 hr, 800°C (*21*).

Fig. IX.1.7 (curves A, B, and C) presents the DTA of cements formed from calcined dolomite obtained at 800°C for 1 hr. The peak due to oxychloride at about 400°C increases with time and is as intense as that obtained for dolomite fired at 750°C (Fig. IX.1.6). There is an indication of a hump at about 500°C especially at 7 days and this might be due to $Ca(OH)_2$ or a complex containing Ca^{++}. The formation of CaO during the decomposition of dolomite at 800°C after 1 hr is perhaps the cause of lower strengths of 48, 139, and 146 kg/cm.^{2} at 3, 7, and 28 days respectively. The sample heated to 800°C for 2 hr also indicates a hump at about 500°C (Fig. IX.1.7, D and E) and this sample also gives lower strengths, the maximum value being

106 gk/cm.²

The above results have shown that strengths depend not only on the amount of lxychloride but also on the deleterious constituents such as CaO. The nature of the filler, especially its particle size, is an equally important factor influencing strengths. Some experiments were hence carried out by mixing precipitated $CaCO_3$ with calcined magnesite in the ratio 3:1.

The curves in Fig. IX.1.8 show that the low temperature endothermic effect decreases with time. Use of precipitated $CaCO_3$ with a large surface area involves use of a large quantity of magnesium chloride solution for standard consistency. The large endothermic peak at 160°C is mainly due to

Fig. IX.1.8: DTA of oxychloride from MgO/precipitated $CaCO_3$=1:3. (A) 2 days, (B) 3 days, (C) 7 days, (D) 28 days (*21*).

$MgCl_2 \cdot 6H_2O$ and this progressively decreases up to 28 days because magnesium chloride is gradually consumed to form oxychloride complex. A large endothermic peak at 250°C on 2- and 3-day samples decreases in intensity up to 28 days. The corresponding peak with sand as aggregate is smaller in intensity (Fig. IX.1.4).

The large endothermic effect is due partly to magnesium chloride and partly to oxychloride and at initial periods is mainly due to magnesium chloride. The intensity of the peak at 400°C even on the 2nd day indicates the formation of a considerable amount of the oxychloride complex. This may be due to a greater availability of magnesium chloride in the mix. Though oxychloride is formed in large quantities, the strengths are only 42, 71, and 106 kg/cm^2 at 3, 7, and 28 days respectively.

The presence of small particles of the filler in the mix prevents accommodation of stresses developed during the formation of complexes. Further, the mechanical binding and inter-penetration of whisker crystals, growth at the sites of crystal contact, and winding leader crystals, which contribute to strength development (*23*), may be hindered by the intervening fine particles of $CaCO_3$. In addition, during setting and drying, more than the normal amount of water is expelled from the large surface of $CaCO_3$ particles, resulting in large shrinkage.

Though oxychlorides are the main products formed in the $MgCl_2$–MgO–H_2O system, under natural conditions of exposure chlorocarbonate of the formula $Mg(OH)_2 \cdot 2MgCO_3 \cdot MgCl_2 \cdot 6H_2O$ is reported to be formed (*24*).

Magnesium Oxysulfate Cement. Magnesium oxysulfate cement, like that of magnesium oxychloride, has very good binding properties. The resistance of oxysulfate to abrasion is about 1.5 times that of Portland cement but only 50% that of magnesium oxychloride. The compressive and transverse strengths of 1:3 mortar from this cement after 28 days are about 4500 and 1500 lb/in.^{2} respectively compared with about 4000 and 1200 lb/in.^{2} respectively for 1:3 Portland cement mortar (*25*). Magnesium oxysulfate is less sensitive to high temperatures than oxychloride and hence finds use particularly in prefabricated structural units requiring processing at elevated temperatures (*26*).

Magnesium oxysulfate is obtained by the reaction of an aqueous solution of magnesium sulfate ($MgSO_4 \cdot 7H_2O$) with magnesium oxide. A comprehensive review on the system MgO–$MgSO_4$–H_2O is to be found in the paper of Demediuk and Cole (*27*). Four oxysulfates are formed in the system at temperatures between 30 and 120°C. They have the compositions, $5Mg(OH)_2 \cdot MgSO_4 \cdot 3$ or $2H_2O$, $3Mg(OH)_2 \cdot MgSO_4 \cdot 8H_2O$, $Mg(OH)_2 \cdot MgSO_4 \cdot 5H_2O$, and $Mg(OH)_2 \cdot 2MgSO_4 \cdot 3H_2O$. Only the 3-form is found to be a stable compound below 35°C.

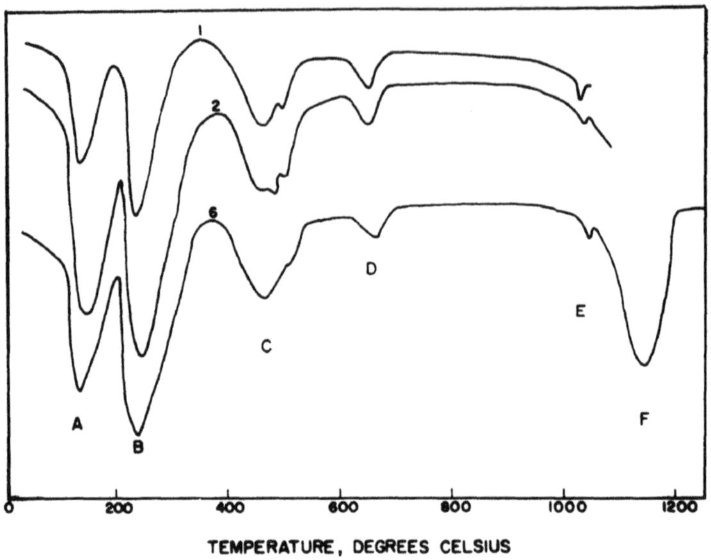

Fig. IX.1.9: Differential thermal analysis of $3Mg(OH)_2 \cdot MgSO_4 \cdot 8H_2O$ (*28*).

Newman (*28*) obtained the DTA of oxysulfate formed by mixing MgO with $MgSO_4$ solution for 2 hr. In Fig. IX.1.9 thermal curves of three preparations are given. Curve 1 refers to a preparation containing MgO=36.47 %, SO_3=19.15% and H_2O=44.38%; curve 2 contains MgO=37.46%, SO_3=18.55%, H_2O=43.99% and curve 3 contains MgO=35.98%, SO_3= 18.66%, H_2O=45.36%. The curves are almost identical and refer to the compound $3Mg(OH)_2 \cdot MgSO_4 \cdot 8H_2O$.

At temperatures 100-140°C four molecules of H_2O are expelled (A). At peak B (205-230°C) three more molecules are expelled. The remaining water is lost at 436-460°C (C). Traces of water are lost at 630-645°C (D). This also corresponds to the decomposition temperature of $MgCO_3$. AT E (1020-1028°C) a reversible transition of $MgSO_4$ takes place. The decomposition of $MgSO_4$ to MgO is attended by an endothermic effect at F (1118-1135°C).

Admoavichyute and co-workers (*15*) studied the hardening and thermal behaviour of magnesium oxysulfate formed from materials having initial concentrations of $MgSO_4$/MgO=0.6, 0.4, or 0.2. Thermograms of the hardened material as well as those of $MgSO_4 \cdot 7H_2O$ and $Mg(OH)_2$ are shown in Fig. IX.1.10.

The product with $MgSO_4$/MgO=0.4 or 0.2 gives $Mg(OH)_2$ as the main product. The mix containing $MgSO_4/MgO \geq 0.6$ forms oxysulfate with a peak at 680°C. It is concluded that at lower $MgSO_4$ concentrations, hydra-

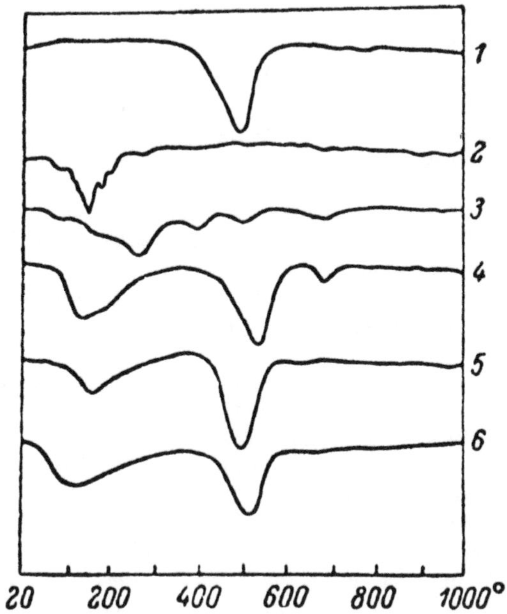

Fig. IX.1.10: Thermograms of hardened cement. (1) $Mg(OH)_2$, (2) $MgSO_4 \cdot 7H_2O$, (3) $MgSO_4 \cdot 3Mg(OH)_2 \cdot 5H_2O$ with $MgSO_4 \cdot MgO$ ratios of (4) 0.6, (5) 0.4, (6) 0.2 (*15*).

tion of MgO leads to hardening of the cement, and at higher concentrations the oxysulfate also contributes to hardening.

REFERENCES

1. Sorel, S.T., *Compt. Rend.*, *65*, 102, 1867.
2. Mohan Rai, *Sorel Cement, Building Materials Note 4*, Central Building Research Institute, India, 1964.
3. Lukens, H.S., *J. Am. Chem. Soc.*, *54*, 2372, 1932.
4. Bury, C.R., and Davies, E.R.H., *J. Chem. Soc.*, *135*, 2008, 1932.
5. Chassevent, L., *14th Congr. Chim. Ind., Paris*, 1934.
6. Feitknecht, W., and Held, F., *Helv. Chim. Acta*, *27*, 1480, 1944.
7. Walter-Levy, L., and de Wolff, P.M., *Compt. Rend.*, *229*, 1077, 1949.
8. de Wolff, P.M., and Walter-Levy, L., *Compt. Rend.*, *229*, 1232, 1949.
9. de Wolff, P.M., and Walter-Levy, L., *Acta. Cryst.*, *6*, 40, 1953.
10. Demediuk, T., Cole, W.F., and Hueber, H.V., *Australian J. Chem.*, *8*, 215, 1955.
11. Cole, W.F., and Demediumk, T., *Australian J. Chem.*, *8*, 234, 1955.
12. Newman, E.S., *J. Res. Nat. Bur. Std.*, *54*, 347, 1955.
13. Vyrodov, I.P., *Zh. Prikl. Khim.*, *34*, 1208, 1961.

14. Berg. L.G., and Ganelina, S.G., *Izv. Kazanskogo FAN-SSSR Ser. Khim-nauk.*, *2*, 91, 1955.
15. Admoavichyute, O.B., Yanitskii, I.V., and Vektaris, B.I., *Zh. Prikl. Khim.*, *35*, 2551, 1962.
16. Kassner, B., *Tonind.-Z. Keram. Rundschau.* *82*, 290, 1958.
17. Berg, L.G., Nikolaev, A.V., and Rode, E.Y., *Bull. Acad. Sci.* (*USSR*), *25*, 88, 1944.
18. Berg, L.G., and Burmistrova, N.P., *Zh. Neorgan. Khim.*, *5*, 676, 1960.
19. Mackenzie, R.C., *Scifax-Differential Thermal Analysis Data Index*, Cleaver-Hume Press Ltd., London, 1962.
20. Behl, W.K., and Gaur, H.C., *Proc. Nat. Inst. Sci.* (*India*), *27*, 33, 1961.
21. Ramachandran, V.S., Kacker, K.P., and Mohan Rai, *Zh. Prikl. Khim.*, *40*, 1687, 1967.
22. Thorpe, J.F., and Whiteley, M.A., *Dictionary of Applied Chemistry*, Longman Green & Co., Vol. 7, p. 448, 1948.
23. Tooper, B., and Cartz., L., *Nature*, *211*, 64, 1966.
24. Walter-Levy, L., and Bianco, Y., *Compt. Rend.*, *232*, 730, 1951.
25. Mohan Rai, and Garg, S.K., *Indian J. Tech.*, *2*, 273, 1964.
26. Kirk, R.E., and Othmer, D.F., *Encyclopaedia of Chemical Technology*, The Interscience Encyclopaedia Inc., New York, Vol. 3, 1953.
27. Demediuk, T., and Cole, W.F., *Australian J. Chem.*, *10*, 287, 1957.
28. Newman, E.S., *J. Res. Nat. Bur. Std.*, *68A*, 645, 1964.

IX.2. GYPSUM PLASTER

Gypsum is a hydrous calcium sulfate mineral represented by the chemical formula $CaSO_4 \cdot 2H_2O$ and is a well known retarding agent in Portland cement. Calcination of gypsum at a temperature of 160-170°C results in the expulsion of three-fourths of the water of crystallization. The resulting product is known as plaster of Paris with the formula $CaSO_4 \cdot \frac{1}{2}H_2O$, and is variously described as wall plaster, molding plaster, finishing plaster, pottery plaster, and dental plaster. The plaster sets to a very hard mass when treated with water.

Types of Gypsum Plaster. British Standard Specification 1191 classifies gypsum plaster into four types, *viz.* plaster of Paris, retarded hemihydrated gypsum plaster, anhydrous gypsum plaster, and Keene's or Parian cement.

Plaster of Paris is obtained in practice by calcining a finely ground gypsum to a temperature of about 160°C in a kettle or in an open iron pan. The product usually contains some overburnt material of composition $CaSO_4$ and unburnt gypsum. Another product of the same composition as plaster of Paris known as α-gypsum is made by steam curing of gypsum at about 120-130°C in presence of salts. This material is of greater density and yields higher strength than plaster of Paris.

Plaster of Paris hardens within 20-30 min after treatment with water. In practice the hardening period is prolonged by an hour or two by the addition of a chemical known as a retarder. The most widely used retarder is an organic compound based on keratin or a glue. Other compounds such as borax, tartaric acid, citric acid, acetic acid, starch, clay, and sugar also act as retarders. Plaster of Paris containing a retarder is termed retarded hemihydrated gypsum plaster.

Gypsum loses two molecules of water when heated to about 200°C and forms the anhydrous calcium sulfate known as soluble anhydrous gypsum plaster. The soluble anhydrous gypsum is also a product of calcination of plaster of Paris and sets in presence of water. The hemihydrate converts to a soluble anhydrite very efficiently by steam curing (*1*).

Keene's cement is made by calcining gypsum at 110-130°C, soaking it in a solution of potash alum, then rebaking at a temperature of 400-500°C. Borax is used in place of potash alum to make Parian cement. Martin's cement uses potassium carbonate. Setting is also accelerated in presence of small amounts of crushed gypsum, sodium chloride, potassium chloride, or ammonium thiocyanate. Keene's cement, Parian cement, and Martin's cement are very hard after setting. The German flooring plaster known as *Estrich Gips* is a product of calcination of gypsum at a temperature of 1100-

TABLE IX.2.1
Transverse Strengths of Hemihydrate Plaster Containing Different Proportions of Sand [Andrews (2)]

Proportions by weight		*Water by wt. of dry materials, %*	*Water by wt. of plaster present, %*	*Modulus of rupture lb/in.*^{2}	
Plaster	*Sand*			*Dry specimens*	*Saturated specimens*
1	0	39.0	39.0	1105	510
1	1	22.0	44.0	865	310
1	2	16.0	48.0	725	225
1	3	14.5	58.0	480	95
1	4	12.5	62.5	290	55
1	6	11.0	77.0	165	25

1200°C. At this temperature a small amount of CaO produced by the dissociation of $CaSO_4$ acts as an accelerator of set.

Uses. Plaster of Paris finds innumerable applications in the ceramic industry, architectural decoration, dental and surgical fields. But it is in the building industry that gypsum finds the widest application. The various purposes for which gypsum is used in building include structural tile, wall plaster, lath, hollow or lightweight blocks, floor finish, precast units, plaster boards, fibrous plaster, reinforced and wood wool slabs. The set gypsum plaster develops adequate strength alone or in combination with aggregates. A Keene's cement may, for example, yield a strength as high as 5000-6000 lb/in.^{2} In Table IX.2.1 are given the strengths of plaster of Paris containing different proportions of sand.

Gypsum is soluble to an extent of about 0.2% in water and hence is not recommended for use under outdoor exposure conditions and at places where it is continuously exposed to moisture. It is known to corrode steel.

Chemistry of Hydration. The mineral gypsum has a layer lattice structure with water separating the sheets of calcium and sulfate ions. Calcination of gypsum at about 125°C produces plaster of Paris and the decomposition reaction can be represented by the equation

$$\underset{\textit{gypsum}}{CaSO_4 \cdot 2H_2O} = \underset{\textit{plaster of Paris}}{CaSO_4 \cdot \tfrac{1}{2}H_2O} + 1\tfrac{1}{2}H_2O$$

The hemihydrate $CaSO_4 \cdot \frac{1}{2}H_2O$ formed under dry heating conditions is not exactly similar to that formed by autoclave treatment or digestion in nitric acid. Though both types of hemihydrate give very similar X-ray diffraction patterns, they differ in their energy contents and behavior with water.

The hemihydrate obtained by dry method, known as the β-form, gives thicker pastes with water, is more soluble, and has a larger surface area than the α-form produced under wet conditions. There is controversy on the exact structure of the hemihydrate. It is not certain whether it is a definite hydrate of the composition $CaSO_4 \cdot \frac{1}{2}H_2O$ or a member of the zeolitic system with the composition ranging from $CaSO_4 \cdot 0H_2O$ to $CaSO_4 \cdot 2/3H_2O$.

Heating of gypsum or plaster of Paris to about 185-200°C yields the soluble anhydrite according to the equations

$$\underset{\textit{gypsum}}{CaSO_4 \cdot 2H_2O} = \underset{\textit{soluble anhydrite}}{CaSO_4 + 2H_2O}$$

or

$$\underset{\textit{plaster of Paris}}{CaSO_4 \cdot \tfrac{1}{2}H_2O} = \underset{\textit{soluble anhydrite}}{CaSO_4 + \tfrac{1}{2}H_2O}$$

At a temperature of 400-500°C gypsum is dehydrated to an insoluble anhydrite of the composition $CaSO_4$. The soluble anhydrite is more reactive and is quickly hydrated by water vapor to $CaSO_4 \cdot \frac{1}{2}H_2O$. Kelly and others (*3*) recognized α and β-forms of soluble anhydrite derived from the corresponding α and β-hemihydrates.

The conversion of plaster of Paris or soluble anhydrite to gypsum is an exothermic process and the reaction can be followed by plotting the temperature rise with time. In Fig. IX.2.1, curve A shows the behavior of plaster of Paris with an initial setting time of 3 min and a final setting time of 4½

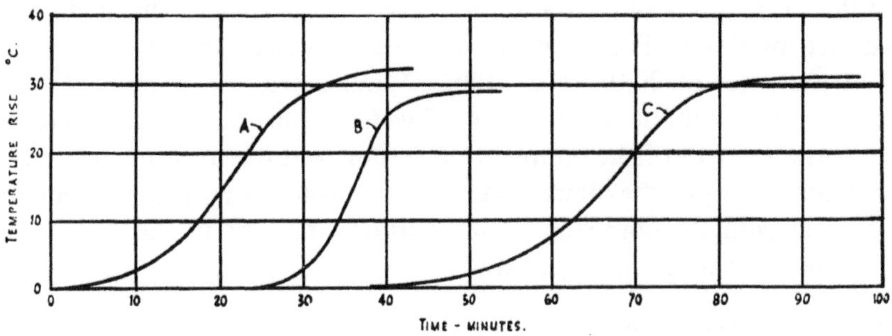

Fig. IX.2.1: The effect of retardation on rate of hydration (*2*).

min. Curve B gives the behavior of plaster of Paris containing 0.1% keratin and gauged in 1% K_2SO_4. There is no temperature rise up to about 25 min, after which the curve rises steeply. In other words, the mix has an induction period of about 25 min in the presence of a retarder. The retarder increases the initial setting time to 21 min and the final setting time to 26 min. With an addition of 0.2% sodium citrate, the induction period is further extended (curve C) and the initial and final setting times are 36 min and 44 min respectively.

Theories of Hydration. Hydration of plaster of Paris to calcium sulfate dihydrate appears simple, but the exact mechanism of hydration is a subject of much speculation. Essentially two theories have been proposed to explain the mechanism of setting and hardening.

Le Chatelier (*4*, *5*) proposed a theory according to which the hemihydrate being more soluble than the dihydrate produces a highly supersaturated solution with respect to gypsum. Subsequently the less soluble dihydrate separates out from the solution as interlocking needle-like crystals. Michaelis (*6*) proposed the colloidal theory for explaining the setting of gypsum plaster. In this postulate also hemihydrate is believed to dissolve first in water but the formation of a colloidal product is thought to be a prelude to the final crystallization to gypsum. There are supporters for both theories over years and these have been discussed in detail in some reviews (*7-11*).

Hansen's modified theory of setting and hardening of gypsum plaster involves initially a direct formation of $CaSO_4 \cdot 2H_2O$ on the surface of $CaSO_4 \cdot \frac{1}{2}H_2O$ or soluble $CaSO_4$. Subsequently, the $CaSO_4 \cdot 2H_2O$ crystallites, acting as nuclei, grow into needle-like crystals of gypsum. The theory explains the dimensional changes in the set plaster and also the role of retarders and accelerators.

One of the techniques developed recently by Sereda and coworkers (*12-14*), known as 'compact technique,' appears to be a promising tool to investigate the mechanism of hydration of plaster of Paris. In this method the unhydrated plaster of Paris powder is made into circular disc-shaped compacts at pressures of 20,000-50,000 lb/in.^{2} and hence the process of setting is accomplished without the addition of water. This technique makes possible a better control of the temperature throughout the mass of the material and permits a study of the dimensional changes from the moment water comes into contact with the compact. The hydration process in the vapor phase can also be studied advantageously.

Differential Thermal Analysis

There are innumerable references to the application of DTA to the study of gypsum and plaster but those who have used this technique as the main

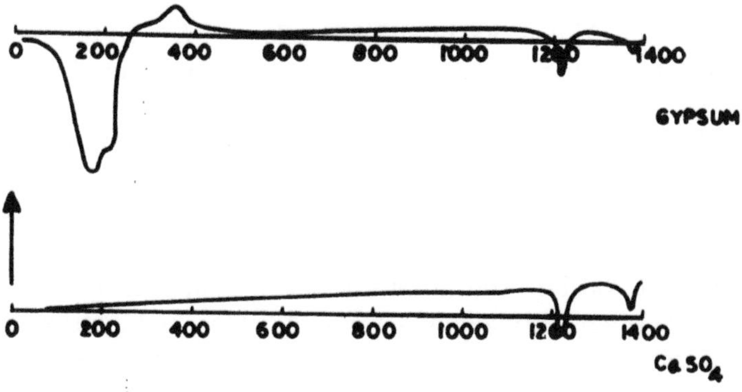

Fig. IX.2.2: DTA of gypsum and anhydrite.

method of investigation are Volzhenskii (*15*), Budnikov and Kosyreva (*16*), Gruver (*17*), Mitsuki (*18*), Yamauchi and Tagai (*19*), West and Sutton (*20*), Kuntze *et al* (*21-23*), Powell (*24-27*), Sekiya *et al* (*28*, *29*), Trzebiatowski and Romotowski (*30*), Sen (*31*), Murakami *et al* (*32*, *33*), Yamada *et al* (*34*), Ramachandran *et al* (*35*, *36*), and Stajadinoric (*37*).

Gypsum. The thermal behavior of natural gypsum, alabaster, selenite, or synthetic $CaSO_4 \cdot 2H_2O$ is very similar, and a representative thermogram of gypsum is shown in Fig. IX.2.2. The first endothermal peak at 180°C represents the removal of 1½ molecules of water from gypsum. The second endothermic effect is less intense and is caused by the removal of the remaining ½ molecule of water. A small endothermal dent has also been reported by some investigators just after the second endothermal effect and is attributed to the expulsion of small amounts of residual water from the hemihydrate.

An exothermal hump at about 355°C is ascribed to the formation of insoluble anhydrite (β-$CaSO_4$) from the soluble anhydrite and the X-ray diffraction results of the sample before and after the temperature of the exothermal effect confirm this observation. The exothermal hump has been missed by some workers.

There is no other thermal effect up to a temperature of about 1200°C, and the endothermal change at 1220°C is caused by the transformation of β-$CaSO_4$ to α-$CaSO_4$ (*38*). The high temperature endothermic peak at 1380°C is due to the melting of the eutectic compound between $CaSO_4$ and CaO. In the cooling curve two exothermal effects are registered at 1350 and 1175°C, due respectively to the freezing of the eutectic compound and the transformation of α-$CaSO_4$ to β-$CaSO_4$. The heat of transformation of α to β-$CaSO_4$ as determined by DTA is 5000$\pm$300 cal/mole.

An artificially prepared anhydrite exhibits two exothermal effects at 1220 and 1380°C, and the intensities of these peaks are larger than those of gypsum. This is because in gypsum containing about 20% water the effective amount of $CaSO_4$ is less.

Murakami and Tanaka (*32*) and Sekiya *et al* (*28*) examined the effect of addition of inorganic salts on the dehydration behavior of gypsum. Murakami found that NaCl or $MgCl_2$ lowered the first endothermic peak at 125°C, while insoluble organic compounds, metallic oxides, and hydroxides increased the peak temperature. Sekiya *et al* (*28*) included for their study 15 salts comprising chlorides, nitrates, sulfates, carbonates, and fluorides. The chlorides lowered the first endothermal peak temperature. With 1.0% $CaCl_2$ or 0.1-1.0% $MgCl_2$ an additional peak resulted at about 230°C, attributed to the formation of α-$CaSO_4 \cdot \frac{1}{2}H_2O$. A similar trend, but to a lesser degree, was found for nitrate additions. The sulfates, carbonates and fluorides had negligible effect on the dehydration peaks.

Setting of Gypsum. The final setting and hardening in gypsum plaster is caused by a progressive growth and interlocking of gypsum crystals. But the nature of the processes before and during the initial setting is a subject of dispute. Fischer (*9*) and Powell (*26*) attempted to explain the mechanism of the initial setting in plasters by adopting the DTA technique.

Fischer chose for his studies an uncontaminated plaster of Paris having an initial setting period of 20 min. In Fig. IX.2.3, curve a of the plaster of Paris exhibits a single endothermal effect at a characteristic temperature of 185°C due to the conversion of $CaSO_4 \cdot \frac{1}{2}H_2O$ to soluble anhydrite. The characteristic temperature (*Tc*) represents the temperature of the onset of the reaction and is determined by drawing a tangent at an angle of 45° to the knee of the endotherm. The dry gypsum dehydrates in two stages, exhibiting two endotherms with *Tc* values at 125 and 190°C.

Figs. IX.2.4, 2.5 and 2.6 give the DTA behavior of plaster of Paris exposed to water for 3 min, 10 min, 30 min, 45 min, and 7 days. In all the preparations the plaster of Paris was stirred for 1 min with water and allowed to stand for the required period and washed with absolute ethyl alcohol and acetone. Subsequently the mix was spread on an adsorptive paper for the acetone to evaporate.

Fig. IX.2.4a is the thermogram of the sample exposed to water for 3 min. The endothermic effect represents the volatilization of some acetone left in the sample and the endotherm B is ascribed to the desorption of water from the large surface and the intermicronic channels of the hemihydrate particles. This water penetrates the interionic interstices and causes hydration of $CaSO_4$

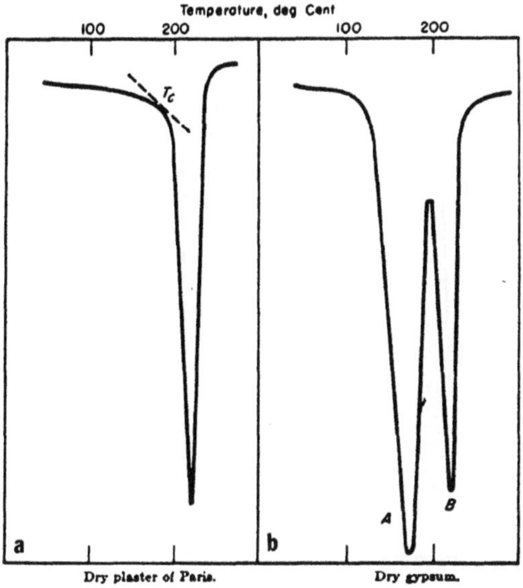

Fig. IX.2.3: Reference samples (*9*).

$\cdot\frac{1}{2}H_2O$ to $CaSO_4\cdot 2H_2O$. The endotherm C at 190°C is due to the expulsion of ½ molecule of water from $CaSO_4\cdot\frac{1}{2}H_2O$. The sample b obtained after 10 min also exhibits three endothermal effects at 65, 100, and 185°C representing respectively volatilization of acetone, desorption of H_2O, and dehydration of plaster of Paris.

In Fig. IX.2.5 is shown the DTA of samples obtained after 30 and 40 min of contact with water. At this period the set has already taken place. The thermogram at 30 min shows an endotherm at 190°C and is caused partly by the dehydration of the plaster of Paris originally present and partly by the hemihydrate produced by a partial dehydration of gypsum formed at this period. The peak B represents the dehydration of some gypsum formed at this stage. At 45 min there is a large endothermic effect at 135°C indicating the presence of a considerable amount of the dihydrate. At 7 days (Fig. IX.2.6) a very large endothermal effect between 100 and 200°C shows a complete conversion of hemihydrate to dihydrate and the second endotherm at 190°C indicates the dehydration of the hemihydrate formed by the partial dehydration of the dihydrate.

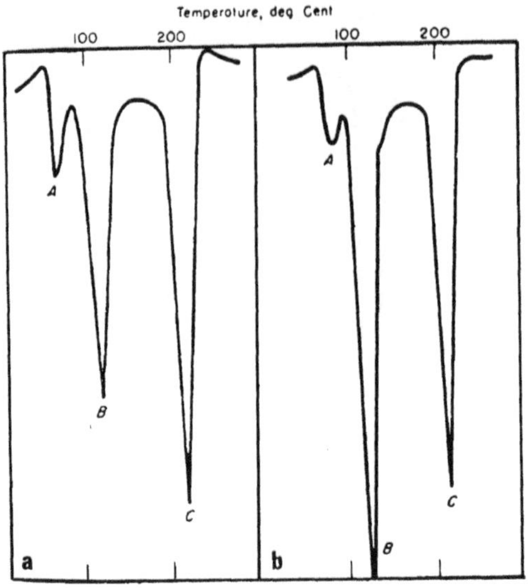

Fig. IX.2.4: Gauged plaster of Paris. (a) 3 min, (b) 10 min (*9*).

From these results Fischer concluded that water initially adsorbed on the large surface of crystals leads to the gelation and initial setting. The next stage is an intermicellar swelling with the chemisorption of water. The chemisorption promotes hydration and the formation of microcrystals that interlock together to cause the final setting. Powell's (*26*) results are contradictory to these findings. He found by DTA that in a commercial plaster with an initial setting period of 33 min no dihydrate was present at 13 min, 4% was present at 25 min, and 80% at about 33 min.

These results would indicate that the setting process is one of crystallization. A more systematic work is needed before a definite conclusion could be reached. The purity of the material, the method of stopping the reaction, the arbitrary means of defining the setting time, the interference of the endothermal reaction of water desorption with that due to the first dehydration effect of gypsum, and the sensitivity of the DTA apparatus are some of the important factors to be considered in such studies.

α and β-hemihydrates The two types of hemihydrate designated as α and β exhibit similar X-ray diffraction patterns and hence this method cannot be

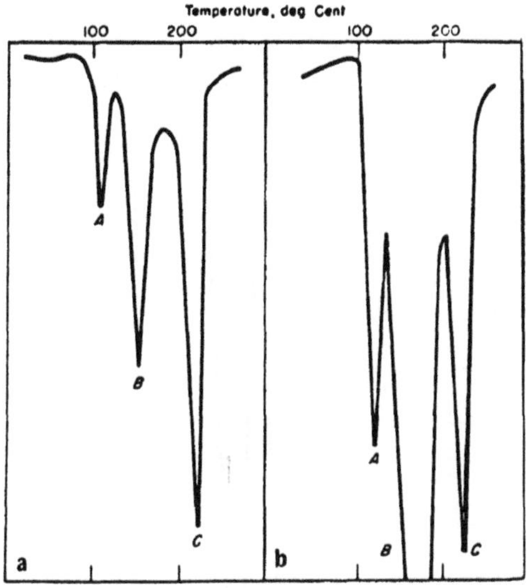

Fig. IX.2.5: Gauged plaster of Paris. (a) 30 min, (b) 45 min (*9*).

applied to differentiate them. Attempts have been made by Powell (*25*), Budnikov and Kosyreva (*16*), and Sekiya (*29*) to differentiate the two forms by the DTA technique. Powell (*24*) prepared the α-form by heating gypsum in nitric acid, magnesium sulfate solution, or in an atmosphere of steam. The β-form was made by heating gypsum at 150, 200, or 240°C. The α and β-varieties obtained by different methods showed similar thermograms and typical results are presented in Fig. IX.2.7. Curve a of the α-hemihydrate shows a sharp exothermic effect at 255°C denoting a rapid transformation of the α-hemihydrate to the insoluble anhydrite. The conversion of the α-hemihydrate to the insoluble anhydrite is not quantitative under these conditions and X-ray diffraction studies reveal the presence of only 60% insoluble anhydrite in the sample calcined at 290°C. There is also an indication that the transformation of the remaining 40% the α-hemihydrate to insoluble anhydrite is gradual and is almost complete at about 550°C. In the thermogram of the β-form (curve 7b) the exothermal peak is shifted to a temperature of 300°C and is not as sharp as that of the α-form.

The DTA method not only differentiates the two forms of hemihydrate but can provide an approximate estimation of one in the other. According to Budnikov and Kosyreva (*16*) the α-form exhibits the endothermic dehydra-

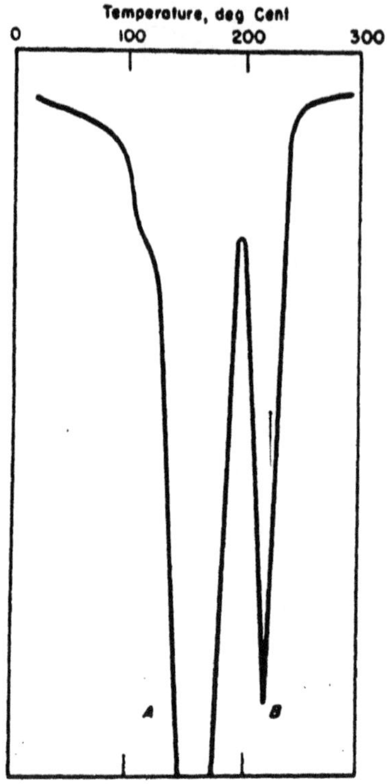

Fig. IX.2.6: Gauged plaster of Paris, aged 7 days (*9*).

tion effect at 190-210°C, while the β-form exhibits the peak at 170-185°C. In their studies only the β-form showed an exothermal dent at 360-370°C. Sekiya (*29*) reported two exothermal peaks in the α-hemihydrate and the temperature of the peak varied, depending on the crystallinity.

Estimation. It is not possible to identify very small amounts of gypsum in the hemihydrate by X-ray methods. The determination of dihydrate in the hemihydrate provides useful information on the role of the dihydrate in setting and on other physical properties. Kuntze (*21*) was able to detect 0.1-1.0% gypsum in $CaSO_4 \cdot 1/2H_2O$ by performing the DTA in a saturated water vapor at 76 cm. Hg. This results are shown in Fig. IX.2.8.

The DTA of gypsum obtained under the ambient conditions is represented by curve A. At a saturated water vapor pressure of 76 cm Hg, the second endothermic effect occurs at a temperature of 187°C, showing thereby a

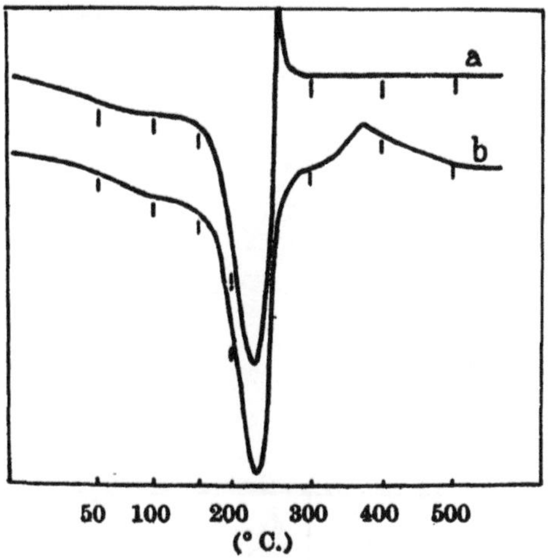

Fig. IX.2.7: Differential thermal analysis curves for calcium sulfate hemihydrate. (a) α-form, (b) β-form (*24*).

temperature shift of about 52°C. The area of the first endothermic effect gives the amount of the dihydrate in the sample. Curves C and D are demonstrative of the efficacy of the method. Curve C is the thermogram of a sample containing less than 2% gypsum and carried out in air. Curve D is the thermogram of a sample containing 0.5% dihydrate but obtained in a saturated water vapor at 76 cm. Hg.

Murakami *et al* (*33*) studied the variation of the DTA peaks in gypsum under an atmosphere of water vapor, ethyl alcohol, or nitrogen at vapor pressures of 1 to 15 atm. The first dehydration peak at 133°C is shifted to 142 and 162°C with saturated water vapor and ethyl alcohol respectively. In a nitrogen atmosphere the second peak shifts according to the equation

$$t = 65 \log p + 192$$

where t is the temperature of the endothermic peak and p the pressure of N_2 gas.

Ramachandran and co-workers (*36*) compared the methods of TGA, DTA, and chemical analysis for estimating the constituents in a white coat plaster containing essentially $CaSO_4 \cdot 2H_2O$, $Mg(OH)_2$, $Ca(OH)_2$, $CaCO_3$, and MgO. Well defined peaks for gypsum, magnesium hydroxide, calcium hydroxide, and calcium carbonate make the DTA technique particularly applicable to the estimation of them in plasters.

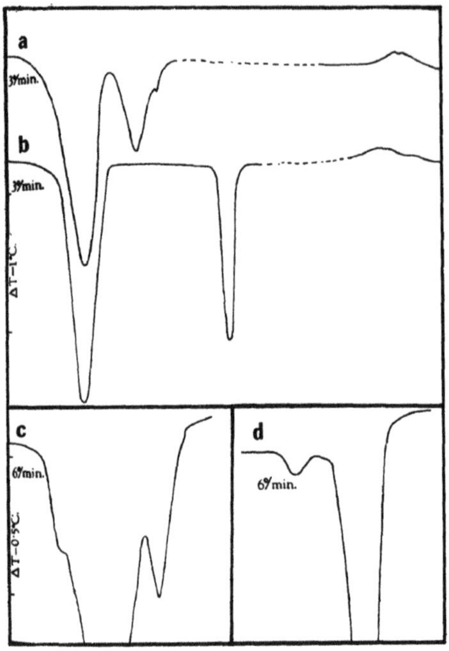

Fig. IX.2.8: DTA of hydrates of $CaSO_4$. (a) Dihydrate uncontrolled atm, (b) dehydrated in saturated water vapor, (c) hemihydrate with 2.5% dihydrate (uncontrolled atm), (d) hemihydrate with 0.5% dihydrate in saturated water vapour (*21*).

The endothermal areas of the four constituents in the sound plaster (a plaster that did not exhibit any failure under normal conditions of exposure) were compared with the corresponding areas obtained for the other three samples designated as failed plaster, autoclaved sound plaster, and autoclaved failed plaster. The relative values calculated by TGA and chemical analysis are shown in Table IX.2.2. The proportions of the individual constituents were calculated by taking the proportions for the sound plaster as unity.

In general, the results of TGA and DTA are in fair agreement with each other. Because of the dynamic nature of the TGA and DTA techniques, the estimated values by these methods are expected to be less than those obtained by chemical analysis and ignition loss. It was concluded that the DTA and TGA methods could be used as negative tests for determining the

TABLE IX.2.2
Proportions of Various Constituents in Plaster as Obtained by Chemical Analysis, TGA, and DTA [Ramachandran, Sereda, and Feldman (36)]

	Gypsum			$Mg(OH)_2$			$Ca(OH)_2$			$CaCO_3$		
Sample	*Chem. anal.*	*TGA*	*DTA*	*Chem. anal.*	*TGA*	*DTA*	*Chem. anal.*	*TGA*	*DTA*	*Chem. anal.*	*TGA*	*DTA*
Sound plaster	1	1	1	1	1	1	1	1	1	1	1	1
Failed plaster	0.96	1.07	0.97	2.08	2.33	2.26	1	1.10	1.03	0.88	0.83	0.90
Autoclaved sound plaster	—	—	—	3.08	3.06	3.12	0.93	1.01	1.03	1	0.98	0.91
Autoclaved failed plaster	—	—	—	2.98	2.94	2.89	1.06	1.12	1.09	0.91	0.89	0.93

suitability of limes.

Deterioration of Plasters. Several investigators have reported the occurrance of 'popping' or 'bulging' in white-coat plaster prepared from a lime putty gauged with plaster of Paris (*35, 39-50*). In these plasters–this term is used here for set plasters—the onset of slight bulging was noticed at periods ranging from 5 to 15 years, with progressive increase and final separation from the wall. Attempts have been made to settle the controversial question of whether the delayed hydration of magnesium oxide is responsible for this type of deterioration. The presence of a great number of constituents such as free water, $CaSO_4 \cdot 2H_2O$, $Ca(OH)_2$, $Mg(OH)_2$, MgO, $CaCO_3$, $CaSO_4$, SiO_2, Al_2O_3, Fe_2O_3, and possibly other compounds of magnesium in the plaster makes identification difficult.

Ramachandran *et al* (*35*) examined a white-coat plaster obtained from the walls of a hospital in which extensive 'bulging' had occurred eight years after the application of the plaster. The DTA technique was applied to investigate the causes leading to the failure.

The main constituents in a gypsum-gauged white-coat plaster are gypsum, formed by the hydration of plaster of Paris (hemihydrate), and calcium hydroxide, formed by the hydration of calcium oxide, provided the lime is obtained from a high-calcium limestone. In a lime obtained by calcining dolomitic limestone the principal constituents before application on the wall are calcium hydroxide, magnesium hydroxide, and considerable amounts of unhydrated magnesium oxide (periclase), especially if the limestone is produced at high temperatures and is not pressure-hydrated. The normal dolomitic hydrated lime (type N) belongs to this category.

Some calcium carbonate is formed by the carbonation of calcium hydroxide and may also be present as an impurity in the gauged plaster or as a core material. Small amounts of calcium sulfate (anhydrite), silica, alumina, and ferric oxide also occur in the plaster. There is little or no evidence of formation of magnesium carbonate.

A gypsum-gauged plaster containing (mainly) plaster of Paris, calcium oxide, and magnesium oxide would transform to gypsum, $Ca(OH)_2$, $CaCO_3$, and $Mg(OH)_2$ after being treated with water and subjected to normal exposure.

Table IX.2.3 gives the molar volumes of various constituents and the theoretical molar volume changes for some of the possible reactions in the white-coat plaster.

On hydration, plaster of Paris converts to $CaSO_4 \cdot 2H_2O$, with an increase in molar volume of 41.13%. The hydration reaction is sufficiently fast as to be almost complete during the setting period. Hence bulging in plaster cannot be attributed to this expansion.

TABLE IX.2.3
Some Physicochemical Constants for the Reactions Occurring in White-Coat Plaster [Ramachandran, Sereda, and Feldman (35)]

Component	*Chemical composition*	*Molecular weight*	*Density, g/ml*[a]	*Molar volume, ml*	*Possible chemical transformation*	*Molar volume change, %*
Plaster of Paris	$CaSO_4 \cdot \frac{1}{2}H_2O$	145.15	2.75[b]	52.796	$CaSO_4 \cdot \frac{1}{2}H_2O \longrightarrow CaSO_4 \cdot 2H_2O$	41.13
Gypsum	$CaSO_4 \cdot 2H_2O$	172.18	2.32	74.218		
Lime (calcia)	CaO	56.08	3.37	16.642		
Hydrated lime	$Ca(OH)_2$	74.10	2.343	31.623	$CaO \longrightarrow Ca(OH)_2$	90.03
Calcite	$CaCO_3$	100.09	2.711	36.92	$Ca(OH)_2 \longrightarrow CaCO_3$	16.75
Periclase	MgO	40.32	3.58	11.264	$CaO \longrightarrow CaCO_3$	121.87
Brucite	$Mg(OH)_2$	58.34	2.385	24.462	$MgO \longrightarrow Mg(OH)_2$	117.22

[a] See Ref. *(48)*.
[b] See Ref. *(49)*.

The volume increase resulting from the formation of $Ca(OH)_2$ from CaO is as high as 90.03%. Because CaO reacts rapidly with water, even before setting and hardening of the plaster, it does not contribute to delayed expansion. Stutterheim and co-workers (*47*) found that in dolomites calcined in the range 600-1400°C calcium oxide hydrated rapidly and completely for all temperatures and times studied.

The formation of $CaCO_3$ from $Ca(OH)_2$ is attended by a volume change of about 16.75%. This reaction is extremely show under normal conditions of exposure. Wells *et al* (*44*) have shown that carbonation of $Ca(OH)_2$ is far form complete even after a number of years. Although the reaction of CaO with CO_2 to form $CaCO_3$ involves a large volume increase of 121.87%, it is improbable because the presence of free CaO is unlikely in the plaster; even if free CaO were present, hydration would be the preferred reaction.

MgO converts to $Mg(OH)_2$ with a volume increase of 117.22%. Magnesia is very likely to be present in an overburnt form in dolomitic limes. In this form MgO hydrates extremely slowly, so that even after the setting of a finish coat there is a considerable quantity of MgO in the plaster. There is every reason to believe that the gradual hydration of MgO is attended by a corresponding expansion. Campbell (*50*) found that MgO prepared from $MgCO_3$ at 800°C hydrated completely in three days, that prepared at 1000-1100°C hydrated in about three months, and that formed at 1200°C required three years. Hydration was not complete even after six years in MgO prepared at 1300°C. Wells and Taylor (*43*) reported that among 27 regularly hydrated dolomitic limes an average of only 15.7% of the total MgO was hydrated in each sample.

The specimens were subjected to differential thermal analysis to identify the components present in the plaster. Fig. IX.2.9 shows the results. Plaster exhibits an endothermic doublet with peaks at 142 and 158°C, due to stepwise dehydration of gypsum. An endothermic peak at 391°C corresponds to the decomposition of $Mg(OH)_2$; the other peak, at 464°C, should be due to $Ca(OH)_2$ decomposition. Pure $Ca(OH)_2$ exhibits an endothermic peak above 500°C, but the temperature of the peak may be lowered by the presence of other constituents. A large endothermic effect with a peak at 800°C is caused by decarbonation of $CaCO_3$ formed over several years by the reaction of $Ca(OH)_2$ with atmospheric CO_2.

The endothermic peak area resulting from $CaCO_3$ decomposition is several times greater than that due to $Ca(OH)_2$ decomposition. A comparison of the areas would not give a direct indication of the relative amounts of $Ca(OH)_2$ and $CaCO_3$ in the specimen, because the heat of reaction per mole of $CaCO_3$ is about three times that of $Ca(OH)_2$ decompostition. The DTA results show that the specimen under question is gypsum-gauged lime, possibly obtained

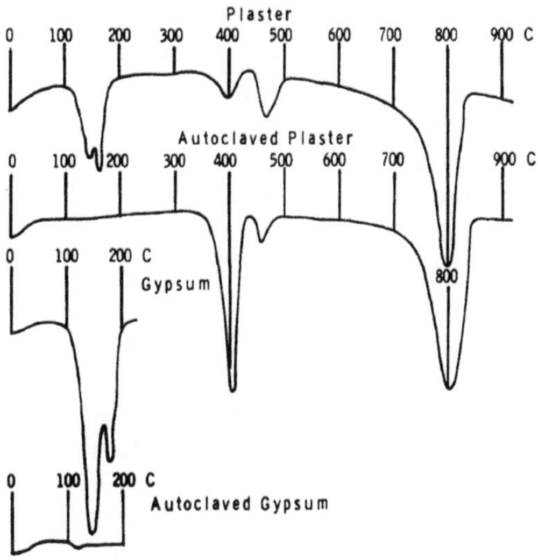

Fig. IX.2.9: Differential thermal curves of plaster and gypsum before and after autoclaving (*35*).

from dolomitic limestone.

The presence of $Ca(OH)_2$ and $Mg(OH)_2$ in the failed plaster indicates the possibility of the presence of unhydrated MgO in the sample. It is difficult to prove its presence by chemical analysis. Another approach is to hydrate the MgO under severe conditions and study the formation of $Mg(OH)_2$ by measuring the dimensional changes or by measuring the intensity of the endothermic effect resulting from $Mg(OH)_2$ decomposition at about 400°C by DTA.

White-coat plaster was ground to pass through a 200-mesh sieve, and circular compacts of a nominal diameter of 1.250 in. were formed at a load of 10,000 or 20,000 lb. The two autoclave-treated compacts showed uniform linear expansions of 12.22 and 14.67% respectively. In order to verify whether any other constituent in the plaster compact contributed to this expansion, compacts were formed from pure $Ca(OH)_2$, $Mg(OH)_2$, $CaSO_4 \cdot 2H_2O$, and $CaCO_3$, and were subjected to autoclave treatment. In none of these compacts could measurable expansion be observed. Hence it was inferred that the expansion was mainly due to the formation of $Mg(OH)_2$ from MgO.

The DTA of the autoclaved sample shows endothermic peaks corresponding to $Ca(OH)_2$, $CaCO_3$, and $Mg(OH)_2$ (Fig. IX.2.9). The endothermic effect due to $Mg(OH)_2$ decomposition, however, is much more intense for the autoclaved than the untreated specimen, showing that the autoclave treatment results in the formation of $Mg(OH)_2$ from unreactive MgO. An autoclaved specimen fails to show endothermic effects for the presence of gypsum.

The effect of autoclave treatment on a compact of precipitated $CaSO_4 \cdot 2H_2O$ is presented in Fig. IX.2.9. The unreacted material shows two endothermic peaks at 153 and 188°C. The autoclaved specimen shows almost complete absence of endothermic effects, indicating that during autoclave treatment gypsum is decomposed to $CaSO_4$. The above results were confirmed by weight loss measurements.

It is well known that repeated calcination of gypsum results in deterioration of the strength of gypsum-based products (*52*). Mtschedlov-Petrosian and co-workers (*53*) investigated the possible causes for this phenomenon by utilizing DTA. Repeated cycles of calcination and hydration bring about a decrease in the endothermal effects of dehydration and hence indicate a diminution of the dehydrated gypsum crystals. In contrast to this the range of temperature of transformation of hemihydrate to soluble anhydrite shifts from 208-390°C to 220-438°C. This shows that the hemihydrate is formed in a more stable form. The repeated calcination promotes a regular crystalline structure, thus causing a loss in activity.

The DTA technique should be applicable as a routine tool in the gypsum industry (*51*). The technique enables detection of unburnt gypsum, and semi-quantitative estimation of impurities such as quartz, clay minerals, organic matter, calcium carbonate, magnesium carbonate, dolomite, and pyrites.

REFERENCES

1. Matsuki, C., and Kita, M., *Nature, 184*, 1481, 1959.
2. Andrews, H., *Building Res. Congr.*, London, pp. 135-144, 1951.
3. Kelly, K.K., Southard, J.C., and Anderson, C.T., *Tech. Paper, Bur. Min., Washington, No. 625*, 1941.
4. Desch, C.H., *Trans. Faraday Soc., 14*, 1, 1919.
5. Desch, C.H., *Trans. Ceram. Soc., 18*, 15, 1918.
6. Michaelis, W., *Kolloid. Zh., 5*, 9, 1909.
7. Ridge, M.J., *Rev., Pure and App. Chem., 10*, 243, 1960.
8. Weiser, H.B., and Moreland, F., *J. Phys. Chem., 36*, 1, 1932.
9. Fischer, H.C., *Am. Soc. Testing Mater. Bull. No. 192*, 43, 1953.
10. Schiller, K., *J. Appl. Chem. (London), 12*, 135, 1962.
11. Hansen, W.C., *Mater. Res. Stad., 3*, 359, 1963.
12. Sereda, P.J., and Feldman, R.F., *J. Appl. Chem. (London), 13*, 150, 1963.
13. Feldman, R.F., and Sereda, P.J., *J. Appl. Chem. (London), 13*, 158, 1963.
14. Sereda, P.J., Feldman, R.F., and Ramachandran, V.S., *Am. Ceram. Soc.*

Bull., *44*, 151, 1965.
15. Volzhenskii, A.V., *Keramika, No. 11*, 64, 1939.
16. Budnikov, P.P., and Kosyreva, Z.S., *Voprosy Petrograf i Mineral, Akad. Nauk SSSR*, *2*, 342, 1953.
17. Gruver, R.M., *J. Am. Ceram. Soc.*, *34*, 353, 1951.
18. Mitsuki, S., *J. Ceram. Assn., Japan*, *60*, 92, 1952.
19. Yamauchi, I.T., and Tagai, H., *J. Ceram. Assn., Japan*, *61*, 600, 1953.
20. West, R.R., and Sutton, W.J., *J. Am. Ceram. Soc.*, *37*, 221, 1954.
21. Kuntze, R.A., *Nature*, *193*, 772, 1962.
22. Kuntze, R.A., *Mater. Res. Std.*, *2*, 640, 1962.
23. Fleck, W.E.P., Jones, M.H., Kuntze, R.A., and McAdie, H.G., *Can. J. Chem.*, *38*, 936, 1960.
24. Powell, D.A., *Nature*, *182*, 792, 1958.
25. Powell, D.A., *J. Sci. Instr.*, *34*, 225, 1957.
26. Powell, D.A., *Nature*, *178*, 428, 1956.
27. Powell, D.A., *Aust. J. Appl. Sci.*, *9*, 24, 1958.
28. Sekiya, M., Sugiyama, Y., and Okamoto, S., *Sekko To Sekkai*, *61*, 263, 1962.
29. Sekiya, M., *Sekko To Sekkai*, *64*, 89, 1963.
30. Trzebiatowski, D.J., and Romotowski, T., *Roczniki Chem.*, *30*, 431, 1956.
31. Sen, S., *Central Glass & Ceram. Res. Inst.*, *Bull.* (*India*), *5*, 93, 1958.
32. Murakami, K., and Tanaka, H., *Sekko To Sekkai*, *24*, 6, 1956.
33. Murakami, K., Shimamura, Y., and Tanaka, H., *Sekko To Sekkai*, *31*, 16, 1957.
34. Yamada, T., Goto, K., Iwata, T., Hiyama, S., and Nagai, S., *Sekko To Sekkai*, *1*, 1529, 1957.
35. Ramachandran, V.S., Sereda, P.J., Feldman, R.F., *Mater. Res. Std.*, *4*, 663, 1964.
36. Ramachandran, V.S., Sereda, P.J., and Feldman, R.F., *Mater. Res. Std.*, *8*, 24, 1968.
37. Stajadinoric, St., *Zement-Kalk-Gips*, *55*, 9, 1966.
38. Newman, E.S., *J. Res. Nat. Bur. Stad.*, *27*, 191, 1941.
39. Murray, J.A., and Fischer, H.C., *Proc. Am. Soc. Testing Mats.*, *51*, 16, 1951.
40. Newman, E.S., *Bull. Am. Ceram. Soc.*, *26*, 117, 1947.
41. Richardson, D.F., *Ind. Eng. Chem.*, *19*, 625, 1927.
42. Wells, L.S., Clarke, W.F., and Levin, *J. Res. Nat. Bur. Stad.*, *41*, 179, 1948.
43. Wells, L.S., and Taylor, K., *J. Res. Nat. Bur. Stad.*, *19*, 215, 1937.
44. Wells, L.S., Clarke, W.F., Newman, E.S., and Bishop, D.L., *Build. Mat. and Struc. Rep. No. 121*, Nat. Bur. Stad., p. 42, 1951.
45. Fox, E.L., and Mathers, F.C., *Proc. Indiana Acad. Sci.*, *43*, 113, 1934.
46. Welch, F.C., *Symp. on Lime, ASTM STP 40, Am. Soc. Testing Mat.*, p. 10, 1939.
47. Stutterheim, N., Webb, T.L., and Uranovsky, B., *Building Res. Cong.*, London, p. 120, 1951.
48. Hodgman, C.D., *Handbook of Chemistry and Physics*, Chemical Rubber Publishing Co., Cleveland, Ohio, 1954.
49. O'Kelly, B.M., *Bull. Am. Soc. Testing Mats.*, *No. 237*, 55, 1959.
50. Campbell, E.D., *Ind. & Eng. Chem.*, *1*, 665, 1909; *10*, 595, 1918.
51. Wiedman, T., *Tonind.-Z. Keram. Rundschau*, *82*, 331, 1958.

52. Budnikov, P.P., *Aufl., Moscow, 3*, 53, 1943.
53. Mtschedlov-Petrosian, O.P., Bunakov, A.G., Govorov, A.A., Latischew, F.A., Lewtschuk, N.A., and Strelkova, I.S., *Silikat Tech.*, *9*, 556, 1958.

IX.3. SAND-LIME MORTAR

Hydrated lime is not a hydraulic cement but becomes hard on exposure to moist air containing CO_2. It has been used as a cementitious material since antiquity.

A sand–lime mortar consists of calcium hydroxide and sand. The hardening of this mortar is different from that of Portland cement or gypsum plaster in requiring the presence of carbon dioxide. The lime in the mortar reacts with carbon dioxide in presence of moisture to form calcium carbonate which binds the sand particles together. A mix of 1:3 hydrated lime:sand mortar (by weight) in presence of carbon dioxide may yield a tensile strength of about 170 $lb/in.^2$ at 28 days (*1*).

Differential Thermal Analysis

Sand–Lime Mortar. In a sand–lime mortar, the extent of carbonation as well as the amount of uncarbonated calcium hydroxide can be estimated semi-quantitatively by the DTA method. Wieden (*2*) studied a sand–lime mortar by exposing it to atmosphere for a period of one year. The DTA of the samples obtained at intervals of weeks indicated initially an endothermal effect for $Ca(OH)_2$ decomposition and another at about 570°C for the $\alpha \rightleftharpoons \beta$ inversion in quartz. Carbonation set in very quickly, and at six weeks all the $Ca(OH)_2$ was carbonated, as indicated by the absence of an endothermic effect corresponding to $Ca(OH)_2$ decomposition.

Wieden's results are interesting because they imply that lime in mortars is completely carbonated within a few weeks. There are several instances quoted in the literature wherein the sand–lime mortar is reported to contain a considerable amount of uncarbonated $Ca(OH)_2$ even after several years of exposure to the atmosphere. The rate of carbonation is very high in the first few weeks and hence there is a possibility that during this period the covering layers may interfere with the carbonation of lime at the core.

Reactivity of Lime. In all the applications of lime it is useful to determine the reactivity of CaO. The conventional calorimetric methods either involve measurements of heat of solution in water or in dilute acids or of the rate of heat evolution during the hydration of CaO. The rate of solution method has been described by Haslem, Adams, and Kean (*3*). The radio-isotope method of Briss and Thorvaldson (*4*) may also be applied for such studies. The surface area values also provide a good indication of the relative reactivities of CaO. Glasson (*5*) has published a series of papers on the application of surface area determination to the investigation of calcined limes. In X-ray diffraction patterns the more reactive limes should show broadening of the

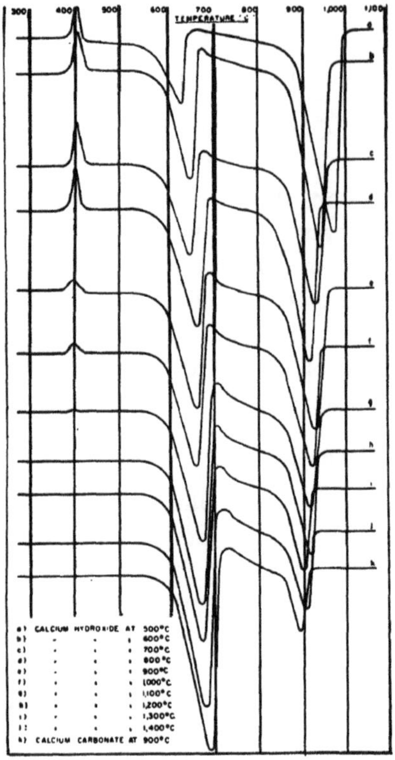

Fig. IX.3.1: Thermograms of equimolar mixtures of magnesite and calcium oxide of origin indicated (*7*).

lines (*6*).

Webb and Van der Walt (*7, 8*) used the DTA technique for the study of the reactivities of CaO by calcining $Ca(OH)_2$ or $CaCO_3$ at different temperatures. Their method is based on the measurement of peak areas of the exothermal reaction between CaO and $MgCO_3$ or between CaO and $Mg(OH)_2$, according to the following equations:

$$CaO + MgCO_3 \longrightarrow CaCO_3 + MgO$$

or

$$CaO + Mg(OH)_2 \longrightarrow Ca(OH)_2 + MgO$$

The extent to which CaO reacts with $MgCO_3$ or $Mg(OH)_2$ depends on its reactivity.

Fig. IX.3.1 shows the thermograms of equimolar mixtures of $MgCO_3$ and CaO obtained with $Ca(OH)_2$ calcined to temperatures in the range 500-1400°C. Certain trends are observed as the activity of the CaO decreases with increase in the calcination temperature. The exothermic effect at about 380°C for mixtures containing $Ca(OH)_2$ calcined up to 1100°C is ascribed to the exothermic reaction between CaO and $MgCO_3$. The endotherm between 600 and 700°C is due to the decarbonation of unreacted $MgCO_3$, and that between 900 and 1000°C is caused by the decomposition of $CaCO_3$ formed by the exchange reaction.

As the temperature of calcination is raised, the peak area of the exothermal effect decreases while the endothermal peak area of the $MgCO_3$ decompo-

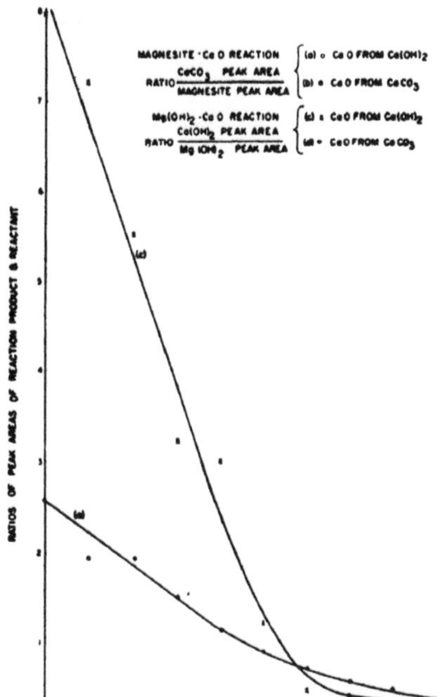

Fig. IX.3.2: Effect of origin and temperature of preparation of calcium oxide on ratio of peak areas of reaction product and reactant (7).

sition increases. The peak area due to calcium carbonate decreases from about 44 units to about 13 units as the calcination temperature of CaO increases from 500 to 1400°C. In the determination of the activities, the ratio of the peak area of the reactant to the reaction product was found to be linear with respect to the temperature up to 1100°C (Fig. IX.3.2).

REFERENCES

1. Lea, F.M., *The Chemistry of Cement and Concrete*, Edward Arnold (Publishers) Ltd., London, 1956.
2. Wieden, P., *Zement-Kalk-Gips*, *55*, 14, 1966.
3. Haslem, R.T., Adams, F.W., and Kean, R.H., *Ind. & Eng. Chem.*, *18*, 19, 1926.
4. Briss, F.W., and Thorvaldson, T., *Can. J. Chem.*, *33*, 881, 1955.
5. Glasson, D.R., *J. Appl. Chem.* (*London*), *8*, 798, 1958; *10*, 38, 1960; *10*, 42, 1960; *11*, 24, 1961; *11*, 28, 1961; *11*, 201, 1961; *13*, 111, 1963; *13*, 119, 1963; *13*, 124, 1963; *14*, 121, 1964.
6. Fischer, H.C., *J. Am. Ceram. Soc.*, *38*, 284, 1955.
7. Webb, T.L., and Van der Walt, T., *South Afr. Ind. Chem.*, *11*, 258, 1957.
8. Webb, T.L., and Van der Walt, T., *Nature*, *181*, 411, 1958.

IX.4. PHOSPHATE CEMENT

Phosphate cements are obtained by heating phosphoric acid with a finely divided oxide such as Al_2O_3. This cement sets under water and is used as an adhesive in cast, pressed, or molded products.

The thermal decomposition of inorganic cements containing the phosphates of aluminum, boron, and zirconium may be followed by DTA (*1*). Heats of reaction have also been determined by using potassium dichromate and calcium phosphate monohydrate as standards. Monoaluminum phosphate exhibits two endotherms with peak temperatures at 200 and 500°C and another with a characteristic temperature at 1500°C.

The first endothermic effect is attributed to the reaction

$$Al(H_2PO_4)_3 \longrightarrow Al(PO_3)_3 + 3H_2O$$

the second to the decomposition

$$Al(PO)_3)_3 \longrightarrow AlPO_4 + P_2O_5$$

the third to

$$2AlPO_4 \longrightarrow Al_2O_3 + P_2O_5$$

The heats of reaction for these reactions are respectively 11.3, 17.4, and 58.5 kcal/g.mole.

Vaporization of boron orthophosphate involves 33.5 kcal/g.mole and the inversion in zirconium pyrophosphate results in an absorption of 0.1 kcal/g.mole. The conversion of zirconium pyrophosphate to zirconyl pyrophosphate is attended by an absorption of 43 kcal/g.mole.

REFERENCES

1. Rickles, R.N., *J. Appl. Chem.* (*London*), *15*, 74, 1965.

X. Aggregates

Aggregates, forming the major portion in concretes and mortars, have to satisfy certain physical and chemical requirements. An aggregate containing minute quantities of deleterious constituents may cause serious failures in concrete. The importance of the role of aggregates in concrete is not realized by many cement chemists because failures due to them may become apparent only several years after making the concrete. In addition, most often the crushed rocks and sands are composed of quartz which does not produce any harmful effect in concrete. Very rightly, Bredsdorff *et al* (*1*) have stressed the inadequancy of the present knowledge on the chemistry of aggregates and their reactions with cements.

Aggregates may either be dense or relatively light. Sands, gravel, and air-cooled slags belong to the denser type, while perlite, pumice, exfoliated vermiculite, sintered fly ash and bloated clay, shale, or slate form the lighter type. Saw dust is used as an organic aggregate with cements but its use has yet to attain popularity. Asbestos fibers are also used with cement to make concrete products of good transverse strength.

Chemistry. Sands contain quartz in addition to felspar and clays. Gravels are composed of flint, chert, quartz, quartzite, granite, etc. Crushed rocks may contain granites, dolerites, and basalts. The air-cooled slag consists mainly of CaO, Al_2O_3, and SiO_2 with minor amounts of MgO, sulfur, and alkalies. In slags the deleterious constituents are C_2S, sulfur-bearing compounds, and iron in the ferrous state.

Volcanic glasses, perlite, and obsidian expand when fired to a temperature of incipient fusion. Vermiculite also expands when fired to temperatures of 600-900°C and forms a very light aggregate. The exfoliation is believed to be due to the creation of force in the interlayer of vermiculite when the interlayer water converts to steam. The cause of exfoliation in vermiculites has been explained by another theory (*2*) according to which the dehydrated vermiculite is subjected to warping and twisting due to unequal strains. These forces being greater than the bonding force between the planes of cleavage promote expansion.

Fly ash and clays bloat when heated rapidly to temperatures of 1000-1300°C. The illitic clays with a longer vitrification range should form a better lightweight aggregate than those containing montmorillonite. Kaolinites, with a higher temperature of fusion, may not be as suitable as illite.

The importance of the mineralogical composition for making lightweight clay aggregates has been discussed by Everhart (*3*). Asbestos is used as an aggregate with cement for making corrugated and plain sheets. The most widely used asbestos belongs to the chrysotile group having the formula $3MgO \cdot 2SiO_2 \cdot 2H_2O$.

An aggregate is not suitable if it contains a material that would deleteriously react with cement or undergoes oxidation, hydration, or carbonation. The sulfide minerals such as pyrite, marcasite, and pyrrhotite are reported to have caused failure in concrete even if present to an extent of only 1%. The sulfide minerals cause failure by forming ettringite and gypsum. Some aggregates containing MgO may also cause failure by slow hydration and expansion.

Blanks and Kennedy (*4*) have reported that an aggregate containing more than 0.25% gypsum may cause expansion and cracking in concrete. Deterioration of concrete may also occur in the presence of montmorillonite. Clays of the montmorillonitic group possess large cation exchange capacity and the Na^+ and K^+ ions of the clay mineral may exchange with Ca^{++}ions of the cement and cause efflorescence and deterioration. Many types of organic compounds are also undesirable constituents of aggregates, as for example humus, wood particles, coal, lignite, and sugar.

Failure caused by the alkali-aggregate reaction is perhaps the most widely studied field in cement-aggregate reactions. This reaction can cause considerable damage to concrete and is due to the reaction between the alkali and lime of the cement with aggregates containing an amorphous or a low crystalline form of silica. The alkali reactive minerals may be opal, opaline shale, some cherts, cristobalite, tridymite, or zeolites. In addition to these there are many instances reported wherein aggregates are known to be responsible for failures but the exact mechanism has not been resolved. One of the well known cases is that of Kingston dolomite of Canada which, used as an aggregate, causes failure by reaction with the alkali in cement, but the mechanism is not an alkali-aggregate type (*5*).

Differential Thermal Analysis

The technique of DTA has not been applied to any significant extent with a view to studying the aggregates and the deleterious constituents present in them. DTA equipment of high sensitivity could be used to examine various types of aggregates and detect the ingredients that may cause deterioration of concrete. The technique could be used with other methods to investigate the mechanism of the reactions leading to the failure of concrete.

Notably, DTA has been applied by Ramachandran, Kacker, and Garg (*6*), Ramachandran, Majumdar, and Patwardhan (*7*), Chopra, Kishan Lal, and Ramachandran (*8*), and Budnikov, Gorshkov, and Titovskaya (*9*) to investigage certain aspects of clay lightweight aggregates.

Materials such as quartz, calcite, dolomite, felspar, saw dust, vermiculite, fly ash, obsidian, and asbestos have been subjected to DTA, and many of them yield characteristic thermograms. In Fig. X.1 the thermograms of some of the aggregates are shown. Curve a refers to quartz which shows a sharp endothermic peak at 573°C due to $\alpha \rightarrow \beta$ transformation. The decomposition of calcium carbonate is indicated by an intense endothermic peak at about 920°C.

Dolomite decomposes in two stages—the first endothermic peak represent-

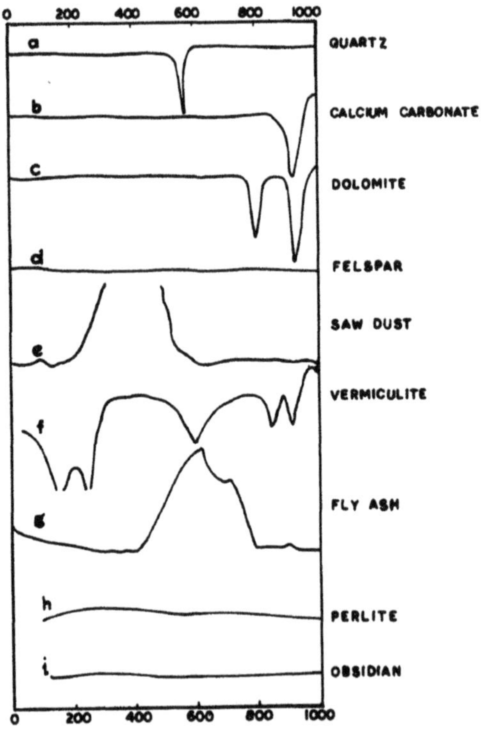

Fig. X.1: DTA of some of the aggregates.

ing decarbonation of $MgCO_3$ at about 800°C and the second endothermic peak at about 925°C representing decarbonation of $CaCO_3$. Thermograms of felspar, perlite, and obsidian are almost flat and do not yield any characteristic heat effects. Saw dust oxidizes over a very wide range with an exothermic effect extending from about 150 to 600°C.

In vermiculite the interlayer water is expelled in two stages below 300°C. Some vermiculites exhibit an endothermic peak at about 600°C and this is possibly caused by the chlorite as an impurity. The two high temperature endothermic peaks may represent the release of (OH) water from the talc-like phase and the exothermic trend may be ascribed to the formation of enstatite. Fly ash shows a very large exothermal effect between 400 and 800°C due to oxidation of carbon.

Harmful Impurities. In Fig. X.2 are given the thermograms of some of

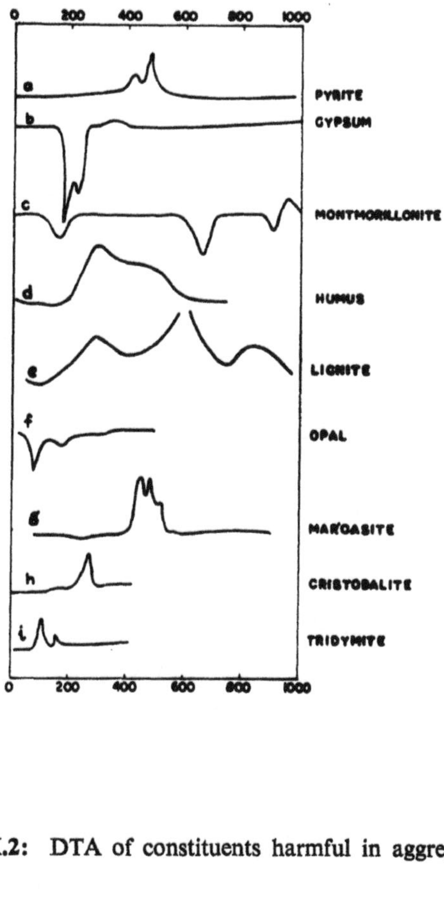

Fig. X.2: DTA of constituents harmful in aggregates.

the harmful impurities that might be present in aggregates. Curve a for pyrite indicates two exothermic effects between 400 and 500°C representing a stepwise oxidation. Gypsum shows two characteristic endothermic peaks at about 180 and 220°C for dehydration in two stages. Montmorillonite is characterized by three endothermic effects at about 160, 660, and 900°C due respectively to the release of interlayer water, dehydroxylation, and destruction of lattice. An exothermic peak at about 950°C is attributed to the formation of spinel.

Humus gives a very strong exothermic effect over a wide range (200-600°C) and lignite exhibits three exothermic peaks caused by oxidation. Opal, in the form of hydrated low cristobalite gives an endothermic peak at about 80°C for the loss of water and another at about 170°C due to a structural transformation. Marcasite manifests three exothermic effects at 450, 470, and

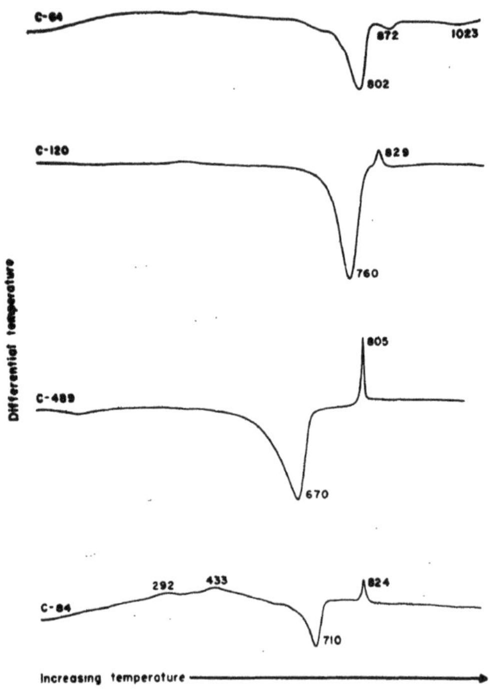

Fig. X.3: Differential thermal analysis curves. C.64—antigorite; C.120—natural mixture of antigorite with some chrysotile; C.489—chrysotile synthetic; C.84—chrysotile (*11*).

520°C. Cristobalite is identified by a peak at about 265°C for $\alpha \rightleftharpoons \beta$ transformation. Tridymite shows two endothermic peaks at about 110 and 160°C respectively for $\alpha_1 \rightleftharpoons \beta_1$ and $\alpha_2 \rightleftharpoons \beta_2$ inversions respectively. The thermograms for cristobalite and tridymite are obtained under cooling conditions.

Asbestos. Asbestos finds many applications with its widest use in in the manufacture of asbestos–cement products. Asbestos is divided into two groups, *viz.* serpentine and amphibole. The principal minerals in the serpentine group are chrysotile, antigorite, and lizardite, all having the general formula $3MgO \cdot 2SiO_2 \cdot 2H_2O$ and differing in their morphology. Chrysotile, the most abundant of asbestos minerals, is fibrous, whereas antigorite is flaky. The amphibole group comprises crocidolite ($Na_2O \cdot Fe_2O_3 \cdot 3FeO \cdot 8SiO_2 \cdot H_2O$), anthophyllite ($7MgO \cdot 8SiO_2 \cdot H_2O$), tremolite ($2CaO \cdot 5MgO \cdot 8SiO_2 \cdot H_2O$), amosite ($5.5FeO \cdot 1.5MgO \cdot 8SiO_2 \cdot H_2O$), and actinolite ($2CaO \cdot 4MgO \cdot FeO \cdot 8SiO_2 \cdot H_2O$).

The utility of asbestos, especially as a reinforcement in cements, is normally determined by its tensile strength. Hodgson (*10*) determined the tensile strengths of various asbestos minerals. Those of chrysotile, amosite, and crocidolite are respectively 31,000, 25,000, and 35,000 kg/cm.^{2} Tensile strengths of anthophllyite, tremolite, and actinolite are less than 5,000 kg/cm.^{2} DTA may be used advantageously for examining the various types of asbestos minerals and their mixtures. Nagy and Faust (*11*) have demonstrated by DTA that antigorite and chrysotile could be identified in natural minerals by their characteristic thermal curves. In Fig. X.3 are shown the thermal curves of antigorite, antigorite–chrysotile mixtures (a natural mixture of 75 % antigorite and 25% chrysotile), synthetic chrysotile, and natural chrysotile. The endotherms in the range 670-824°C represent dehydroxylation and the exothermic peaks in the range 805-829°C correspond to the crystallization of forsterite.

Wittels (*12*, *13*) and Vermaas (*14*) were the first to investigate amphibole group by DTA. Hodgson *et al* (*10*, *15-17*, *24*) have made an extensive investigation of various types of asbestos minerals in atmospheres of oxygen and argon. Typical thermograms of amphibole asbestos minerals, crocidolite, amosite, anthophyllite, tremolite, and prieskaite (fibrous actinolite) obtained by Hodgson are shown in Fig. X.4 (*24*) Amphiboles generally exhibit thermal effects corresponding to the loss of physically held water, loss of chemically bound water, and breakdown of the product into simpler units, consisting mainly of pyroxene.

Asbestos–cement products are either cured at ordinary temperatures or in high pressure steam. Kalousek and co-workers (*18*) applied the methods of DTA and X-ray and identified a lattice-substituted tobermorite with small amounts of CSH_n in autoclaved asbestos-cement products. DTA was also

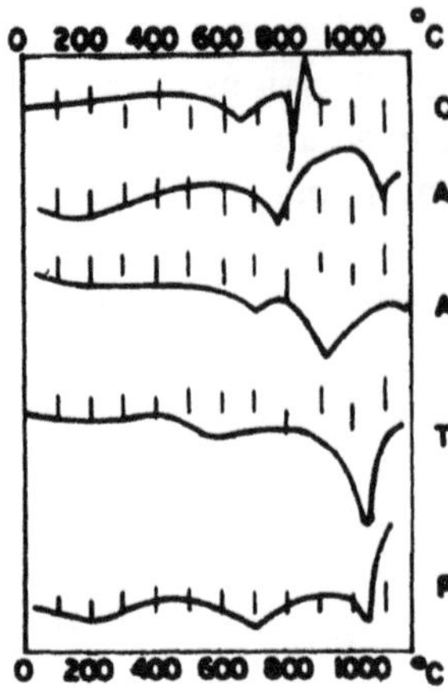

Fig. X.4: Typical DTA curves of amphibole asbestos in an inert atmosphere (*24*).

used to estimate quantitatively the uncombined $Ca(OH)_2$ and this led to the conclusion that the glycerol method in the ASTM method extracts more $Ca(OH)_2$ from normal-cured specimens than that indicated by DTA.

Bloated Clay Aggregates. Clays form an important group of raw materials for the production of lightweight aggregates. It is generally recognized that a clay will bloat if two conditions are satisfied. First, the material should fuse to a viscosity high enough to entrap the gas that is formed, otherwise the gas will bubble through the melt. Second, some material or combination of minerals must be present or formed during the heating that will dissociate and liberate a gas at the same time as the mass of the clay forms a viscous melt. In practice, pellets are fired to temperatures of the order of 1000-1300°C.

The bloating phenomenon has been known for several decades, but only recently has some attention been directed to the study of its mechanism. Various materals piresent in the natural clay are reported to be responsible. The gas producing agents are variously given as hematite, pyrite, carbonates, sulfates, alkali earths, graphite, lignite, and micaceous minerals, the gases responsible for bloating being O_2, S, SO_2, SO_3, CO_2, CO, H_2O, hydrocarbons, and entrapped air either alone or in combination. Most of the natural clays contain several possible minerals as gas producers, and consequently substances reported by some workers as excellent bloating agents were rejected as non-gas producing agents by others.

Ramachandran *et al* (*6*) selected 21 substances as gas-producing agents and studied the mechanism of bloating in clays by DTA. These agents were chosen on the following considerations: (i) Many of them are present in natural clays, (ii) some of them are chemicals that could possibly be added to a non-bloating clay for producing bloating, (iii) the substances evolve gases over a wide range of temperature, (iv) they have varying fluxing action, and (v) they evolve different types of gases.

Two clay compositions were chosen, one corresponding to a dehydrated illite with the composition Al_2O_3:SiO_2:Na_2, $K_2(O)$=15:75:10 (formed with a mixture of kaolinite, silica, and microcline) and the second corresponding to montmorillonite. The gas-producing agents consisted of natural minerals and a few chemicals such as hematite, calcite, pyrite, gypsum, siderite, magnesite, dolomite, graphite, sodium chloride, anhydrous ferric sulfate, sodium sulfate, ferrous chloride, sodium carbonate, ferrous sulfate, molasses, saw dust, fly ash, purified illite, pyrophyllite, talc and muscovite.

Some of the gases expected to be liberated by these gas-producing agents are oxygen, carbon dioxide, sulfur dioxide, sulfur trioxide, chlorine, water vapor, and hydrocarbons.

TABLE X.1
Efficiency of Various Gas-Producing Agents on Bloating of Montmorillonite and Illite [Ramachandran et al (6)]

		Montmorillonite		*Illite*		
No.	*Bloating agent*	*1100 °C*	*1150 °C*	*1200 °C*	*1250 °C*	*1300 °C*
1	Hematite	Pb	S	S	S	S
2	Calcite	Gb	Vg	Gb	Pb	S
3	Pyrite	Gb	Vg	Gb	Pb	S
4	Gypsum	Pb	Pb	Sh	S	S
5	Siderite	Gb	Vg	Gb	Gb	Gb
6	Magnesite	Pb	Vg	Gb	Pb	Pb
7	Dolomite	Gb	Vg	Gb	Gb	Pb
8	Graphite	Gb	Vg	Pb	Pb	Pb
9	Sodium chloride	Gb	Gb	Sh	S	S
10	Ferric sulfate	Nb	Gb	Nb	S	Pb
11	Sodium sulfate	Sh	Gb	Pb	Pb	Gb
12	Ferrous chloride	S	Gb	Nb	S	S
13	Sodium carbonate	Gb	Vg	Pb	Pb	Pb
14	Ferrous sulfate	Pb	Gb	Nb	S	S
15	Molasses	Vg	Vg	Gb	Gb	Pb
16	Sawdust	Vg	Vg	Gb	Pb	Pb
17	Fly ash	Gb	Gb	Gb	Pb	Pb
18	Illite (purified)	Pb	Pb	Gb	Gb	Gb
19	Pyrophyllite	Pb	Gb	Gb	Gb	Pb
20	Talc	Pb	Pb	Pb	Sh	S
21	Muscovite	Gb	Pb	Gb	Gb	Gb

Montmorillonite was fired at 1100 and 1150°C because treatment at higher temperatures resulted in complete fusion and sagging. The firing temperatures for illites were fixed at 1200, 1250, and 1300°C. Below 1200°C there was no appreciable fusion to entrap a gas and above 1300°C the mixture slagged. All the samples were fired at a retention time of 5 min.

Table X.1 summarizes the effect of various additives on the bloating behavior of montmorillonite. The bloating effects are arbitrarily divided into six categories for comparison:

a) Vg = very good bloating (expansion of or over 100%)
b) Gb = good bloating (expansion 30-99%)
c) Pb = poor bloating (expansion up to 29%)
d) Nb = no bloating (expansion is nil)
e) Sh = shrinkage (pellets shrink without fusion)
f) S = slagging down due to fusion

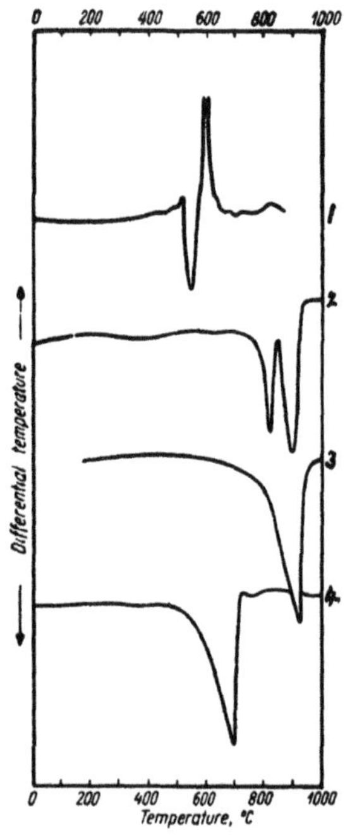

Fig. X.5: Thermograms of some gas producing agents. (1) siderite, (2) dolomite, (3) calcite, (4) magnesite (*6*).

Differential thermal curves of 21 gas-producing agents are presented in Figs. X.5-X.9.

Ferric oxide in the form of hematite does not effect bloating either in montmorillonite or illite. Fe_2O_3 induces early fusion and slagging of the pellets. The decrease in volume is as high as 22.54% for montmorillonite and 26.11% for illite. At 1100°C montmorillonite vitrified to a viscous melt but failed to bloat. Hostetter and Roberts (*19*) found that Fe_2O_3 liberated oxygen at 1300°C. Failure to produce bloating at 1100°C is perhaps due to non-liberation of any gas from Fe_2O_3. At higher temperatures the pellets slagged down and could not entrap the gas. DTA does not indicate any effect for the oxidation of Fe_2O_3 below 1000°C (Fig. X.8, curve 1).

Contrary to the observations of Riley (*20*) calcite behaves as a good bloating agent. In montmorillonite the bloating increased from 53.93% at 1100°C to 102.25% at 1150°C. In illites the bloating was 42.27% at 1200°C. Riley (*20*) and Johnston (*21*) believe that at 1200°C no calcite should survive to liberate CO_2. DTA shows that CO_2 is evolved at a temperature of 800°C.

Pyrite produces a better bloating effect in montmorillonite than calcite. The maximum bloating is 113.1%. In illite there is a small bloating effect.

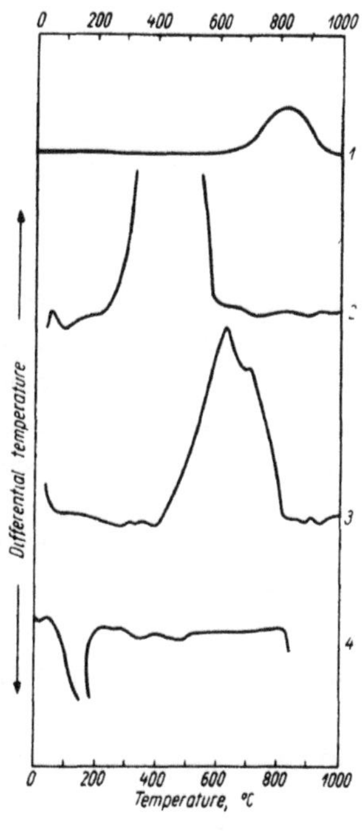

Fig. X.6: Thermograms of gas-producing agents. (1) graphite, (2) saw dust, (3) fly ash, (4) sodium carbonate (*6*).

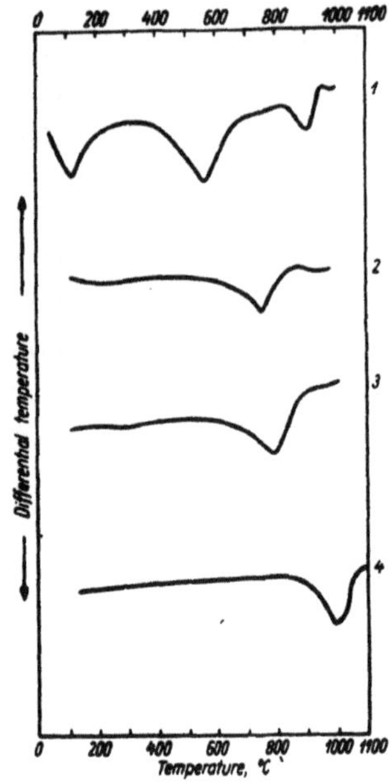

Fig. X.7: Thermograms of some gas-producing agents. (1) illite, (2) muscovite, (3) pyrophyllite, (4) talc (*6*).

It is surprising that though pyrite is oxidized at a low temperature of 400-500°C (Fig. X.8, curve 4) it acts as a good gas producing agent at 1150°C. This indicates that in flash firing, independent of the temperature of liberation of gas by the gas producing agent, bloating takes place if the rate of formation of the viscous layer is quicker than the gas evolution.

Gypsum is a poor additive for producing any expansion in the fired pellets. In montmorillonite, bloating is only 22.51% and in illite it acts as a good fluxing agent and produces slagging. Gypsum dehydrates at very low temperatures and liberates small amounts of SO_3 at about 1200°C (Fig. X.9, curve 1).

Addition of siderite results in good bloating in montmorillonite. Bloating is 138.52% at 1150°C. In illite it has comparatively less effect. $FeCO_3$ decomposes to FeO at 545°C and oxidizes to Fe_2O_3 at 600°C (Fig. X.5, curve 1).

Magnesite behaves similarly to calcite in producing a good expansion at 1150°C. Its influence in illite is comparatively less. Magnesite gives an

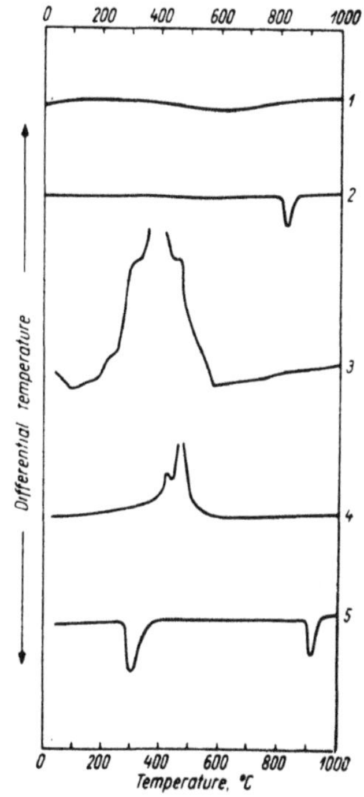

Fig. X.8: Thermograms of some gas-producing agents. (1) anhydrous hematite, (2) sodium chloride, (3) molasses, (4) pyrites, (5) sodium sulfate (*6*).

endothermic peak at about 700°C liberating CO_2.

It is interesting to note that dolomite showed a better bloating effect than $CaCO_3$ or $MgCO_3$. In montmorillonite the expansion was as high as 284.5 % at 1150°C. Even in illite, dolomite exhibits a considerable bloating effect. It is possible that a molar mixture of $CaCO_3$ and $MgCO_3$ may provide the optimum viscosity at 1150°C for gas entrapment.

Graphite oxidizes over a temperature range of about 700-950°C (Fig. X.6, curve 1) and is a better bloating agent than $CaCO_3$ at 1150°C.

Sodium chloride induces fusion and slagging in illite at 1250 and 1300°C. However, at 1200°C there is only shrinkage of the pellet. In montmorillonite there is an expansion of 80.76% at 1150°C. Sodium chloride melts at 816°C (Fig. X.8, curve 2). Ferric sulfate produces bloating at 1150°C. At other temperatures there is negligible bloating. The sample of ferric sulfate shows a very large endothermic effect at about 750°C.

Addition of sodium sulfate results in sufficient bloating at 1150 and 1300°C. It is possible that proper conditions of viscosity were unavailable for entrapment of any gas at 1200 and 1250°C.

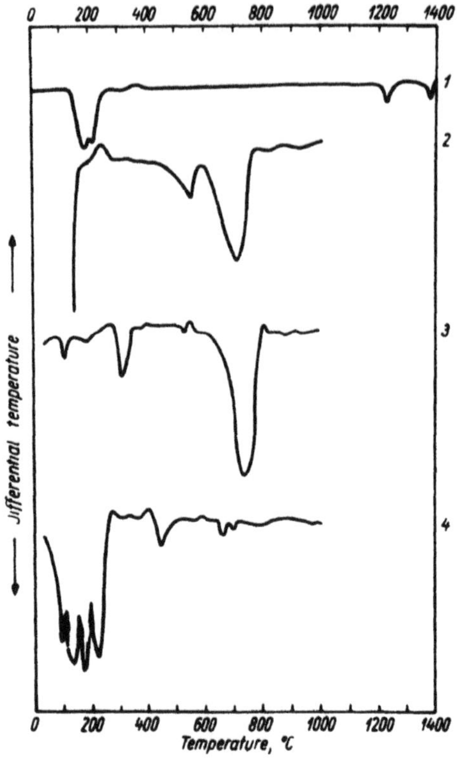

Fig. X.9: Thermograms of some-gas producing agents. (1) gypsum, (2) ferrous sulfate, (3) anhydrous ferric sulfate, (4) ferrous chloride (*6*).

Ferrous chloride shows a peculiar behavior. At 1100°C there is fusion and slagging of montmorillonite pellets. But at 1150°C there is a good bloating effect in the order of 65%. No explanation could be offered, but it is probable that there is a sudden evolution and entrapment of gas at 1150°C. In illites, ferrous chloride acts as a good flux and exhibits no bloating effect.

In the mixture containing sodium carbonate there is a maximum bloating effect at 1150°C, the bloating being 258%. There is negligible bloating in illite. It is known that decomposition of Na_2CO_3 takes place above 1200°C. The evolution of gas at 1150°C could be due to a catalytic decomposition of Na_2CO_3 in the presence of the constituents of montmorillonite.

Ferrous sulfate behaves like ferrous chloride in producing slagging in the illite sample. In montmorillonite the effect is about the same as found for ferrous chloride.

Molasses is oxidized (with volatilization) over a low temperature range 200-600°C (Fig. X.8, curve 3). It makes a very good admixture in producing a bloating of 149.7% and 129.3% in montmorillonite at 1100 and 1150°C respectively. In illite the effect is comparatively less.

Saw dust is the second best bloating additive studied. In montmorillonite the bloating is 225.1% and 246% at 1100 and 1150°C respectively. In illites the effect is negligible, particularly at higher temperatures. Saw dust shows a very wide exothermal effect in the range 100-600°C (Fig. X.6, curve 2).

Fly ash containing about 7% carbon shows uniformly good bloating effect at 1100, 1150, and 1200°C. Its effect at 1250 and 1300°C is less. The oxidation of carbon in fly ash occurs in the range 400-800°C (Fig. X.6, curve 3).

Purified illite in admixture with montmorillonite has only a small effect. With illitic composition the bloating is 46.03%, 52.07%, and 58.89% at 1200, 1250, and 1300°C respectively. Pure illites have been known to show bloating effect. This could perhaps be explained if it is assumed that at the surface there is quick dehydration and fusion. The expansion would then result from the pressure developed by steam evolved by dehydroxylation and expelled after an impermeable surface layer is formed.

Pyrophyllite proves a good bloating agent at 1150, 1200, and 1250°C. Unlike purified illite it is a good agent for montmorillonite as well as for illite. Pyrophyllite exhibits an endothermic hump with a peak at about 800°C (Fig. X.7, curve 3).

Talc has a small bloating effect at 1100, 1150, and 1200°C. At 1250 and 1300°C there is a shrinkage. Talc shows an endothermic peak at about 1000°C (Fig. X.7, curve 4).

Muscovite seems to be a good bloating agent for montmorillonite at 1100 and illite at 1200°C.

From the DTA and bloating results described above, Ramachandran and co-workers (*6*) concluded as follows:

Montmorillonite bloats to a greater extent than illite. Almost all substances studied, except hematite, induce bloating in montmorillonite either at 1100 or at 1150°C. The high percentage of hematite in montmorillonite may induce quick fusion and shrinkage of the pellets.

Hematite, gypsum, sodium chloride, ferrous chloride, and ferrous sulfate are practically ineffective as bloating agents in admixture with illite. On the addition of these agents either the volume remains practically the same or there is a shrinkage of the pellets.

No correlation could be found between the temperature of gas evolution and the efficacy of the gas-producing agent at different temperatures of firing. For example, molasses, which evolves gases at a low temperature of 200-600°C and saw dust with the oxidation and volatilization temperature at 100-600°C, are excellent bloating agents. A substance may produce gas at a low temperature and still bloat, since under the conditions of flash firing the rate of gas evolution may be lower than the rate at which the pellet fuses and forms a hard coating. Budnikov and others (*9*) showed by means of DTA that an endothermic peak for desorbed water at 160°C obtained at a rate of about 10°C/min may shift to 560°C if the rate is as high as 100-300°C/min.

At high temperatures of flash firing the mass fuses and there is also an evolution of gas depending on the quantity of the gas producing agent and the temperature and rate of the gas-producing reaction. If the rate of gas liberation is too high the gas escapes before fusion takes place and there is no bloating. If there is no gas liberation the pellets shrink and slag. If the mass forms a melt of proper viscosity and coats the surface before there is considerable gas liberation, bloating results.

Substances liberating gases over a wider range of temperature would also liberate gases over a longer period and promote better gas-entrapment. Such behavior is shown by saw dust, fly ash, and molasses.

Many of the additives have a dual role to play—they provide a source for gases and also act as fluxes. If an additive is too good a flux, bloating would not be expected to take place because no gas would be entrapped. For example, gypsum and sodium chloride seem to act as good fluxing agents for illite, inducing only shrinkage.

A substance may evolve a gas and produce bloating at a temperature lower than the temperature at which it would by itself normally give off a gas. In such cases the components of a clay may be acting as catalysts. Such an effect seems to offer an explanation for the bloating behavior of montmorillonite treated with sodium carbonate.

Fig. X.10: DTA. (P_1) original clay, (P_2) clay fraction, (P_3) non-clay fraction (*8*).

Chopra and co-workers (*8*) found that the DTA technique can advantageously be applied to examine the possible constituents responsible for bloating in clays. The clay selected for their investigation was a silty clay known as 'Palta silt,' formed as a result of sedimentation in tanks at a water works.

Differential thermal curves of Palta clay, clay fraction, non-clay fraction, and variously treated samples are presented in Figs. X.10-12. DTA of Palta clay (P_1 in Fig. X.10) shows a low temperature endothermic peak at about 110°C due to the expulsion of adsorbed water. An exothermic effect over a wide range with two peaks at about 330 and 430°C results from the oxidation of organic matter.

Many clays are known to exhibit a stepwise oxidation of organic matter (*22*). Chemical analysis shows that the clay contains organic matter corresponding to 0.36% carbon. An endothermic peak at 520°C is caused by the dehydroxylation of the clay mineral, and this effect is shown by clay minerals such as kaolinite, illite and nontronite.

The high cation exchange capacity (c.e.c.) of 60.6 m. equiv./100 g rules out the possibility of kaolinite as the major mineral in the clay fraction. The low SiO_2/Al_2O_3 ratio of 2.57, Fe_2O_3 content, and c.e.c. indicate the absence of nontronite as the predominant clay mineral.

The presence of chlorite could not be confirmed for the following reasons: Chlorite does not exhibit a low temperature endothermic peak for the loss of interlayer water. In addition, chlorite generally registers two endothermic peaks in the temperature range 500-800°C and an exothermic peak above

Fig. X.11: DTA. Clay treated with cold 1:9 HCl (P_4), boiling 1:1 HCl (P_5), H_2O_2 at room temperature (P_6), H_2O_2 at 80°C (P_7) (*8*).

Fig. X.12: DTA. Clay heated for 3½ hr at 300°C (P_8), at 400°C (P_9), 500°C (P_{10}), 600°C (P_{11}), 700°C (P_{12}), 800°C (P_{13}) (*8*).

800°C. Chlorites hould normally have high MgO or FeO contents, but the Pelta clay fraction has only 1.35% of MgO and very low FeO.

From the above facts it was concluded that the clay contains illite as the predominant clay mineral. The c.e.c. values of illite normally vary in the range 10-40 m.equiv./100 g. A high value of 60.6 m.equiv./100 g. indicates that this illite is poorly crystallized. X-ray diffraction data also show strong lines characteristic of illite. A sharp endothermic effect at about 750°C is due to decarbonation of calcium carbonate. The calcium carbonate contents as found by direct chemical analysis and CO_2 estimation are respectively 2.84 and 3.18%. Pure calcium carbonate shows an endothermic peak at about 850-900°C, but in presence of impurities the temperature of the peak is known to decrease by 100°C. The presence of calcium carbonate could also be confirmed indirectly.

P_4 and P_5 (Fig. X.11) refer to the thermograms of the clay treated with hydrochloric acid. The acid-treated clay does not exhibit an endothermic peak as a result of the decomposition of calcium carbonate. After the completion of the high-temperature endothermic peak, there is a considerable base line shift. The clay progressively fuses and forms a hard mass on the thermocouple bead. Hence, in all the DTA studies the clay was mixed with an inert material, α-alumina.

The clay fraction (P_2, Fig. X.10) shows an endothermic peak at about 130°C due to the evolution of adsorbed water. The exothermic peak is more intense than that for the original clay (P_1, Fig. X.10), showing thereby that a substantial quantity of organic matter is adsorbed on the clay mineral. Chemical analysis shows that the clay fraction contains about 0.5% carbon, whereas the original clay contains 0.36% carbon.

The endothermic effect occurring due to dehydration is obviously more pronounced because the clay mineral is concentrated in the clay fraction. The endothermic effect for decomposition of $CaCO_3$ is of very low intensity because the carbonate predominates in the non-clay fraction. A slight endothermic dent at about 910°C is characteristic of illitic mineral and is attributed to the destruction of the lattice.

The DTA of the non-clay fraction (P_3, Fig. X.10) shows a smaller water absorption effect, followed by an exothermal dent with two inflections. A sharp endothermic peak at 575°C is due to the transformation of α to β-quartz and the endothermic peak at 750°C corresponds to the decomposition of calcium carbonate. DTA indicates that most of the calcium carbonate is present in the non-clay fraction.

The above results suggest that Palta silt comprises mainly an illitic clay in the clay fraction and quartz in the non-clay fraction. In addition, the results lead to the conclusion that the clay contains accessory materials such as organic matter, calcium carbonate, and ferric oxide. The clay shows good bloating characteristics, the density of the fired pellet being as low as 0.61 g/cm.^{3} The possible sources of gas evolution are one or more of the following: (i) organic matter, (ii) clay mineral itself (contributing adsorbed as well as chemically held water), (iii) calcium carbonate, (iv) ferric oxide, and (v) entrapped air.

In order to study all the factors that might induce bloating, the clay was systematically subjected to various treatments and its bloating and thermal behavior were studied.

The role of organic matter as a bloating agent was studied by firing the clay from which organic matter was removed either by hydrogen peroxide treatment or by heating the clay. Organic matter is generally oxidized by hydrogen peroxide in the cold but certain types of lignin–humus complexes are resistant to such treatment. Hence the clay was also treated with hydrogen peroxide at 80°C for several hours. Both treatments seem to be effective in destroying most of the organic matter, as is evident by the absence of exothermal effects in curves P_6 and P_7 (Fig. X.11).

The clay treated with hydrogen peroxide showed bloating to a lesser extent than the untreated sample. The densities of the treated and untreated samples were respectively 1.37 and 0.61 g/cm.^{3} Heat treatment at 300°C for 3½ hr does not destroy the organic matter (P_8, Fig. X.12) and bloating occurs to nearly the same extent (0.625 g/cm^3) as with the original sample. Heat treatment at 400°C for 3½ hr causes a decrease (1.38 g/cm^3) in the bloating capacity and there is no exothermal effect in the DTA curve (P_9, Fig. X.12).

These results indicate that organic matter is partly responsible for bloating. Although organic matter decomposes at lower temperatures, it is retained even at high temperatures as a consequence of rapid firing.

A second series of experiments was carried out to investigate the role of calcium carbonate in the bloating of Palta clay. The carbonate could either be destroyed with hydrochloric acid or decomposed at about 700-800°C. The clay was treated with both 1:9 and 1:1 HCl and the resultant clay washed free of chloride and its bloating behavior studied. Both treatments were effective in destroying $CaCO_3$ as is evident from the absence of endothermic peaks in the thermograms at high temperatures (P_4 and P_5 in Fig. X.11 and P_{12} and P_{13} in Fig. X.12).

The sample treated with dilute HCl (1:9) bloated (0.998 g/cm.^{3}), although to a lesser extent than the untreated clay. This showed that $CaCO_3$ is partly responsible for bloating. Several investigators have found that clays containing $CaCO_3$ showed good bloating characteristics. The presence of $CaCO_3$ beyond 1000°C during flash firing should explain its effective role in the bloating of clays. Many substances that decompose at very low temperatures have been found to be good gas producing agents.

The sample treated with concentrated and hot HCl (1:1) showed no bloating at all, and the fired pellet did not even exhibit vitrification. Treatment with conc. HCl and subsequent washing would bring about the to a large extent removal of the fluxes CaO and Fe_2O_3. If the conditions of proper viscosity of the melt and of gas evolution are not fulfilled, the clay is not expected to bloat.

Samples heated to 500 and 600°C retain the high temperature endothermic peak, indicating thereby that calcium carbonate is not decomposed at these temperatures (P_{10} and P_{11}, Fig. X.12), but when subjected to flash firing they failed to show any bloating. Treatment at 500°C and above results in the expulsion of the chemically held water. The removal of this water creates numerous open pores that would provide free access to gases that otherwise would be entrapped. Mielenz and King (*23*) also found that the structure of the original clay is a significant factor in the bloating process—a porous, open structure permitting ready burning out of carbon without producing any bloating.

The clay in which both organic matter and calcium carbonate were decomposed by treatment with hydrogen peroxide and dilute HCl respectively showed no bloating and had a density as high as 2.09 g/cm.^{3} This obviously should mean that (OH) water, entrapped air in the pores, and Fe_2O_3 do not significantly affect bloating. Since both $CaCO_3$ and organic matter are gas-producing agents, the bloated clay may be expected to contain CO_2 as the entrapped gas. Hence, a quantitative determination of the entrapped gas would directly confirm this fact. The gas chromatographic analysis of the entrapped gases produced by crushing the bloated clays showed a direct confirmation of the presence of CO_2.

REFERENCES

1. Bredsdorff, P., Idorn, G.M., Kjaer, A., Plum, N.M., and Poulsen E., *Proc. 4th Intern. Symp. Chem. Cements*, Washington, pp. 749-806, 1960.
2. Byers, W.B., *U.S. Pat.*, 2,030,239, Feb. 11, 1936.
3. Everhart, J.O., *Brick and Clay Record*, 134, 58, 1959.
4. Blanks, R.F., and Kennedy, H.L., *The Technology of Cement and Concrete*, John Wiley & Sons Inc., New York, Chapman Hall Ltd., London, p. 422, 1955.
5. Gillot, J.E., *Geol. Soc. Am. Bull.*, *74*, 759, 1963.
6. Ramachandran, V.S., Kacker, K.P., and Garg, S.P., *Proc. 7th Conf. Silicate Industry*, Budapest, pp. 183-197, 1963.
7. Ramachandran, V.S., Majumdar, N.C., and Patwardhan, N.K., *Indian Concrete J.*, *35*, 453, 1961.
8. Chopra, S.K., Kishan Lal and Ramachandran, V.S., *J. Appl. Chem. (London)*, *14*, 181, 1964.
9. Budnikov, P.P., Gorshkov, V.S., and Titovskaya, V.T., *Stoitel Materialy*, *6*, 30, 1960.
10. Hodgson, A.A., *Lecture Series No. 4*, The Royal Institute of Chemistry, London, 1964.
11. Nagy, B., and Faust, T., *Am. Mineralologist*, *41*, 817, 1956.
12. Wittels, M., *Am. Mineralologist*, *36*, 851, 1951.
13. Wittels, M., *Am. Mineralologist*, *37*, 28, 1952.
14. Vermaas, F.H.S., *Trans. Geol. Soc. S. Africa*, *55*, 199, 1952.
15. Hodgson, A.A., Freeman, A.G., and Taylor, H.F.W., *Mineral. Mag.*, *35*, 5, 1965.
16. Hodgson, A.A., Freeman, A.G., and Taylor, H.F.W., *Mineral. Mag.*, *35*, 445, 1965.
17. Hodgson, A.A., *Mineral. Mag.*, *35*, 291, 1965.
18. Kalousek, G.L., Camarda, F.M., Kopanda, J.E., and Jugovic, Z.T., *Mater. Res. Std.*, *6*, 169, 1966.
19. Hostetter, J.C., and Roberts, H.S., *J. Am. Ceram. Soc.*, *4*, 927, 1921.
20. Riley, C.M., *J. Am. Ceram. Soc.*, *34*, 121, 1951.
21. Johnston, J., *J. Am. Ceram. Soc.*, *30*, 1357, 1908.
22. Ramachandran, V.S., Ahmad, F.U., and Jain, L.C., *Proc. Indian Acad. Sci.*, *50A*, 314, 1959.
23. Mielenz, R.C., and King, M.E., *Calif. Div. Mines Bull.*, *No. 169*, 196, 1955.
24. Hodgson, A.A., Ph.D. Thesis, London, 1963.

Subject Index

Author Index

www.ingramcontent.com/pod-product-compliance
Ingram Content Group UK Ltd.
Pitfield, Milton Keynes, MK11 3LW, UK
UKHW041130071026
15587UKWH00009B/407

9 780820 600246